AF694287

Developments in Applied Phycology 8

Series editor

Michael A. Borowitzka, Algae R&D Centre, School of Veterinary and Life Sciences, Murdoch University, Murdoch, WA, Australia

Aims and Scope

Applied Phycology, the practical use of algae, encompasses a diverse range of fields including algal culture and seaweed farming, the use of algae to produce commercial products such as hydrocolloids, carotenoids and pharmaceuticals, algae as biofertilizers and soil conditioners, the application of algae in wastewater treatment, renewable energy production, algae as environmental indicators, environmental bioremediation and the management of algal blooms. The commercial production of seaweeds and microalgae and products derived there from is a large and well established industry and new algal species, products and processes are being continuously developed.

The aim of this book series, Developments in Applied Phycology, is to present state-of-the-art syntheses of research and development in the field. Volumes of the series will consist of reference books, subject-specific monographs, peer reviewed contributions from conferences, comprehensive evaluations of large-scale projects, and other book-length contributions to the science and practice of applied phycology.

Prospective authors and/or editors should consult the Series Editor or Publishing Editor for more details. Series Editor: Michael A. Borowitzka - M.Borowitzka@murdoch.edu.au Publishing Editor: Alexandrine Cheronet - alexandrine.cheronet@springer.com

More information about this series at http://www.springer.com/series/7591

Katarzyna Chojnacka • Piotr Pawel Wieczorek
Grzegorz Schroeder • Izabela Michalak
Editors

Algae Biomass: Characteristics and Applications

Towards Algae-based Products

Springer

Editors
Katarzyna Chojnacka
Faculty of Chemistry, Department of Advanced Material Technologies
Wrocław University of Science and Technology
Wrocław, Poland

Grzegorz Schroeder
Faculty of Chemistry
Adam Mickiewicz University in Poznan
Poznan, Poland

Piotr Pawel Wieczorek
Faculty of Chemistry, Department of Analytical and Ecological Chemistry
Opole University
Opole, Poland

Izabela Michalak
Faculty of Chemistry, Department of Advanced Material Technologies
Wrocław University of Science and Technology
Wrocław, Poland

Developments in Applied Phycology
ISBN 978-3-030-09059-3 ISBN 978-3-319-74703-3 (eBook)
https://doi.org/10.1007/978-3-319-74703-3

© Springer International Publishing AG, part of Springer Nature 2018
Softcover re-print of the Hardcover 1st edition 2018
This work is subject to copyright. All rights are reserved by the Publisher, whether the whole or part of the material is concerned, specifically the rights of translation, reprinting, reuse of illustrations, recitation, broadcasting, reproduction on microfilms or in any other physical way, and transmission or information storage and retrieval, electronic adaptation, computer software, or by similar or dissimilar methodology now known or hereafter developed.
The use of general descriptive names, registered names, trademarks, service marks, etc. in this publication does not imply, even in the absence of a specific statement, that such names are exempt from the relevant protective laws and regulations and therefore free for general use.
The publisher, the authors and the editors are safe to assume that the advice and information in this book are believed to be true and accurate at the date of publication. Neither the publisher nor the authors or the editors give a warranty, express or implied, with respect to the material contained herein or for any errors or omissions that may have been made. The publisher remains neutral with regard to jurisdictional claims in published maps and institutional affiliations.

Printed on acid-free paper

This Springer imprint is published by the registered company Springer International Publishing AG part of Springer Nature.
The registered company address is: Gewerbestrasse 11, 6330 Cham, Switzerland

Preface

The chemical industry surrounding algae products has grown enormously in the last 50 years. Today, algae biomass is not merely a food ingredient or an animal feed additive but a rich source of bioactive chemical compounds. These natural products, which can be isolated from algae biomass, have attracted the attention of biologists and chemists around the world. The demand for and popularity of these products has grown significantly among ecologically aware consumers, as the chemical compounds of which they are comprised originated from natural sources and were not created as a result of chemical synthesis. Consumers' requirements and expectations are increasing, as they yearn for their products or materials to be "bio," "natural," and "organic." Recently, novel technologies have been developed that can convert algae biomass into valuable products and allow for the extraction of biologically active compounds from algae in an efficient and environmentally friendly manner. The development of an algae industry centered on the industrial application of naturally grown algae biomass contributes to sustainable development, as it preserves the given environment so it can be later used for active recreation and leisure purposes. Algae biomass and natural products obtained from algae biomass have played a very large part in the world's history, and the trade in natural products is an increasingly significant part of the world's economy.

An interdisciplinary 24-person team of biologists, pharmacists, and agricultural professionals present their monographs on the development of algae biomass in the European climate zone. The aim of this book is the presentation of knowledge about the possible use of algae biomass in various branches of industry and agriculture.

The 12 chapters of the book expose the reader to various aspects of key issues surrounding the industrial and agricultural application of algae biomass. In Chaps. 1 and 2, the authors present the environmental benefits of the algae biomass industry. Algae are considered a potential feedstock for many products, such as food, feed, biofuels, biofertilizers, and cosmetics. Additionally, the product of atmospheric carbon dioxide and sunlight conversion, i.e., biomass, can be transferred via various chemical processes into valuable materials. The biologists in Chap. 3 describe cosmopolitan taxa of freshwater macroalgae occurring en masse in inland waters with respect to their biology and the factors influencing the fluctuations of macroalgal populations (e.g., *Cladophora*, *Oedogonium*). Chapter 4 is composed of four short parts, two of which illustrate algae as a source of valuable selected chemicals (phycobiliproteins, polyphenols), while the other two are dedicated to the biocatalytic abilities of those organisms that allow them to protect ecosystems against organic pollutants and transitional metal ions. Chapter 5 was prepared by specialists in industrial technology. Dealing with biomass processing by extraction on a daily basis, they discuss how to obtain and extract algae for potential industrial use. In food and animal feed, algae biomass can be employed in different forms – whole seaweed meal, powder, extract, homogenate, or fermented. In Chaps. 6 and 7, the advantages (e.g., nutritional value, accessibility, etc.) and disadvantages (e.g., toxic metals, sensory perception, etc.) of the application of seaweeds as a component of food are discussed. Close attention was paid to their application in cereals, dairy, and meat products. Chapter 8 is devoted to the issues behind using freshwater algae biomass in the cosmetics industry. It shows that the presence of bioactive substances in their thalli determines the potential use of freshwater algae biomass in the production of cosmetics. The use of algae and algae extracts in agriculture is presented in Chaps. 9, 10, and 11. Agricultural formulations containing algae extracts stimulate the growth and yield of plants in a very efficient way, because of their efficacy at low concentrations. Concentration of phytohormones depends mostly on the botanical origin of the obtained biomass, time and place of its collection, and the method of extraction of the active compounds. Currently, legislation restricts the use of mineral fertilizers and pesticides and thus forces new approaches for reducing the use of chemical products, either by parallel application or partial replacement with enhanced formulations. Among natural materials of such capability, there are algae that contain a variety of biologically active compounds verified to have a beneficial influence on plants. Depending on their formulation, algae-based products may show the functionality of organic fertilizers or components of organo-mineral fertilizers, soil amendments (improvers), (bio)stimulants, and pesticides. Consumers expect that with the increase in food products, their quality will also improve. This is especially evident in the livestock sector, as the demand for animal protein is systematically increasing. This situation led to a search for innovative products of natural origin that could be used in animal husbandry and breeding. That product could be algae, which contain such ingredients in its biomass that have a positive impact on animal and human organisms. Not only can algae-based feed additives improve production parameters and animal health, but they can also affect the quality of animal products. In Chap. 12, the authors present the use of biomass, as well as extracts derived from said biomass, from the economic and life cycle assessment perspectives. The economic aspects of obtaining algae biomass in the product life cycle are discussed with regard to the bioproducts industry.

Processing algae biomass begins in the environment, is dictated by the variability of species, and ends with the use of the product, whether it is biodiesel, algae cream, growth stimulant, or animal feed ingredient.

We hope this book will be useful for academia, industry, and agriculture, as well as for those who work in sustainable development and ecology and are responsible for shaping the environment for recreation and leisure. The economic and environmental aspects of the processing of algae biomass should be of interest not only to scientists but also to politicians, environmentalists, and business representatives.

Wrocław, Poland Katarzyna Chojnacka
Opole, Poland Piotr Pawel Wieczorek
Poznan, Poland Grzegorz Schroeder
Wrocław, Poland Izabela Michalak

Contents

Introduction: Toward Algae-Based Products

1

Izabela Michalak and Katarzyna Chojnacka

Abstract

This book provides an overview of the importance of different forms of algae (microalgae, as well as marine and freshwater macroalgae) in different applications. Algae as a renewable biomass can be used as a component in the production of a wide array of products, including food, pharmaceuticals, health-related products, nutraceuticals, cosmetics, fine chemicals (dyes and colorants), feed components, feed additives, aquaculture products, and agriculture products (fertilizers, biostimulants for plant growth). Recently, novel and emerging technologies have been developed to convert algal biomass into valuable products and extract biologically active compounds from algae in an efficient and environmentally friendly manner. This book describes the characteristic features of algae cultivation, identification, and production, as well as its subsequent applications. Algae-based products may play an important role in a sustainable future. The development of an algal sector in industry could help to solve many of the problems that modern society is currently facing, including the security of energy and water, food supplies, and climate protection.

Keywords

Algae-based products · Extraction · Bioactive compounds · Utilitarian properties

I. Michalak (✉) · K. Chojnacka
Faculty of Chemistry, Department of Advanced Material Technologies, Wrocław University of Science and Technology, Wrocław, Poland
e-mail: izabela.michalak@pwr.edu.pl

1.1 Introduction

Consumers have developed great interest in algal biomass and its applications in many algae-based products. The ongoing research on the cultivation, harvesting, characterization, processing, and applications of algal biomass in many branches of industry has led to the popularization of algae-based products worldwide (e.g., Priyadarshani and Rath 2012; Enzing et al. 2014; Kim and Chojnacka 2015; Ruiz et al. 2016; Bux and Chisti 2016). Some species of algae are edible and can be consumed by humans. They also constitute an ingredient in animal feed and can be converted into organic fertilizers. Extraction of biologically active compounds from algae offers a new range of products, which can be used in the food, pharmaceutical, cosmetic, and agricultural industries (Fig. 1.1).

The interest in algae is continuously increasing, due to their unique composition. They are known to be a rich source of biologically active compounds, such as oils, fats, polyunsaturated fatty acids, proteins, carbohydrates, minerals, antioxidants (e.g., polyphenols, vitamin C, tocopherols, mycosporine-like amino acids), and pigments (Borowitzka 1995; Michalak and Chojnacka 2015).

As photoautotrophic organisms, algae play the role of primary producers in the biosphere (Bellinger and Sigee 2015). Since algae synthesize their biomass in the process of photosynthesis, they could become a promising, cheap raw material for industry, as well as a renewable source of energy. Algae are characterized by quick growth and a relatively high ability to fix carbon dioxide. This book covers both freshwater and marine macroalgae, as well as microalgae. Their wide range of tolerance is based on efficient adaptation to biochemical processes, as well as their specific cellular structure, which predisposes these biota to growth and development under laboratory and industrial conditions (e.g., Andersen 2005; Demirbas and Demirbas 2010; Enzing et al. 2014; Kim and Chojnacka 2015; Bux and Chisti 2016 etc.).

© Springer International Publishing AG, part of Springer Nature 2018
K. Chojnacka et al. (eds.), *Algae Biomass: Characteristics and Applications*, Developments in Applied Phycology 8,
https://doi.org/10.1007/978-3-319-74703-3_1

The chapters in this book present current aspects of various applications of algal biomass derived from natural and artificial environments. This book is intended for a wide audience interested in new methods of conversion of algal biomass into valuable products and in a variety of goods based on bioactive substances extracted from algal biomass.

1.2 Algae Biomass: Characteristics and Applications

Many environmental benefits stemming from the use of algal biomass in different branches of industry (Chap. 2 "The Environmental Benefits Arising from the Algae Biomass Industry") have arisen. The replacement of nonrenewable raw materials with renewables is a strategy that has gained much attention, due to high energy consumption and increasing CO_2 emissions. Algae are considered potential feedstock candidates, not only for the production of biofuels but also for other products, such as food, feed, biofertilizers, etc., in a biorefinery system (e.g., Demirbas and Demirbas 2010; Das 2015). High value-added products obtained from algal biomass are considered sustainable, renewable, and environmentally friendly. Due to the photosynthesis performed by algae, oxygen and high-energy carbonaceous compounds (biomass) are produced and may be processed into valuable products.

Another advantage of algae is the possibility of their cultivation locally, on nonarable lands and at higher productivity rates than for terrestrial biomass. Additionally, wastewaters and wastes rich in organic and inorganic nutrients may be used for their cultivation. In this way, water and nutrients are preserved and wastewater is treated (Ruiz et al. 2016; Enzing et al. 2014; Das 2015).

In the literature, less attention is paid to freshwater algae than to marine seaweeds and microalgae. A survey of filamentous freshwater macroalgae, their growth forms, morphology, and competitive interactions is presented in Chap. 3 "Biology of Freshwater Macroalgae and Their Distribution." This chapter describes cosmopolitan taxa of freshwater macroalgae (e.g., *Cladophora*, *Oedogonium*) that occur in masses in inland waters with respect to their biology. This abundant occurrence of algae is caused by an increase in water fertility, which itself is caused by the biogenic compounds of phosphorus and nitrogen, stagnant waters with low flow rate, good light conditions, etc. (Smith 2003). Knowledge concerning the factors that influence the fluctuations of macroalgal populations and their metabolism and composition can inform us about the possibilities for their use in various branches of industry (Smith 2003; Bellinger and Sigee 2015; Kim and Chojnacka 2015).

The natural and unique characteristics of algae have opened the door wide for their multidirectional biotechnological use, examples of which are described in Chap. 4 "Algae in Biotechnological Processes." Among the variety of examples, there are two main areas of activity in which algae are involved. The first involves the use of intact algal biomass or manufactured algal extracts that are a rich source of biological substances of required quality (e.g., phycobiliproteins, polyphenols). The second area involves the biochemical metabolism of algae that are able to produce, de novo, a vast array of organic compounds but that are also able to transform existing compounds. This approach allows us to use algae as effective biocatalysts. These properties of algae help to protect ecosystems against organic pollutants and toxic metal ions. Some other examples of the biotechnological potential of algae have been described in previous books,

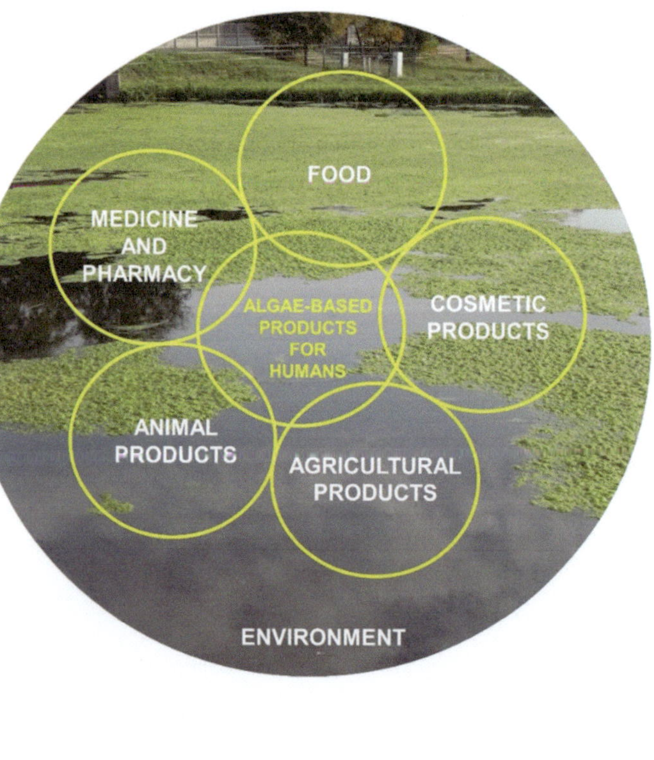

Fig. 1.1 Valorization of algae into high value-added products

e.g., Chen and Jiang (2001), Kim and Chojnacka (2015), and Bux and Chisti (2016).

A chapter on technological issues, Chap. 5 "The Methods of Algal Biomass Extraction Towards to Use of the Extract," includes a description of extraction techniques that are used for the isolation of biologically active compounds from the algal biomass. Problems with algae transport and storage (especially fresh seaweeds) led to the development of techniques that enable biomolecule isolation, including extraction. Various approaches have been successfully applied in extracting biologically active compounds from the algal biomass, among which solvent, pressure, and temperature treatment are the most common. Recently, supercritical fluid extraction, as well as ultrasound-, microwave-, and enzyme-assisted extractions, has also been reported (Hayes 2012; Michalak and Chojnacka 2014; Kim and Chojnacka 2015; Grosso et al. 2015; Bux and Chisti 2016). The methods differ in efficacy, selectivity, and purity of extraction, determining the usability of the final product in either bulk manufacturing or as a high-value material.

1.3 Algae-Based Products

This section is dedicated to a wide array of products that can be produced from algal biomass. In Chap. 6 "Seaweeds As a Component of Human Diet," many examples concerning the application of algae in the human diet are detailed. Algae can be served in different forms – whole seaweed meal, powder, extract, homogenate, or fermented. In this chapter, both the advantages (e.g., nutritional value, accessibility, etc.) and disadvantages (e.g., toxic metals, sensory perception, etc.) of the application of seaweeds as a component of food are discussed. Special attention is paid to their application in cereals, dairy, and meat products. Seaweeds can be also used as a bio-based salt, as well as a source of hydrocolloids, especially in the confectionery industry. Macroalgae can deliver biologically active compounds to food products of either plant or animal origin indirectly, through their use in plant cultivation and animal feed. It is predicted that seaweed as food and seaweed-derived food flavors, colors, and nutrients will continue to attract considerable commercial attention (Chacón-Lee and González-Mariño 2010; Priyadarshani and Rath 2012; Enzing et al. 2014; Das 2015; Bux and Chisti 2016; Kim and Chojnacka 2015; Wells et al. 2017).

In the last few decades, knowledge about the influence of algae in the diet on health and well-being has improved. Scientific research has resolutely confirmed that algae are an inestimable and renewable source of hundreds of chemical compounds, characterized by a wide spectrum of biological activity that can also find application in medical treatment. Chapter 7 "Algae and Their Extracts in Medical Treatment" deals with biologically active compounds, such as polysaccharides, phenolic compounds and their derivatives, pigments, proteins, lipids, and fatty acids, that possess antimicrobial, antiviral, antifungal, antilipidemic, antitumor, antidiabetic, anticoagulant, antioxidant, and antiallergic properties. These compounds can be successfully utilized for human needs in the pharmaceutical (e.g., nutraceuticals, wound healing/dressing, drug delivery, and controlled release) and medicinal fields (e.g., support of the nervous and circulatory system). Biodegradability into environmentally harmless by-products, excellent biocompatibility, the lack of toxicity, and physiological indifference are their main advantages. Although the healing properties of seaweeds have been proved by many researches, as well as in clinical trials, there are still various possibilities for using biological compounds from seaweeds to improve human health and wellness (Hayes 2012; Das 2015; Kim and Chojnacka 2015; Michalak and Chojnacka 2015; Yan et al. 2016).

Besides the application of algae in food and pharmacy, they have also been found to be useful in cosmetic products – Chap. 8 "Application of Algae Biomass and Algal Extracts in Cosmetic Formulations." In the literature, mainly marine macroalgae are described as a source of ingredients for cosmetic purposes. However, as is shown in this chapter, interdisciplinary research has proven that freshwater macroscopic green algae species (e.g., *Chara fragilis*, *Cladophora glomerata*, *Ulva flexuosa*) may also turn out to be a rich source of macro- and micronutrients and other bioactive substances, such as fatty acids, polysaccharides, pigments, polyphenols, etc., that can be used in cosmetics. Freshwater macroalgae are a rare object of study and are practically nonexistent within the cosmetics market, despite their high potential (Priyadarshani and Rath 2012; Wang et al. 2015; Kim and Chojnacka 2015).

Beside their use in products for humans, algae and algal extracts can also be used in agriculture – in plant cultivation as fertilizers, biostimulants for plant growth, and in animal feed as additives. Currently, agricultural legislation limits the use of mineral fertilizers and plant protection products and enforces a new approach to decreasing the use of conventional preparations by applying treatment-enhancing formulations. The application of algal extracts in agricultural production is described in Chap. 9 "The Biomass of Algae and Algal Extracts in Agricultural Production" and Chap. 10 "Algae As Fertilizers, Biostimulants and Regulators of Plant Growth." There are a few forms of algal preparation suitable for agricultural use, including raw biomass, compost, dried meal, and extracts. Fertilizers and plant growth regulators (PGRs) based on seaweeds are commonly known, as they have already been produced and applied for a number of years. Algal-based materials have been confirmed to provide crops with nutrients (as fertilizers), increase biomass production, and activate the natural ability of plants to respond properly to stress agents (such as biostimulants and bioregulators). Agricultural formulations containing algal extracts stimulate the growth and yield of plants in a very efficient way, because of their action at low concentrations. Concentration of phytohormones depends mainly on the botanical origin of the obtained biomass, the time and place of its collection, and the method of extraction of the active compounds. Plant growth regulators are mainly obtained from brown algae, due to the relatively high content of those substances and their long accessibility during the year. Algal extracts are often enriched with such substances as urea, humic acids, ammonium phosphate, potassium sulfate, and additional doses of growth hormones. Depending on the species and living environment of an alga and the treatment method involved, the effect of biomolecules on crops varies. The composition of such products may influence not only growth and development processes in plants but also indirect factors, such as soil fertility or the presence of soil microorganisms (Khan et al. 2009; Craigie 2011; Kim and Chojnacka 2015).

Similarly as in plant cultivation, in regard to animal production, consumers expect that, with the increase in food products, their quality will also improve. This is particularly important in the livestock sector, as the demand for animal protein is systematically increasing. This situation led to a search for innovative products of natural origin that could be used in animal husbandry and breeding. Such a product could be algae containing ingredients in their biomass that have a positive impact on animal and human organisms – Chap. 11 "The Algae Biomass in Animal Production." Algae-based feed additives not only improve production parameters and animal health but also affect the quality of animal products. Several studies have been done to develop algae in feed for poultry, pigs, cattle, and horses. These studies have shown that the use of algae as feed additives can bring many benefits, due to their unique properties (Evans and Critchley 2014; Enzing et al. 2014; Kim and Chojnacka 2015; Bux and Chisti 2016).

Another distinctive approach concerning the economic aspects of the utilization of algal biomass in bio-based industry is presented in Chap. 12 "Economic Aspects of Algal Biomass Harvesting for Industrial Purposes: The Life Cycle of the Product." Special attention is paid to the chemical composition of the algae biomass obtained from the natural environment, cultivated under natural conditions or in bioreactors, which determine their application as energy or fuel, or in the cosmetic industry. The use of biomass, as well as extracts derived from said biomass, is discussed in regard to the economic aspect and assessment of the life cycle.

1.4 Conclusions

The ongoing researches concerning the cultivation, harvesting, pretreatment, and conversion of algal biomass have fostered the development of new algal-based products. Additionally, new algae species, especially from natural environments with unique properties, are still being identified. Moreover, advanced research techniques are enabling the identification of biologically active compounds in algae. All of these factors have combined to make the number of algal products currently on the market very large, and only getting larger. In the present book, the latest trends in the use of algae in many formulations – functional food, pharmaceuticals, cosmetics, agricultural preparations for both plants and animals – are presented.

References

Andersen RA (2005) Algal culturing techniques. Elsevier Academic Press, London

Bellinger EG, Sigee DC (2015) Freshwater algae: Identification, enumeration and use as bioindicators, 2nd edn, Wiley-Blackwell, Hoboken

Borowitzka MA (1995) Microalgae as sources of pharmaceuticals and other biologically active compounds. J Appl Phycol 7:3–15

Bux F, Chisti Y (2016) Algae biotechnology: products and processes. Springer, Cham

Chacón-Lee TL, González-Mariño GE (2010) Microalgae for "healthy" foods – possibilities and challenges. Compr Rev Food Sci Food Saf 9:655–675

Chen F, Jiang Y (2001) In: Chen F, Jiang Y (eds) Algae and their biotechnological potential, Proceedings of the 4th Asia-Pacific Conference on Algal Biotechnology, 3–6 July 2000, Hong Kong. Kluwer Academic Publications, Dordrecht

Craigie JS (2011) Seaweed extract stimuli in plant science and agriculture. J Appl Phycol 23(3):371–393

Das D (2015) Algal Biorefinery: an integrated approach. Capital Publishing Company, Springer, Cham

Demirbas A, Demirbas MF (2010) Algae energy – algae as a new source of biodiesel. Springer, London

Enzing C, Ploeg M, Barbosa M et al (2014) Microalgae-based products for the food and feed sector: an outlook for Europe, EUR – Scientific and Technical Research Reports, Joint Research Centre. Publications Office of the European Union, Luxembourg. https://doi.org/10.2791/3339,5-73. Available from: http://publications.jrc.ec.europa.eu/repository/bitstream/JRC85709/final%20version%20online%20ipts%20jrc%2085709.pdf. Accessed 24 Aug 2017

Evans FD, Critchley AT (2014) Seaweeds for animal production use. J Appl Phycol 26:891–899

Grosso C, Valentão P, Ferreres F, Andrade PB (2015) Alternative and efficient extraction methods for marine-derived compounds. Mar Drugs 13(5):3182–3230

Hayes M (2012) Marine bioactive compounds: sources, characterization and applications. Springer, New York

Khan W, Rayirath UP, Subramanian S et al (2009) Seaweed extracts as biostimulants of plant growth and development. J Plant Growth Regul 28:386–399

Kim S-K, Chojnacka K (2015) Marine algae extracts: processes, products, and applications. Wiley-VCH Verlag GmbH & Co. KGaA, Weinheim

Michalak I, Chojnacka K (2014) Algal extracts: technology and advances. Eng Life Sci 14:581–591

Michalak I, Chojnacka K (2015) Algae as production systems of bioactive compounds. Eng Life Sci 15:160–176

Priyadarshani I, Rath B (2012) Commercial and industrial applications of micro algae – a review. J Algal Biomass Util 3(4):89–100

Ruiz J, Olivieri G, de Vree J et al (2016) Towards industrial products from microalgae. Energy Environ Sci 9:3036–3043

Smith VH (2003) Eutrophication of freshwater and coastal marine ecosystems a global problem. Environ Sci Pollut Res 10(1):1–14

Wang HM, Chen CC, Huynh P, Chang JS (2015) Exploring the potential of using algae in cosmetics. Bioresour Technol 184:355–362

Wells ML, Potin P, Craigie JS, Raven JA et al (2017) Algae as nutritional and functional food sources: revisiting our understanding. J Appl Phycol 29(2):949–982

Yan N, Fan C, Chen Y, Hu Z (2016) The potential for microalgae as bioreactors to produce pharmaceuticals. Int J Mol Sci 17(6):962. 1–24

2 The Environmental Benefits Arising from the Use of Algae Biomass in Industry

Małgorzata Mironiuk and Katarzyna Chojnacka

Abstract

The replacement of nonrenewable raw materials with renewables is a strategy that has gained much attention in the face of high-energy consumption and increasing CO_2 emissions. The introduction of algae biomass into industry could contribute to solving some of the main challenges that modern society faces: energy security, precarious water and food supplies, and climate change. Algae are considered as potential feedstock candidates for many products, such as food, feed, biofuels, biofertilizers, and cosmetics. Goods obtained from algae biomass are considered sustainable, renewable, and environmentally friendly, as they are generally formed through photosynthesis and use atmospheric CO_2 and sunlight to produce oxygen and high-energy carbonaceous compounds (i.e., biomass) that can be transformed into valuable products. They can be produced locally on non-arable lands. An additional benefit of the application of algae is their productivity rates, which are higher than those of terrestrial biomass, such as corn. It has been suggested that wastewaters and wastes rich in organic and inorganic nutrients may be used in place of freshwater and fertilizers in algae cultivation. Thus, the utilization of waste and wastewaters to cultivate algae could simultaneously solve the problems of freshwater demand, the high cost of nutrients, and the need to remediate waste.

Keywords

Renewable energy · Wastewater treatment · Climate change · Algae industry · Environmentally friendly technologies

M. Mironiuk (✉) · K. Chojnacka
Faculty of Chemistry, Department of Advanced Material Technologies, Wrocław University of Science and Technology, Wrocław, Poland
e-mail: malgorzata.mironiuk@pwr.edu.pl

2.1 Introduction

Environmental pollution is a result of an increase in population, urbanization, technological advancement, and the rapid growth of industrialization. All of these factors have created negative impacts on each component of the environment. A number of pollutants, such as fertilizers, pesticides, and heavy metals, can have serious effects on human beings (Bharti 2012).

Algae are a class of photosynthetic organisms that exhibit very high biological diversity and metabolic elasticity. Compared to terrestrial plants, they are far better able to adapt their biochemical metabolic pathways and cellular composition in response to external conditions (Chen et al. 2015; Laurens et al. 2017). Many scientific publications report that the microalgae efficiency of CO_2 fixation is 10–50 times higher compared to terrestrial plants (Bhola et al. 2014; Chen et al. 2015). In a realistic approach, the maximum photosynthetic efficiency of microalgae is estimated at 4.5% (Walker 2009).

Their growth is strongly influenced by several cultural parameters: nutrient quality and quantitative profiles (especially carbon, nitrogen, and phosphorus), pH, temperature, light supply, dissolved oxygen and CO_2, and the presence of toxic elements in the medium (Pires 2017). Commercial production of microalgae on a large scale costs more than traditional crop production, as a result of the consumption of nutrients and water essential for algae growth (Chen et al. 2015).

There are two different types of alga, namely, micro- and macroalgae (seaweeds) (Anastopoulus and Kyzas 2015). Both macro- and microalgae are used as a source of food, for production of biofuels (oil, biodiesel, bioethanol, biohydrogen, and biogas), as stabilizing agents (agar, carrageen), as a biofertilizer, as a source of nutrients, and for controlling pollution (wastewater treatment, reduction of CO_2 emissions) (Bixler and Porse 2011; Ibañez and Cifuentes 2013; Singh and Sharma 2012; Singh and Singh 2014).

© Springer International Publishing AG, part of Springer Nature 2018

K. Chojnacka et al. (eds.), *Algae Biomass: Characteristics and Applications*, Developments in Applied Phycology 8,
https://doi.org/10.1007/978-3-319-74703-3_2

Considering the high biodiversity of microalgae, the development of genetic and metabolic engineering, and the development of detailed culture and screening techniques, this group of organisms is one of the most promising sources for new products and applications (Harun et al. 2010).

2.2 Capturing of CO_2

One of the biggest threats of the modern world is climate change, which is associated with increasing emissions of greenhouse gases, especially carbon dioxide. The increase in CO_2 emissions over the last few decades is shown in Fig. 2.1.

The increase of atmospheric CO_2 concentrations causes global warming and ocean acidification, which has a negative impact on biodiversity (Brierley and Kingsford 2009). Climate change is manifested in a rise of temperatures and sea levels or more extreme weather events and a change of precipitation patterns. All of this has a negative impact on agriculture, water supplies, food production, ecosystems, energy security, and infrastructure. Therefore, climate change demands action, and said action is one of the Sustainable Development Goals. An increase in the supply of clean energy and reduction of emissions of greenhouse gases are necessary to avoid most environmental problems (Nascimento et al. 2015; World Bank 2017).

Among the methods of CO_2 removal are land use management, the use of biomass as a carbon neutral energy source, enhancement of natural weathering processes to capture atmospheric CO_2, direct engineered capture (physicochemical processes), and enhancement of oceanic CO_2 uptake. Biological fixation of CO_2 is the only process that ensures both clean energy and a reduction in atmospheric CO_2 at the same time (Pires et al. 2014).

Photosynthesis is the practical form of CO_2 capturing. One option that has high potential to be an important sink of CO_2 is photoautotrophic algae (Pires 2017). Algae are responsible for 50% of the photosynthetic processes that take place on Earth (Singh and Singh 2014). Their ability to uptake and transport bicarbonate into cells makes them well suited to capturing carbon dioxide. Thanks to their photosynthetic activity, CO_2 may be fixed from the atmosphere, industrial exhaust gases, and soluble carbonate. Captured carbon is transported into algal cells by bicarbonate transporters present in the plasma membrane and chloroplast envelope. Inside the chloroplast, bicarbonate is converted into CO_2. In *Cyanobacteria* and eukaryotic microalgae, the assimilation of inorganic carbon into organic matter microalgae is based on the Calvin cycle (photosynthetic carbon reduction cycle) and enzyme Rubisco (ribulose-1,5-bisphosphate carboxylase oxygenase), which catalyzes the fixation of CO_2 to the acceptor molecule RuBP (ribulose-1,5-bisphosphate) and the oxygenation of that same molecule. To convert CO_2 into carbohydrates, lipids, and proteins, algae use solar energy. Because of their greater abilities to capture light and convert it into usable chemical energy, algae's photosynthetic efficiency is up to ten times higher than that of terrestrial plants. Each kilogram of microalgae corresponds to about 1.83 kg fixed CO_2 (Jorquera et al. 2010; Sayre 2010; Cai et al. 2013; Pires et al. 2014; Beardall and Raven 2016; Pires 2017).

Algae growth is divided into two main classes: autotrophic and heterotrophic. Autotrophic growth directly consumes CO_2 as a carbon source in the presence of light as an energy source. For large-scale production, CO_2 can be supplied from an upstream power plant or another emission source, which provides effective capture and recycle of carbon. In a heterotrophic pathway, algae use organic substances as a source of carbon and energy used for growth. At a large scale, heterotrophic cultivation is cheaper and easier than autotrophic (Davis et al. 2011; Perez-Garcia et al. 2011; Morales-Sánchez et al. 2016). Some algae are mixotrophic, able to utilize both organic and inorganic carbon substrates. The possibility of utilizing both processes at the same time leads to the increased productivity of algae (Bhatnagar et al. 2011).

The efficiency of CO_2 capture is related to the physiology of the algae, temperature, and the amount of CO_2 and other

Fig. 2.1 Annual worldwide emission of CO_2. (http://www.data.worldbank.org; searched on 20 June 2017)

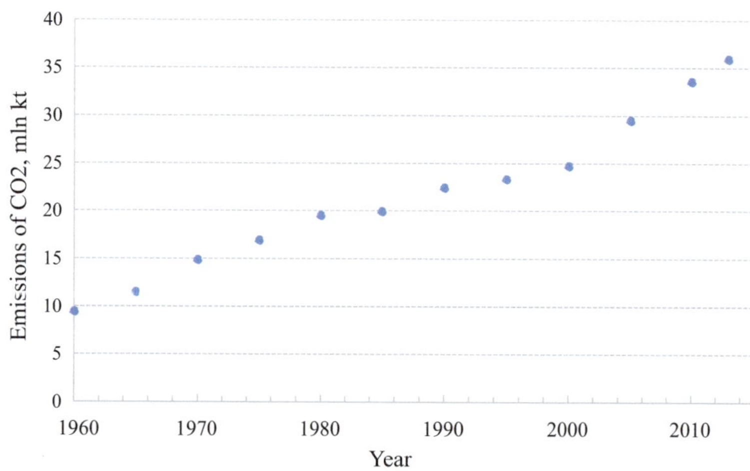

gases, such as SO_x and NO_x. Efficiency at the level of 80–99% can be achieved under optimal conditions and with gas residence times as short as 2 s (Sayre 2010; Bhola et al. 2014).

Some microalgae species are able to grow at very high CO_2 concentrations and have a high efficiency of CO_2 capturing (Singh and Singh 2014). The captured carbon is incorporated into carbohydrates and lipids and therefore stores energy and produces chemicals and foods (Chen et al. 2015). Commercial production of algae biomass has great potential to contribute to world energy supplies and control emissions of CO_2 (Beema Jainab et al. 2014).

2.3 Wastewater Treatment

Global annual freshwater consumption was estimated at almost 3986 billion m^3 during 2014 (World Bank Open Data). Most of the consumed water is turned into wastewater of different chemical compositions, depending on the source. For example, it can be rich in ammonium and organic nitrogen (wastewater from animal farms) or contain less nitrogen and phosphorus but more heavy metals (municipal wastewater). For the discharge and reuse of wastewater, the nutrients and toxic metal ions must be removed to acceptable limits (Junzhuo et al. 2016; Luo et al. 2017).

The current approach to the removal of metal ions and other inorganic compounds in wastewater is not simply about the recovery of high-value by-products but also about reducing health concerns and environmental pollution caused by their disposal (WWAP 2017). Improving water quality and reducing the proportion of untreated wastewater are necessary to achieve the Sustainable Development Goals (World Bank 2016).

The methods of wastewater treatment have been classified into three categories: physical, chemical, and biological. They can be combined into a variety of systems (primary, secondary, and tertiary wastewater treatment) to achieve different levels of contaminant removal. Chemical methods have serious long-term environmental effects and are expensive. The most environmentally friendly and least expensive method of wastewater treatment is the class of biological treatment processes, which use microorganisms to break down the chemicals in wastewater and valorize the residues through the production of added-value compounds (Rawat et al. 2011; Bhattacharjee and Siemann 2015). It has been shown that microalgae are efficient in removing nitrogen, phosphorus, and toxic metals from a wide variety of wastewaters (municipal, agricultural, and industrial), unlike the physical and chemical methods, which are not economical for the treatment of agricultural wastewater (Luo et al. 2017). The efficiency of nitrogen and phosphorus removal from wastewater is affected by the N/P ratio and initial nutrient concentration (Boonchai et al. 2012).

Because of the small size of microalgae and quite high dilution of the cultures, one of the limitations of the development of microalgae-based systems of wastewater treatment is a problem with the harvest and separation of the biomass at the end of the process. Current algae harvesting techniques (chemical, mechanical, electrical, and biological) require a great deal of both chemicals and energy, which limits their use in large-scale algae cultivation. The use of an immobilized-cell system over a cell-free system could make this process easier (Ruiz-Marin et al. 2010; Cai et al. 2013).

Microalgae, which have a high capacity for inorganic nutrient uptake, can be proposed as an alternative to the biological treatment of wastewater. It has been proven that microalgae have the potential to efficiently accumulate nutrients and metal ions when the source of organic carbon is not present (Ruiz-Marin et al. 2010; Abdel-Raouf et al. 2012). The biomass can be used for the production of various other value-added products, such as bioethanol or biomethane, or for liquid fuel production after thermal conversion (Rawat et al. 2011). Algae in wastewater treatment have been used and described in the literature for many years (Oswald et al. 1957; Bartsch 1961; Oswald 1988). This approach is environmentally friendly and is in the forefront of sustainable development (Rawat et al. 2011).

2.3.1 Recovery of Nutrients from Wastewater

The total concentration of nutrients in wastewater, especially nitrogen and phosphorus, can be very high. Biogenic compounds mainly occur in wastewater in the form of ammonia, nitrite, nitrate, organic nitrogen, and orthophosphate, and their concentration in some industrial and agricultural wastewater can be higher by as much as three orders of magnitude in comparison with natural water. Without the proper treatment, wastewater may overload the bio-system, having a negative impact on the natural processes of photosynthesis, respiration, nitrogen fixation, evaporation, and precipitation. Nutrients can be released into aquatic systems. This leads to eutrophication and ecosystem damage through stimulation of the growth of unwanted plants, such as algae and aquatic macrophytes. Furthermore, non-ionized ammonia is toxic to fish and other aquatic organisms (Abdel-Raouf et al. 2012; Kumar et al. 2013; Chen et al. 2015).

Removal and recovery of nutrients from wastewater mostly involves cost-effective treatment technologies. Many methods for nutrient removal and recovery from wastewater, such as membrane-based technologies, chemical preparation, and bioelectrochemical systems (BES), have been described (Zhang et al. 2014).

Eukaryotic algae play a key role in removing nutrients from wastewater through an assimilation process. The history of their use in removing nitrogen and phosphorus is quite extensive, said use having first been proposed over 50 years ago (Abdel-Raouf et al. 2012). The first research on the use of microalgae assumed that their productivity using recycled wastewater nutrients would be several times greater than that of conventional food plants, particularly in regard to efficiency. Although the productivity of algae turned out to be much lower than originally expected, they are still commonly used in wastewater treatment (Oswald 2003).

Microalgae are able to remove nutrients effectively because they require high amounts of nitrogen and phosphorus for protein, nucleic acid, and phospholipid synthesis (Rawat et al. 2011). The mechanisms for removing nitrogen can be based on nitrogen uptake by algae biomass and the decomposition of ammonia resulting from an increase in pH during algae growth. Phosphorus can be removed from wastewater as a result of bioassimilation, adsorption, and chemical precipitation processes above pH 8 (Wang et al. 2014). The oxygen released by the microalgae can be consumed by bacteria so as to decompose the organic matter in wastewater, giving out carbon dioxide, ammonia, and phosphate, which the microalgae subsequently assimilate and convert into biomass (Choi and Lee 2012; Cuellar-Bermudez et al. 2017).

There are two main commercial systems for algae cultivation: open raceway ponds – simple, low cost and the most common technology for large scales – and closed photobioreactors, which are mostly used for inocula or high-value products (Singh and Sharma 2012; Prussi et al. 2014). Cultivation of algae on a large scale demands energy, water, and significant amounts of nutrients, e.g., algae need 6–8 tons·ha^{-1}·year^{-1} of nitrate as a nitrogen source, which is 55–111 times more than field crops. Nutrients are normally derived from chemical or organic fertilizers, which is a major reason for the high production cost of algae (Chen et al. 2015, Kligerman and Bouwer 2015; Ravindran et al. 2016). A promising way to reduce the costs of algae cultivation is to replace the expensive synthetic chemicals with cheap resources (Lu et al. 2015). In wastewater, the majority of nutrients required for algae cultivation are already present, which makes it an attractive and economical alternative source for growing algae. Algae have a high potential to remove nutrients, mainly nitrogen and phosphorus, from wastewater and to accumulate biomass, which makes the wastewater treatment and algae biomass production process economically viable and cost-effective (Chen et al. 2015; Unc et al. 2017).

The effectiveness of algae use for removing nutrients from wastewater has been confirmed in many studies (Abdel-Raouf et al. 2012). The research shows that *Scenedesmus*, *Chlorella*, and even *Arthrospira*, which is a marine form, could grow in wastewater and produce significant amounts of biomass, with the water after algae removal containing far fewer nutrients than the supplied water. Use of algae allowed for removing more than 90% of nitrogen and more than 50% of phosphorus from wastewater (Bhattacharjee and Siemann 2015). In the study by Choi and Lee, the effectiveness of nitrogen and phosphorus removal by *Chlorella vulgaris* was on the level of 81–85% and 32–36%, respectively (Choi and Lee 2012). Henkanatte-Gedera et al. (2017) have shown the ability of *Galdieria sulphuraria* to remove ammonia nitrogen and phosphorus with efficiencies ranging from 63–89% to 71–95%, respectively. Wang et al. (2014) have shown that the green algae *Micractinium* sp. can be used for nutrient removal with good efficiency as well. The final concentrations of soluble phosphorus and nitrogen were reduced by more than 95% and 94%, respectively.

The bioassimilation process can be affected by many factors, such as light condition, temperature, availability of CO_2 purging, pH, salinity, and the initial concentration of nutrients, which can be very important factors for regulation of the growth characteristics of algae and determination of the subsequent removal kinetics. Each strain of microalgae prefers different conditions, nutrient forms, and N/P ratios suitable for growth, showing varied efficiency in nutrient removal and different resistance to predation. Potentially, the filamentous algae with large cells or colony size and indigestible cell walls, such as *Cladophora*, *Oedogonium*, and *Spirogyra*, have an advantage in wastewater treatment over unicellular species. They are more effective in biomass production and nutrient removal. They also have the ability to grow while attached to substrate and are easy to harvest. Defined physical parameters, selection of appropriate algae strains, and good understanding of the ecology of algae communities are essential for the treatment of a certain type of wastewater and for efficient production of biomass (Junzhuo et al. 2016; Wang et al. 2014).

From resources of nutrients dissolved in wastewater, microalgae are able to produce such high-value products as transportation biofuels, biodegradable bioplastics, biochemicals, nutrition supplements for humans and animals, antioxidants, and cosmetic ingredients (Luo et al. 2017) (Fig. 2.2).

The concept of producing biofuel is based on the conversion of nutrients in wastewater into microalgae biomass, which can be further transformed into biofuel. The algae are able to produce almost 300 times more oil per acre than traditional crops. The basis of the integrated wastewater treatment and biofuel production system is the selection of appropriate species of algae, the requirement of modulated illuminating conditions, high rates of nutrient removal, enhanced biomass productivity, and the capability of efficient harvesting of biomass. This approach can be used for wastewater treatment, capturing CO_2 and producing alternative sustainable energy, and could exceed the cost of produc-

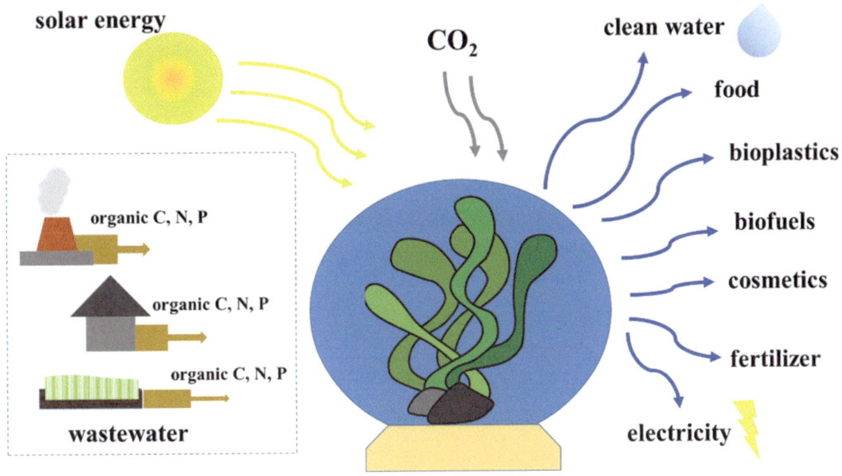

Fig. 2.2 Cultivation of algae – raw materials and products. (On the basis of Luo et al. 2017)

tion by at least 10% (Bhattacharjee and Siemann 2015; Kligerman and Bouwer 2015; WWAP 2017).

Biodegradable plastics, which are produced from wastewater algae, have the potential to replace traditional petroleum-based plastics at lower costs. Algae-derived oils are also useful in the production of cosmetics and medical products (WWAP 2017).

2.3.2 Biosorption of Heavy Metals

Pollution of the environment with heavy metals has become a serious problem. In the last 50 years, human exposure to heavy metals has risen dramatically as a result of an exponential increase in the use of such metals in industrial processes and products. Anthropogenic activities such as mining and smelting operations, industrial sources (metal processing in refineries, coal burning in power plants, petroleum combustion, etc.), domestic and agricultural use of metals and their compounds, atmospheric deposition, soil erosion, weathering, and volcanic eruptions have been reported to significantly contribute to heavy-metal pollution. This has led to an increase in the level of water contamination (Aziz et al. 2008; Bharti 2012; Tchounwou et al. 2012; He and Chen 2014; Kumar et al. 2015).

The most common pollutants found in contaminated water are arsenic, cadmium, chromium, copper, lead, mercury, nickel, and zinc ions. Metals are unique among pollutants, because they cause adverse health effects, occur naturally, and are ubiquitous in the environment. Some of them, like copper, zinc, manganese, nickel, and cobalt, are important for the biochemical and physiological functioning of plants and animals (essential heavy metals), but for many of them, there is a very narrow range of dose between beneficial and toxic effects. Other metals, e.g., aluminum, antimony, arsenic, cadmium, lead, mercury, tin, or titanium, called nonessential metals, have unknown biological functions (Aziz et al. 2008; Bharti 2012; Tchounwou et al. 2012; He and Chen 2014; Kumar et al. 2015). Due to their toxicity to higher life forms and their mobility in aquatic ecosystems, heavy metals are considered the most significant inorganic contaminants in the environment (Kumar et al. 2015).

Removal of heavy metals from the environment is difficult because they cannot be degraded chemically or biologically and their discharge into surface waters has a negative impact on the aquatic system, causing destruction of its ability for self-treatment (Kumar et al. 2015). Unlike organic pollutants, heavy metals are not biodegradable and tend to accumulate in living organisms (Fu and Wang 2011). Arsenic, cadmium, chromium, lead, and mercury rank first among the metals that are of great significance for human health because of their high degree of toxicity. They are all toxicants that are known to induce multiple organ damage, even at lower levels of exposure. They may enter the human body along with food, through water and air, or by absorption through the skin (Bharti 2012), interacting with cell components such as DNA and nuclear proteins and causing damage to DNA structure and conformational changes that may lead to cell-cycle modulation, carcinogenesis, or apoptosis (Tchounwou et al. 2012).

Many technologies have been developed for the removal of heavy metal ions, such as chemical precipitation and coagulation, ion exchange, electrochemical methods, adsorption, membrane filtration, electrodialysis, and photocatalysis. However, very often, the conventional methods are ineffective or very expensive (requiring large amounts of reagents and energy) and produce toxic sludge or other waste products; subsequently, they are generally only used for low concentrations of heavy metals (Nguyen et al. 2013; Kumar et al. 2015; Wang et al. 2017).

Biotechnological methods, such as biosorption that uses the ability of some microorganisms to tolerate and interact with metal ions, are becoming attractive alternatives for removing toxicants from wastewater (Dixit and Singh 2014).

Microorganisms, including bacteria, algae, fungi, and yeast, are capable of accumulating heavy metals with high efficiency (Javanbakht et al. 2014). Algae, as a renewable biomass with different affinities toward different metal ions, are one of the most important biosorbent materials (Anastopulus and Kyzas 2015). Biosorption is a low-cost and high-efficiency process, and therefore has been found to be an economically feasible alternative and one of the most promising methods for heavy metal ion removal. The biosorption mechanism involves ion exchange, wherein ions such as sodium, magnesium, or calcium become displaced by heavy metal ions and complexation between metal ions and functional groups takes place. The process of biosorption of heavy metal ions occurs in two stages. In the first stage (passive removal), rapid uptake takes place due to surface adsorption on the cell wall components. This is a non-metabolic process, essentially reversible and occurring in both living and nonliving cells. The second phase (active uptake) involves slow uptake due to membrane transport of metal ions to the cytoplasm of the cells and subsequent accumulation inside the cell. This is a metabolism-dependent process occurring within the cell, and it is slow, restricted to living cells only, and usually irreversible. The carboxyl, hydroxyl, sulfate, and amino groups in algal cell-wall polysaccharides act as binding sites for metals (Gupta and Rastogi 2008a; Lesmana et al. 2009; Shalaby 2011; Dixit and Singh 2014; Kumar et al. 2015). Initial concentration of heavy metal ions in the solution, contact time, pH, biomass concentration, and temperature affect the course and efficiency of the biosorption process (Kanchana et al. 2014). Many different types of alga have the properties for heavy metal ion removal. Research proves that various species of algae biomass, such as *Arthrospira*, *Nostoc*, *Gelidium*, *Ulva*, *Oedogonium* sp., *Cystoseira*, *Sargassum*, etc., are good sorbents of heavy metals (Gupta et al. 2015).

Richards and Mullins (2013) showed that the mixture of microalga species, *Microchloropsis gaditana*, *Diacronema lutheri*, *Tetraselmis chui*, and *Chaetoceros muelleri*, is able to remove over 95% of the metal ions (aluminum, iron, manganese, barium, cerium, lanthanum) from leachate after 10 days.

Dried *Spirogyra* was found to be a good sorbent for bioremediation of chromium and copper. Under optimal conditions, the alga was able to bind 78% of chromium and even 78–84% of copper ions (Chatterjee and Abraham 2015; Kushwah and Srivastav 2015). *Sargassum fusiforme* was able to adsorb 90% of copper and 72% of mercury from the aquatic solutions during the first 60 min. The adsorption was pH dependent, and the best results were achieved at pH 8 and 10 for both metals (Huang and Lin 2015).

Gupta and Rastogi (2008a, b) showed that the biomass of two dried freshwater algae, *Oedogonium* sp. and *Nostoc* sp., has the ability to remove lead ions from aqueous solutions. In addition, the biomass of green algae *Oedogonium* sp. has the potential to be used as an efficient and economic biosorbent material for the removal of cadmium ions from water as well.

Other studies by Gupta and Rastogi (2009) showed the ability of raw and acid-treated *Oedogonium* sp. in the biosorption of hexavalent chromium with an efficiency similar to that of *Palmaria palmata* (red algae) and *Ulva* spp. (green algae) and greater than that of *Fucus spiralis* (brown algae), *Chlamydomonas reinhardtii*, and *Spirogyra* sp.

The study by Dixit and Singh (2014) showed that cyanobacterium *Nostoc muscorum* has a high capability for cadmium and lead ion removal and that it can be used as an efficient biosorbent for the removal of heavy metal ions from wastewater. They showed that 85% of lead and 78% of cadmium remained adsorbed onto a cell's surface and only a small fraction of metal ions was taken up through the metal transport system into the intracellular compartment. It confirms that living biomass is capable of binding a large fraction of metal ions onto cellular surfaces, and therefore can be recovered by using an appropriate desorbing agent so that algae biomass can be reused again (Dixit and Singh 2014).

In Table 2.1, the capacities for biosorption of heavy metal ions from wastewater by different species of alga are shown.

Biosorption of heavy metal ions from wastewater could be carried out using living algal cells, which need minimal nutrients and environmental conditions, or dead biomass of algae, which does not require specific nutrients or oxygen. It is possible to use algae in both aerobic and anaerobic systems (Kumar et al. 2015). Biosorption using the algae biomass is a reversible process, and the algae biomass can be used repeatedly for the removal of heavy metal ions from aqueous solutions (Gupta and Rastogi 2008a). In comparison with the conventional methods, the technology of biosorption has many advantages, such as high efficiency in the detoxification of heavy metals, low operating costs, a lower amount of spent biosorbent for final disposal, and no nutrient requirements (He and Chen 2014).

2.3.3 Algae As a Bioindicator of Aquatic Pollution

Establishing the level of pollutants in an environment is necessary for evaluating their possible toxicity at different levels of the trophic chain (Volterra and Conti 2000). Photosynthetic organisms have intra- and extracellular mechanisms for metal ion detoxification, some of which are able to play the role of biomarkers of exposure, i.e., quantitative measures of changes in the biological system that may be caused by exposure to the toxic effects of environmental chemicals (Shalaby 2011; Ismail et al. 2017). Biomonitoring of environmental conditions can be carried out in two ways.

Table 2.1 The capacities for biosorption of heavy metal ions by dry biomass of different alga species

Algae	Metal	Capacity (mg/g)	References
Arthrospira	Cd	357	Solisio et al. (2008)
Nostoc sp.	Pb	94	Gupta and Rastogi (2008a)
Nostoc muscorum	Cr	23	Gupta and Rastogi (2008d)
Gelidium corneum	Cd	18	Vilar et al. (2006)
Palmaria palmata	Cd	4.8	Prasher et al. (2004)
	Cr	34	Abdi and Kazemi (2015)
	Cu	6.6	Prasher et al. (2004)
	Ni	3.0	
	Pb	15	
	Zn	28	
Chlorella vulgaris	Ni	60	Aksu (2002)
	Zn	17	Abdi and Kazemi (2015)
Parachlorella kessleri	Cu	37	Horvathova et al. (2009)
	Zn	40	
Cladophora spp.	Pb	46	Abdi and Kazemi (2015)
Ulva lactuca	Cr	28	Abdi and Kazemi (2015)
Ulva ohnoi	Cd	62	Suzuki et al. (2005)
	Zn	75	
Oedogonium sp.	Cd	88	Gupta and Rastogi (2008b)
	Pb	145	Gupta and Rastogi (2008a)
Oedogonium hatei	Cr	31	Gupta and Rastogi (2009)
	Ni	41	Gupta et al. (2010)
		41–44	Vijayaraghavan et al. (2006)
Spirogyra	Pb	140	Gupta and Rastogi (2008c)
	Cu	133	Gupta et al. (2006)
	Cr	15	Gupta et al. (2001)
Fucus ceranoides	Cd	90	Herrero et al. (2006)
Fucus evanescens	Zn	52	Abdi and Kazemi (2015)
Fucus spiralis	Cd	115	Romera et al. (2007)
	Cu	115	
	Ni	50	
	Pb	204	
	Zn	53	
Cystoseira baccata	Hg	329	Herrero et al. (2005)
Saccharina japonica	Zn	92	Abdi and Kazemi (2015)
Sargassum fusiforme	Cu	31	Huang and Lin (2015)
	Hg	7.7	
	Pb	266–303	Chen and Yang (2005)
	Cu	87	
	Cr	53	Abdi and Kazemi (2015)
Sargassum muticum	Cr	196	Abdi and Kazemi (2015)
Sargassum siliquosum	Cr	66	

Environmental monitoring programs are mainly based on the use of native organisms (active approach). In the second way (passive approach), the organisms are translocated to a given environment for a specified period of time (Farias et al. 2017). Algae may play the role of bioindicators of structural components of ecological integrity and functional integrity (Stancheva and Sheath 2016). Algae are directly affected by physical and chemical factors, they accumulate pollutants easily, and their metabolism is sensitive to the variation of environmental and natural fluctuations (Omar 2010). Because of their natural and widespread occurrence along worldwide seashores, algae could be useful for a transfer of the ecosystem response to exposure to toxic compounds (Torres et al. 2008).

The biomonitoring methods that use algae are based on the assumption that a healthy environment is typified by a greater diversity of organisms than that found in degraded environments (Omar 2010). The algae response can be carried out at the level of a biomarker, short-term indicators of exposure to environmental stress, or a bioindicator, long-term indicators that reflect the health status of an aquatic system. Biomarkers that give quick answers are able to act as an early warning system for the monitoring of environmental changes, while bioindicators have a high ecological relevance and the ability to analyze environmental samples at any time after collection (Belinger and Sigee 2015). Both macro- and microalgae are important tools for monitoring heavy metal ion pollution, persistent organic pollutants, polycyclic aromatic hydrocarbons, pesticides, and polychlorinated biphenyls from the euphotic zone by direct sinking of the cells (Torres et al. 2008). One of the requirements of the Water Framework Directive (WFD) is the need to monitor the ecological quality of water on the basis of biological quality elements (Directive 2000/60/EC).

In response to environmental factors, several investigations into the use of algae as indicative organisms of water quality have been described over the last three decades (Abdel-Raouf et al. 2012). Because of algae's comparatively large surface-to-volume ratio and their ability to absorb dissolved nutrients, they are very sensitive to changes in nutrient levels in the environment, making them a good indicator of biogenic compounds, especially nitrogen and phosphorus, which are responsible for eutrophication of aquatic systems (Rosset et al. 2015). The algal indication system is mainly based on diatoms, which are used for the development of a number of indices that assess the degree of saprobization, eutrophication, salinity, or acidity of water (Fetscher et al. 2014; Bielczyńska 2015). Diatom composition can be used to develop an index for monitoring eutrophication of rivers in England and Wales (Kelly and Whitton 1995). In Poland, the multimetric Diatom Index for Lakes, which is based on phytobenthic diatoms, is used for lake environmental evaluation (Bielczyńska 2015). Non-diatoms (soft algae) can be used as indicators as well, but due to a highly variable morphology, their efficacy may be lower than that of diatoms (Blinn and Herbst 2003; Fetscher et al. 2014). In comparison with diatoms, they exhibit a stronger geographical specialization as well, and therefore their use is limited to more localized instances (Stancheva and Sheath 2016).

Macroalgae, along with marine angiosperms, benthic invertebrate fauna, and phytoplankton, are one of the quality elements in the evaluation of the ecological status of a water body (Directive 2000/60/EC; Panayotidis et al. 2004). An environment's nitrogen exposure has a huge impact on species composition and the biodiversity of macroalgae, and their response to nitrogen enrichment is dependent on the algal group (Kim et al. 2014).

The study by Farias et al. (2017) showed that *Ulva australis* is able to accumulate arsenic, copper, lead, and zinc over the short term and has the potential to be a useful tool for the monitoring and management of metal pollution.

Algae can be used for monitoring gas pollution as well. The study showed that *Trebouxia* sp. can be a good bioindicator for rising concentrations of CO_2. The density of this alga inside the FACE (Free-Air Carbon Dioxide Enrichment) system in which CO_2 gas molecules were injected was much higher in comparison with the control site (Ismail et al. 2017).

2.4 Conclusions

The need for novel environmentally friendly technologies and the development of innovative mass production makes the biotechnology of microalgae worthy of attention. In summary, the commercial cultivation of microalgae has many advantages. They grow rapidly, and they can be cultivated both in brackish water and on non-arable land. Macro and microalgae are important indicators of aquatic pollution. They have an ability to absorb pollutants from the aquatic environment, biotransform organic compounds, and immobilize inorganic elements, making them less toxic. Their ability to remove nutrients and heavy metals from most wastewater makes them an alternative for wastewater treatment methods. They are able to capture carbon dioxide, thus reducing greenhouse gas emissions. Because of rapid biomass growth, absorption of significant amounts of CO_2 during the photosynthetic process, and the high content of oil in biomass, algae can be used as a renewable energy source, e.g., for the production of biodiesel. Algae biomass residue, after lipid extraction, can be used as a source of nitrogen for such products as protein-rich animal feed and fertilizer.

The ability to connect the technology of the production of biomass with carbon mitigation and bioremediation makes the commercial production of algae, especially microalgae, an important area for biotechnological development.

References

Abdel-Raouf N, Al-Homaidan AA, Ibraheem IBM (2012) Microalgae and wastewater treatment. Saudi J Biol Sci 19(3):257–275

Abdi O, Kazemi M (2015) A review study of biosorption of heavy metals and comparison between different biosorbents. J Mater Environ Sci 6(5):1386–1399

Aksu Z (2002) Determination of the equilibrium, kinetic and thermodynamic parameters of the batch biosorption of nickel (II) ions onto *Chlorella vulgaris*. Process Biochem 38(1):89–99

Anastopoulus I, Kyzas GZ (2015) Progress in batch biosorption of heavy metals onto algae. J Mol Liq 209:77–86

Aziz HA, Adlan MN, Ariffin KS (2008) Heavy metals (Cd, Pb, Zn, Ni, Cu and Cr(III)) removal from water in Malaysia: post treatment by high quality limestone. Bioresour Technol 99(6):1578–1583

Bartsch AF (1961) Algae as a source of oxygen in waste treatment. J Water Pollut Control Fed 33(3):239–249

Beardall J, Raven JA (2016) Carbon acquisition by microalgae. In: Borowitzka MA, Beardall J, Raven JA (eds) The physiology of microalgae. Springer, Dordrecht, pp 89–90

Beema Jainab SI, Subramanian VV, Sivasubramanian V (2014) Study of bicarbonate tolerance by micro algae and possible application in CO_2 sequestration. Int J Curr Microbiol Appl Sci 3(7):1062–1071

Bellinger EG, Sigee DC (2015) Freshwater algae: identification, enumeration and use as bioindicators. Wiley, Chichester

Bharti PK (2012) Heavy metals in environment. Lambert Academic Publishing, Saarbrucken

Bhatnagar A, Chinnasamy S, Singh M, Das KC (2011) Renewable biomass production by mixotrophic algae in the presence of various carbon sources and wastewaters. App Energy 88:3425–3431

Bhattacharjee M, Siemann E (2015) Low algal diversity systems are a promising method for biodiesel production in wastewater fed open reactors. Algae 30(1):67–79

Bhola V, Swalaha, Kumar RR, Singh M, Bux F (2014) Overview of the potential of microalgae for CO_2 sequestration. Int J Environ Sci Technol 11:2103–2118

Bielczyńska A (2015) Bioindication on the basis of benthic diatoms: advantages and disadvantages of the Polish phytobenthos lake assessment method (IOJ – the diatom index for lakes). Environ Prot Nat Res. 26 4(66):48–55

Bixler HJ, Porse H (2011) A decade of change in the seaweed hydrocolloids industry. J Appl Phycol 23:321–335

Blinn DW, Herbst DB (2003) Use of diatoms and soft algae as indicators of environmental determinants in the Lahontan Basin, USA. Annual report for California State water resources board contract agreement 704558.01.CT766, p 1–25

Boonchai R, Seo GT, Park DR, Seong CY (2012) Microalgae photobioreactor for nitrogen and phosphorus removal from wastewater of sewage treatment plant. Int J Biosci Biochem Bioinforma 2(6):407–410

Brierley AS, Kingsford MJ (2009) Impacts of climate change on marine organisms and ecosystems. Curr Biol 19:R602–R614

Cai T, Park SY, Li Y (2013) Nutrient recovery from wastewater streams by microalgae: status and prospects. Renew Sustain Energy Rev 19:360–369

Chatterjee A, Abraham J (2015) Biosorption capacity of dried spirogyra on heavy metals. Int J ChemTech Res 8(9):387–392

Chen JP, Yang L (2005) Chemical modification of *Sargassum* sp. for prevention of organic leaching and enhancement of uptake during metal biosorption. Ind Eng Chem Res 44:9931–9942

Chen G, Zhao L, Qi Y (2015) Enhancing the productivity of microalgae cultivated in wastewater toward biofuel production: a critical review. Appl Energy 137(1):282–291

Choi HJ, Lee SM (2012) Effects of microalgae on the removal of nutrients from wastewater: various concentrations of *Chlorella vulgaris*. Environ Eng Res 17(S1):3–8

Cuellar-Bermudez SP, Aleman-Nava GS, Chandra R, Garcia-Perez SG, Contreras-Angulo JR, Markou G, Muylaert K, Rittmann BE, Parra-Saldivar (2017) Nutrients utilization and contaminants removal. A review of two approaches of algae and cyanobacteria in wastewater. Algal Res 24:438–449

Davis R, Aden A, Pienkos PT (2011) Techno-economic analysis of autotrophic microalgae for fuel production. Appl Energy 88(10):3524–3531

Directive 2000/60/EC of the European Parliament and of the Council of 23 October 2000 establishing a framework for Community action in the field of water policy. Off J Eur Communities L 327/1 of 22.12.2000, p 72

Dixit S, Singh P (2014) An evaluation of phycoremediation potential of cyanobacterium *Nostoc muscorum*: characterization of heavy metal removal efficiency. J Appl Phycol 26:1331–1342

Farias DR, Hurd CL, Eriksen RS, Simioni C, Schmidt E, Bouzon ZL, Macleod CK (2017) In situ assessment of *Ulva australis* as a monitoring and management tool for metal pollution. J Appl Phycol 29(5):2489–2502

Fetscher AE, Stancheva R, Kociolek JP, Sheath RG, Stein ED, Mazor RD, Ode PR, Busse LB (2014) Development and comparison of stream indices of biotic integrity using diatoms vs. non-diatom algae vs. a combination. J Appl Phycol 26(1):433–450

Fu F, Wang Q (2011) Removal of heavy metal ions from wastewaters: a review. J Environ Manag 92(3):407–418

Gupta VK, Rastogi A (2008a) Biosorption of lead(II) from aqueous solutions by non-living algal biomass *Oedogonium* sp. and *Nostoc* sp.—a comparative study. Colloid Surf B 64(2):170–178

Gupta VK, Rastogi A (2008b) Equilibrium and kinetic modelling of cadmium(II) biosorption by nonliving algal biomass *Oedogonium* sp. from aqueous phase. J Hazard Mater 153(1–2):759–766

Gupta VK, Rastogi A (2008c) Biosorption of lead from aqueous solutions by green algae *Spirogyra* species: kinetics and equilibrium studies. J Hazard Mater 152:407–414

Gupta VK, Rastogi A (2008d) Sorption and desorption studies of chromium (VI) from nonviable cyanobacterium *Nostoc muscorum* biomass. J Hazard Mater 154:347–354

Gupta VK, Rastogi A (2009) Biosorption of hexavalent chromium by raw and acid-treated green alga *Oedogonium hatei* from aqueous solutions. J Hazard Mater 163:396–402

Gupta VK, Srivastava AK, Jain N (2001) Biosorption of chromium (VI) from aqueous solutions by green algae *Spirogyra* species. Water Res 35:4079–4085

Gupta VK, Rastogi A, Saini VK, Jain N (2006) Biosorption of copper (II) from aqueous solutions by algae *Spirogyra* species. J Colloid Interface Sci 96:59–63

Gupta VK, Rastogi A, Nayak A (2010) Biosorption of nickel onto treated alga (*Oedogonium hatei*): application of isotherm and kinetic models. J Colloid Interface Sci 342:533–539

Gupta VK, Nayak A, Agarwal S (2015) Bioadsorbents for remediation of heavy metals: current status and their future prospects. Environ Eng Res 20(1):1–18

Harun R, Singh M, Forde GM, Danquah MK (2010) Bioprocess engineering of microalgae to produce a variety of consumer products. Renew Sustain Energy Rev 14(3):1037–1047

He J, Chen JP (2014) A comprehensive review on biosorption of heavy metals by algal biomass: materials, performances, chemistry, and modeling simulation tools. Bioresour Technol 160:67–78

Henkanatte-Gedera SM, Selvaratnam T, Karbakhshravari M, Myint M, Nirmalakhandan N, Van Voorhies W, Lammers PJ (2017) Removal of dissolved organic carbon and nutrients from urban wastewaters by *Galdieria sulphuraria*: laboratory to field scale demonstration. Algal Res 24:450–456

Herrero R, Lodeiro P, Rey-Castro C, Vilariño T, Sastre de Vicente ME (2005) Removal of inorganic mercury from aqueous solutions by biomass of the marine macroalga *Cystoseira baccata*. Water Res 239:3199–3210

Herrero R, Cordero B, Lodeiro P, Rey-Castro C, Sastre de Vicente ME (2006) Interaction of cadmium (II) and protons with dead biomass of marine algae *Fucus* sp. Mar Chem 99:106–116

Horvathova H, Kadukova J, Stofko M (2009) Biosorption of Cu^{2+} and Zn^{2+} by immobilized algae biomass of chlorella kessleri. Acta Metall Slovaca 15(4):255–263

Huang S, Lin G (2015) Biosorption of Hg(II) and Cu(II) by biomass of dried *Sargassum fusiforme* in aquatic solution. J Environ Health Sci Eng 13:21–28

Ibañez E, Cifuentes A (2013) Benefits of using algae as natural sources of functional ingredients. J Sci Food Agric 93:703–709

Ismail A, Marzuki SD, Yusof NBM, Buyong F, Said MNM, Sigh HR, Zulkifli AR (2017) Epiphytic terrestrial algae (*Trebouxia* sp.) as a biomarker using the free-air-carbon dioxide-enrichment (FACE) system. Biol (Basel) 6(1):19–25

Javanbakht V, Alavi SA, Zilouei H (2014) Mechanisms of heavy metal removal using microorganisms as biosorbent. Water Sci Technol 69(9):1775–1787

Jorquera O, Kiperstok A, Sales EA, Embiruçu M, Ghirardi ML (2010) Comparative energy life-cycle analyses of microalgal biomass production in open ponds and photobioreactors. Bioresour Technol 101(4):1406–1413

Junzhuo L, Danneels B, Vanormelingen P, Vyverman W (2016) Nutrient removal from horticultural wastewater by benthic filamentous algae *Klebsormidium* sp., *Stigeoclonium* spp. and their communities: from laboratory flask to outdoor Algal Turf Scrubber (ATS). Water Res 92:61–68

Kanchana S, Jeyanthi J, Kathiravan R, Suganya K (2014) Biosorption of heavy metals using algae: a review. Int J of Pharm Med Biol Sci 3(2):1–9

Kelly MG, Whitton BA (1995) The trophic diatom index: a new index for monitoring eutrophication in rivers. J Appl Phycol 7(4):433–444

Kim S, Park SR, Kang YH, Kim G-Y, Lee K-S, Lee HJ, Won N-I, Kil H-J (2014) Usefulness of tissue nitrogen content and macroalgal community structure as indicators of water eutrophication. J Appl Phycol 26(2):1149–1158

Kligerman DC, Bouwer EJ (2015) Prospects for biodiesel production from algae-based wastewater treatment in Brazil: a review. Renew Sust Energ Rev 52:1834–1846

Kumar A, Das A, Goel M, Kumar KR, Subramanyam B, Sudarsan JS (2013) Recovery of nutrients from wastewater by struvite crystallization. Nat Environ Pollut Techol 12(3):479–482

Kumar KS, Dahms HU, Won EJ, Lee JS, Shin KH (2015) Microalgae – a promising tool for heavy metal remediation. Ecotox Environ Safe 113:329–352

Kushwah A, Srivastav JK (2015) Biosorption of copper ions by green algae spirogyra. Int J Chem Stud 3(3):36–38

Laurens LML, Chen-Glasser M, McMillan JD (2017) A perspective on renewable bioenergy from photosynthetic algae as feedstock for biofuels and bioproducts. Algal Res 24:261–264

Lesmana SO, Febriana N, Soetaredjo FE, Sunarso J, Ismadji S (2009) Studies on potential applications of biomass for the separation of heavy metals from water and wastewater. Biochem Eng J 44(1):19–41

Lu Q, Zhou W, Min M, Ma X, Chandra C, Doan Y, Ma Y, Zheng H, Cheng S, Griffith R, Chen P, Chen C, Urriola PE, Shurson GC, Gislerød HR, Ruan R (2015) Growing *Chlorella* sp. on meat processing wastewater for nutrient removal and biomass production. Bioresour Technol 198:189–197

Luo S, Berges JA, He Z, Young EB (2017) Algal-microbial community collaboration for energy recovery and nutrient remediation from wastewater in integrated photobioelectrochemical systems. Algal Res 24(B):527–539

Morales-Sánchez D, Martinez-Rodriguez OA, Martinez A (2016) Heterotrophic cultivation of microalgae: production of metabolites of commercial interest. J Chem Technol Biotechnol 92(5):925–936

Nascimento IA, Dominguez Cabanelas IT, Nunes dos Santos J, Nascimento MA, Sousa L, Sansone G (2015) Biodiesel yields and fuel quality as criteria for algal-feedstock selection: effects of CO_2-supplementation and nutrient levels in cultures. Algal Res 8:53–60

Nguyen TAH, Ngo HH, Guo WS, Zhang J, Liang S, Yue QY, Li Q, Nguyen TV (2013) Applicability of agricultural waste and by-products for adsorptive removal of heavy metals from wastewater. Bioresour Technol 148:574–585

Omar WMW (2010) Perspectives on the use of algae as biological indicators for monitoring and protecting aquatic environments, with special reference to Malaysian freshwater ecosystems. Trop Life Sci Res 21(2):51–67

Oswald WJ (1988) Micro-algae and waste-water treatment. In: Borowitzka MA, Borowitzka LJ (eds) Micro-algal biotechnology. Cambridge University Press, Cambridge, pp 305–328

Oswald WJ (2003) My sixty years in applied algology. J Appl Phycol 15(2):99–106

Oswald WJ, Gotaas HB, Golueke CG, Kellen WR, Gloyna EF, Hermann ER (1957) Algae in waste treatment. Sew Ind Waste 29(4):437–457

Panayotidis P, Montesanto B, Orfanidis S (2004) Use of low-budget monitoring of macroalgae to implement the European Water Framework Directive. J Appl Phycol 16(1):49–59

Perez-Garcia O, Escalante FME, de Bashan LE, Bashan Y (2011) Heterotrophic cultures of microalgae: metabolism and potential products. Water Res 45:11–36

Pires JCM (2017) COP21: the algae opportunity? Renew Sustain Energy Rev 79:867–877

Pires JCM, Gonçalves AL, Martins FG, Alvim-Ferraz MCM, Simões M (2014) Effect of light supply on CO_2 capture from atmosphere by *Chlorella vulgaris* and *Pseudokirchneriella subcapitata*. Mitig Adapt Strat Gl 19:1109–1117

Prasher S, Hawri J, Bera P, Patel RM, Kim SH (2004) Biosorption of heavy metals by red algae (*Palmaria palmata*). Environ Technol 25(10):1097–1106

Prussi M, Buffi M, Casini D, Chiaramonti D, Martelli F, Carnevale M, Tredici MR, Rodolfi L (2014) Experimental and numerical investigations of mixing in raceway ponds for algae cultivation. Biomass Bioenergy 67:390–400

Ravindran B, Gupta SK, Cho WM, Kim JK, Lee SR, Jeong KH, Lee DJ, Choi HC (2016) Microalgae potential and multiple roles—current progress and future prospects—an overview. Sustainability 8(12):1215. https://doi.org/10.3390/su8121215

Rawat I, Kumar R, Mutanda T, Bux F (2011) Dual role of microalgae: Phycoremediation of domestic wastewater and biomass production for sustainable biofuels production. Appl Energy 88(10):3411–3424

Richards RG, Mullins BJ (2013) Using microalgae for combined lipid production and heavy metal removal from leachate. Ecol Model 249:59–67

Romera E, Gonzalez F, Ballester A, Blázquez ML, Muñoz JA (2007) Comparative study of biosorption of heavy metals using different types of algae. Bioresour Technol 98(17):3344–3353

Rosset S, A'Angelo C, Wiedenmann J (2015) Ultrastructural biomarkers in symbiotic algae reflect the availability of dissolved inorganic nutrients and particulate food to the reef coral Holobiont. Front Mar Sci 2:103. https://doi.org/10.3389/fmars.2015.00103

Ruiz-Marin A, Mendoza-Espinosa LG, Stephenson T (2010) Growth and nutrient removal in free and immobilized green algae in batch and semi-continuous cultures treating real wastewater. Bioresour Technol 101(1):58–64

Sayre R (2010) Microalgae: the potential for carbon capture. Bioscience 60(9):722–727

Shalaby E (2011) Algae as promising organisms for environment and health. Plant Signal Behav 6(9):1338–1350

Singh RN, Sharma S (2012) Development of suitable photobioreactor for algae production – a review. Renew Sustain Energy Rev 16(4):2347–2353

Singh SP, Singh P (2014) Effect of CO_2 concentration on algal growth: a review. Renew Sustain Energy Rev 38:172–179

Solisio C, Lodi A, Soletto D, Converti A (2008) Cadmium biosorption on *Spirulina* platensis biomass. Bioresour Technol 99:5933–5937

Stancheva R, Sheath RG (2016) Benthic soft-bodied algae as bioindicators of stream water quality. Knowl Manag Aquat Ecosyst 417:15. https://doi.org/10.1051/kmae/2016002

Suzuki Y, Kametani T, Maruyama T (2005) Removal of heavy metals from aqueous solution by nonliving *Ulva* seaweed as biosorbent. Water Res 39:1803–1808

Tchounwou PB, Yedjou CG, Patlolla AK, Sutton DJ (2012) Heavy metal toxicity and the environment. In: Luch A (ed) Molecular, clinical and environmental toxicology, Environmental Toxicology, vol 3. Springer, Heidelberg, pp 133–164

Torres MA, Barros MP, Campos SCG, Pinto E, Rajamani S, Sayre RT, Colepicolo P (2008) Biochemical biomarkers in algae and marine pollution: a review. Ecotoxicol Environ Saf 71(1):1–15

Unc A, Camargo-Valero MA, Smith SR (2017) Algal research, special issue editorial: wastewater and algae; risk, biofuels and long-term sustainability. Algal Res 24(B):A1. https://doi.org/10.1016/j.algal.2017.05.004

Vijayaraghavan K, Padmesh TVN, Palanivelu K, Velan M (2006) Biosorption of nickel(II) ions onto *Sargassum wightii*: application of two-parameter and three-parameter isotherm models. J Hazard Mater 133:304–308

Vilar VJP, Botelho CMS, Boaventura RAR (2006) Equilibrium and kinetic modeling of Cd(II) biosorption by algae *Gelidium* and agar extraction algal waste. Water Res 40:291–302

Volterra L, Conti ME (2000) Algae as biomarkers, bioaccumulators and toxin producers. Int J Environ Pollut 13:1–6

Walker DA (2009) Biofuels, facts, fantasy and feasibility. J Appl Phycol 21:508–517

Wang M, Kuo-Dahab WC, Dolan S, Park C (2014) Kinetics of nutrient removal and expression of extracellular polymeric substances of the microalgae, *Chlorella* sp. and *Micractinium* sp., in wastewater treatment. Bioresour Technol 154:131–137

Wang B, Wen JL, Sun SL, Wang HM, Wang SF, Liu QY, Charlton A, Sun RC (2017) Chemosynthesis and structural characterization of a novel lignin-based bio-sorbent and its strong adsorption for Pb(II). Ind Crop Prod 108:72–80

World Bank (2016) World development indicators 2016. International Bank for Reconstruction and Development/The World Bank, Washington, DC. 20433, pp 1–54

World Bank (2017) World development indicators 2017. International Bank for Reconstruction and Development/The World Bank, Washington, DC. 20433, pp 1–146

World Bank Open Data (2017) http://www.data.worldbank.org. Searched on 20 June 2017

WWAP (United Nations World Water Assessment Programme) (2017) The United Nations world water development report 2017, Wastewater: The Untapped Resource. UNESCO, Paris

Zhang F, Li J, He ZA (2014) A new method for nutrients removal and recovery from wastewater using a bioelectrochemical system. Bioresour Technol 166:630–634

Biology of Freshwater Macroalgae and Their Distribution

3

Beata Messyasz, Marta Pikosz, and Ewa Treska

Abstract

Cosmopolitan taxa of freshwater macroalgae occurring in inland waters in mass are described with respect to their biology. One of their components is autecology, which concerns the relations of individual organisms to the various factors of their environment. The factors influencing the fluctuations of macroalgal populations (e.g., *Cladophora*, *Oedogonium*) and the formations of their life strategies as primary producers can inform us about the possibility of their use in various branches of industry.

Keywords

Filamentous algae · Macroalgae · Growth forms · Morphology · Competitive interactions

3.1 Introduction

Macroalgae are multicellular algae of thalli-like structure (Van den Hoek et al. 1995; Blomster et al. 2000; Yoshii et al. 2004; Sakayama 2008; Sakayama et al. 2009; Leliaert et al. 2000, 2012). They can attach themselves to solid substrates underwater (e.g., stones, mollusks' shells, or macrophytes) or float freely in the water. In freshwater ecosystems, macroalgal mats are mainly built by species of green algae (Chlorophyta), such as *Spirogyra*, *Cladophora*, *Mougeotia*, and *Ulva*, but also by stonewort (Charales) attached to the sandy bottom (Duarte 1995; Messyasz et al. 2015a; Pikosz and Messyasz 2016; Pikosz et al. 2017).

The big thalli of the algae play a significant role as producers in the water ecosystem – they distribute organic substances, create significant biodiversity, diversify the environment, and indicate the ecological state of the water (Hutchinson 1957; Hoffmann and Graham 1984; Dodds and Gudder 1992; Ensminger et al. 2005; Borja et al. 2008; Messyasz and Rybak 2011; Montoya-Moreno and Aguirre-Ramirez 2013; Green and Fong 2015). The most common places where macroalgae can be found are water bodies that are used commercially and recreationally – breeding fishponds, water bodies used for bathing or swimming, and newly created reservoirs.

Massive occurrence of macroalgae is mainly caused by:

- An increase of water fertility caused by nutrient intake, e.g., phosphorus, nitrogen (Herbst 1969; Freeman 1986; Krause-Jensen et al. 1999; Bricker et al. 2008; Pikosz and Messyasz 2016)
- Stagnant waters with low flow rates (Chudyba 1965; Whitton 1970; Hard 1992; Thybo-Christesen and Blackburn 1993; Krause-Jensen et al. 1996; Stewart and Carpenter 2003; Pikosz and Messyasz 2015a)
- Good light conditions (Lester et al. 1988; Berner et al. 1989; Krause-Jensen et al. 1996; Vergara et al. 1997; Dere et al. 1998; Ensminger et al. 2000, 2001; Lenzi et al. 2013)
- Stoneworts (Charales) – low concentration of nutrients, high water hardness (Pełechaty and Gąbka 2003; Gąbka 2007, 2009; Pełechaty 2005)

In water ecosystems with elevated nutrient concentrations and proper physicochemical parameters (basic pH, good light conditions, high water temperature), diversified communities of filamentous green algae grow intensively, creating large mats. Research suggests that there is a correlation between water depth and transparency and the range of filamentous algae in the vertical transect of water (Lichtenthaler and Wellburn 1983; McGlathery et al. 1997; Hainz et al. 2009; Garcia and Aboal 2014; Pikosz et al. 2017). It influences the occurrence of macroscopic green algae from genera: *Cladophora*, *Oedogonium*, *Ulva*, *Rhizoclonium*, *Spirogyra*, and *Mougotia*. The presence of mats created by these species at the water surface leads to ecological changes in water ecosystems, including floristic and faunistic changes (Philips 1990; Fletcher 1996; Chemello and Milazzo 2002; Farina et al. 2003; Das et al. 2014).

The developmental optimum of biomass for the analyzed group of algae is generally considered to take place during

B. Messyasz (✉) · M. Pikosz · E. Treska
Faculty of Biology, Department of Hydrobiology, Adam Mickiewicz University in Poznan, Poznan, Poland
e-mail: messyasz@amu.edu.pl

© Springer International Publishing AG, part of Springer Nature 2018

K. Chojnacka et al. (eds.), *Algae Biomass: Characteristics and Applications*, Developments in Applied Phycology 8,
https://doi.org/10.1007/978-3-319-74703-3_3

the summer period, in July and August. However, two other optima can occur as well: during spring and autumn, for example, in the case of *Cladophora glomerata* (L.) Kütz. (Pikosz et al. 2017). Unfavorable habitat conditions cause changes in the morphometric structure of thalli, the seasonality of their occurrence, and the production of significant amounts of filamentous algae biomass.

For a dozen or so years, in Polish waters in the Wielkopolska region, a constant growth of green algae biomass has been observed (Messyasz and Rybak 2011; Messyasz et al. 2015a; Pikosz and Messyasz 2016). The green algae mats cover a significant portion of the water's surface. In smaller bodies of water, these mats can even cover up to 100% of the surface. The presence of massive blooms of filamentous green algae in water ecosystems, from an ecological point of view, is a negative effect of an increase in water trophy. In the Wielkopolska region, the quality of most water bodies is classified as eutrophic, which is conducive to increasing algal biomass. The biggest increase in biomass of filamentous green algae is observed during summer, at the peak of the vegetative season. A stocktaking of green algae from the genera *Cladophora* and *Ulva* has been underway in recent years in Wielkopolska (Messyasz and Rybak 2011; Pikosz and Messyasz 2016). Their presence was confirmed at several dozen stations. In 2012, the presence of a multispecies mat (consisting of *Cladophora glomerata*, *Rhizoclonium* sp., *Stigeoclonium* sp., and *Microspora* sp.) was observed in the Mogilnica River, a single-species mat in the Nielba and Samica Stęszewska Rivers (consisting of various forms of *Cladophora glomerata*) and a large mat (covering 90% of the water's surface) of the *Oedogonium* species in a field pond (Pikosz and Messyasz 2015a, b). At the beginning of April 2013, the monitoring and sampling of filamentous green algae that were creating mats had commenced in the water ecosystems being studied. Changes in the species composition of filamentous green algae have been observed (Pikosz and Messyasz 2016).

The consequences of massive occurrences of macroalgae on the environment differ depending on the form of the biomass created. They are as follows:

In cases of floating macroalgae:

- Deterioration of light conditions for submerged macrophytes and other organisms
- Fast decomposition – supplying the nutrient pool, oxygen deficits, decay
- Progressive eutrophication of the water body

In cases of macroalgae attached to the bottom sediments:

- Limitation of bottom sediment mobilization
- Acceleration of sedimentation
- Creation of periphytic habitats and shelters for water organisms
- Displacement of submerged macrophytes, originally occurring in such habitats
- Stoneworts – improvement of environmental conditions

Macroalgae are becoming more abundant in freshwater reservoirs, but their effects on other groups of organisms remain poorly described. Favorable conditions for algal growth are created by the increased eutrophication resulting from the unsustainable use of direct catchment areas. As macroalgae increase in abundance, they can actively overgrow the bottom or passively take over a space after coming to the surface of the water. Compared to other algal groups, such as microalgae, macroalgae occupy available space slightly more slowly, but they grow faster (e.g., *Cladophora*, *Ulva*) and are less vulnerable to grazing and water turbulence. Macroalgal mats are dense, mono-, or multispecies and thus are a more complex functional group. Thorough knowledge of the structure and function of such mats will allow for better understanding of the complex factors responsible for the coexistence of species, not only at the macroalgae level but also at the microalgae/macroalgae level. To describe the general characteristics of competitive interactions between certain species of macroalgae and phytoplankton, we presented macroalgal growth forms and the types of mat they create.

3.2 Surveys of Macroalgae Occurrences in Polish Waters

Filamentous algae are common and visible macroscopically, but they are nonetheless underexamined, and their presence in water ecosystems is hardly ever mentioned in literature. Taxonomic identification of filamentous algae is difficult, as these algae grow in various habitats and environments (Bellinger and Sigee 2010; Guiry and Guiry 2017). Filamentous green algae show morphological and phenotypical plasticity (Leliaert et al. 2012). This variability is one of the reasons for problems in regard to taxonomic identification; the systematics is based on morphological features (such as thalli color, cell shape, presence or absence and type of branch). The most common in freshwater ecosystems in Poland are macroscopic green algae of the genus *Cladophora* (30% of all sites in Poland), *Oedogonium* (11%), *Rhizoclonium* (3%), *Ulothrix* (8%), *Spirogyra* (16%), *Vaucheria* (7%), and *Mougotia* (7%) (Messyasz et al. 2015a; Pikosz and Messyasz 2016). A quick guide to their identification has been prepared (Fig. 3.1). Moreover, if *Cladophora glomerata* occurs in a given body of water, it always predominates over the other filamentous algae in that water.

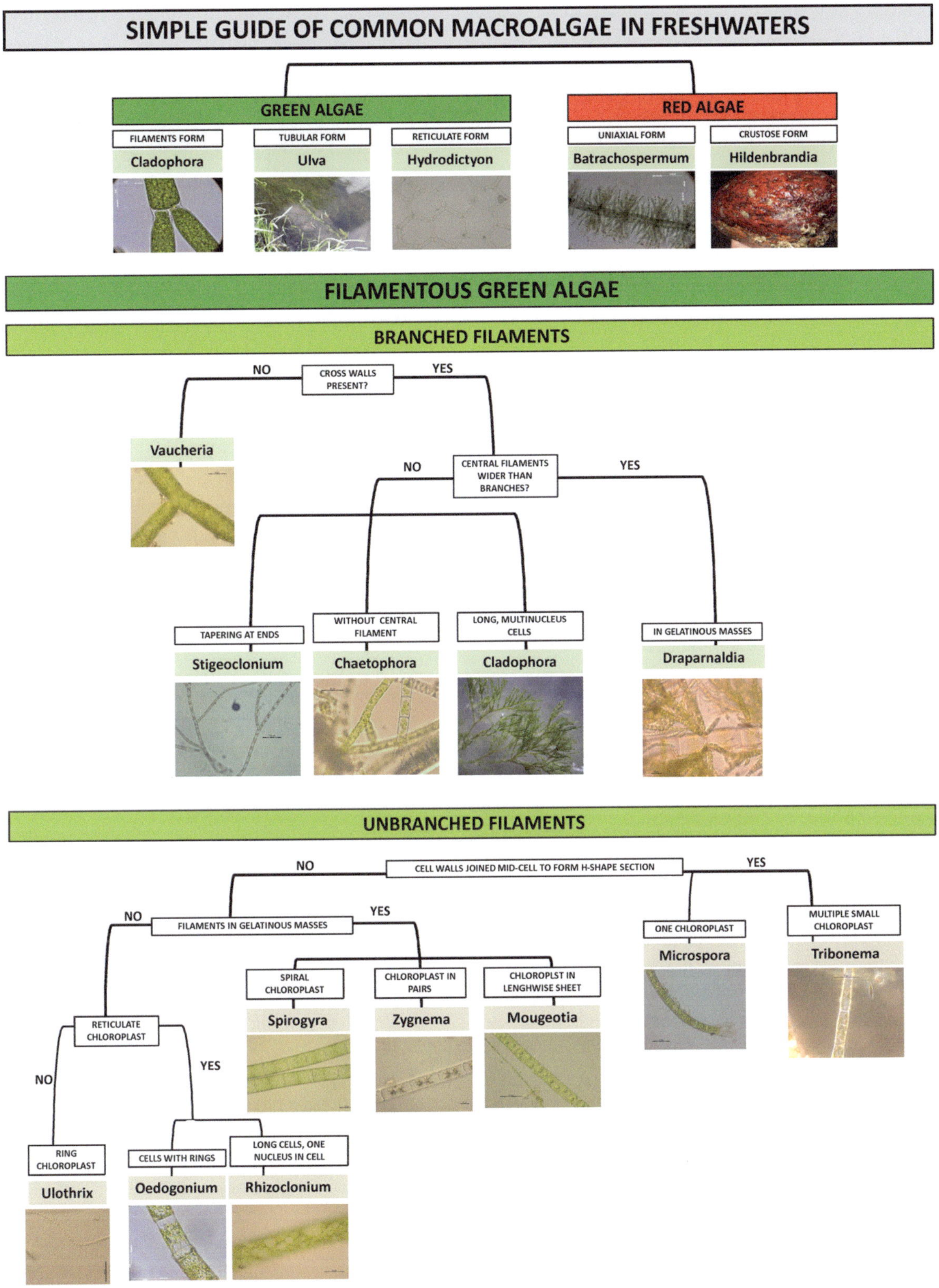

Fig. 3.1 Guide to common macroscopic alga in Polish freshwaters

There are plenty of floristic data describing the presence of filamentous green algae in lakes and lowland rivers; however, in most of those cases, the data do not take into consideration the structure of the algal mats or any other ecological data. Therefore, descriptions of the most important macroalgal genera are summarized below.

1. *Coleochaete* – single-row thalli when they are young. In later periods, they are able to grow into mass threads, which have a tendency to merge into a bigger mass. Their characteristic feature is the ability to spread widely as an epiphyte on hydromacrophytes and other submerged biotic substrates (Guiry and Guiry 2017).
2. *Aphanochaete* – green algae growing on plant substrate, possessing the ability to crawl. This is a broadly spread type of green alga, often seen in waters rich in nutrients as an epiphyte on submerged macrophytes and larger filamentous green algae (Guiry and Guiry 2017).
3. *Bulbochaete* – representative of Oedogoniales. They have branched threads and are easily identified because of the protruding threadlike outgrowths in their bubble-like swollen cells near the basis of branching (Guiry and Guiry 2017). They occur widely, growing attached to submerged plants or stones. They tolerate a high range of pH and substrates (Van den Hoek et al. 1995).
4. *Chaetophora* – threads often attached to stones and submerged plants, they can be very long, up to 10 cm. They are spread broadly along the edges of shallow flowing waters or in deeper parts attached to a substrate (McGlathery et al. 1997; Guiry and Guiry 2017).
5. *Draparnaldia* – straightened (standing) threads are attached to a substrate with rhizoids. Threads are set in soft mucus. Spread broadly, they often occur in waters poor in phosphorous (Van den Hoek et al. 1995; Guiry and Guiry 2017).
6. *Stigeoclonium* – filamentous algae, usually attached to various kinds of substrate with basal cells. Broadly spread, they are usually attached to rocks and stones, but with connections that are easily broken. They can be found floating freely in water (Guiry and Guiry 2017). The most common is *Stigeoclonium tenue* (Agardh) Kütz., used as an indicator of eutrophic waters, often contaminated with organic matter (it shows tolerance for heavy metal contamination) (Van den Hoek et al. 1995).
7. *Microthamnion* – attached, softly branched threads of algae, of constant width throughout their length. Broadly spread, they often occur in acidic waters rich in organic substances with high contents of Fe and Mg. They attach to substrate with hooklike formations. They often occur in communities with *Microspora*. Representatives of this genus can also be found in wet soil (Van den Hoek et al. 1995; Guiry and Guiry 2017).
8. *Rhizoclonium* – creates thick, rough threads with small branching of rhizoids (not always visible). These are common in basic, shallow waters, where they can form dense mats. Often found (and confused) with *Cladophora* (Van den Hoek et al. 1995; Messyasz et al. 2015a; Guiry and Guiry 2017).
9. *Cladophora* – typical filamentous green algae with a high level of branching, but in slow moving waters, they can be scarce or even not form at all. They can float freely in water or attach themselves to a substrate with hooklike formations. Branching is dichotomic or even trichotomic. They create dense, thick mats. The representative for this genus is often present in basic waters, especially those with inflow of sewage. *Cladophora* can grow intensively near filter systems of water treatment plants, such as slow sand filters in which serious problems may occur precisely because of the algal threads (Painter and Kamaitis 1987; Van den Hoek et al. 1995; Messyasz et al. 2015a; Guiry and Guiry 2017).
10. *Microspora* – creating non-branching threads, they attach themselves to a substrate with hooklike formations and are common in small bodies of water, especially during colder seasons (Van den Hoek et al. 1995; Guiry and Guiry 2017).
11. *Spirogyra* and *Mougeotia* – common in many environments. They have straight, non-branched threads (Simons and Beem 1990; Van den Hoek et al. 1995; Guiry and Guiry 2017).
12. *Oedogonium* – cells create long, non-branched threads. Very common in flowing waters where they can create dense mats. *Oedogonium* usually occurs in small, seasonally flooded water bodies; they can even develop in puddles. Small surface, high concentration of nutrients, warmth, and stagnation of water create the perfect conditions for *Oedogonium*'s development. Rarely seen in lakes or rivers. Temperature, pH, light, and certain salts dissolved in water (Ca, Fe, Mn, Mg) have significant influence on their development (Van den Hoek et al. 1995; Messyasz et al. 2015a; Pikosz and Messyasz 2015b; Guiry and Guiry 2017).
13. *Ulva* – typical tubular green algae in freshwaters, attached to substratum or floating on the water's surface. Their mats can be found across a wide range of nutrient conditions but are uncommon in clean waters. Found across a wide range of conductivity (Malta et al. 1999; Blomster et al. 2000; Bischof et al. 2002; Messyasz and Rybak 2009; Pikosz and Messyasz 2015b; Guiry and Guiry 2017).
14. **Stoneworts** (e.g., *Chara*, *Nitella*) – the best known group of macroalgae in the Wielkopolska region. Detailed research on their ecology and morphometric diversity has been conducted since the nineteenth century (Dąmbska et Karpiński 1954). Stoneworts occur

mainly in lakes, creating dense underwater "stonewort meadows," as they used to be called. In Poland, 34 species of these macroalgae occur. Stoneworts are present mainly in lakes with clear waters, reaching up to 2 m. They are considered a very good indicator of aquatic environmental pollution. Stoneworts are attached to sandy or silty bottoms with rhizoids (Pełechaty and Gąbka 2003; Gąbka 2007, 2009; Pełechaty 2005).

Advantageous habitat conditions (i.e., low water depth, high trophy) and a collection of adaption features cause massive occurrences of filamentous green algae in water ecosystems. This, in turn, has serious consequences for water systems: changes in light and oxygen conditions, as well as nutrient circulation. Algae blooms are most often the result of eutrophication (Khanum 1982; Ozimek 1990; Higgins et al. 2008). These blooms create single- or multispecies mats that are a model research subject for researching competition relations. So far, it has been observed that the species' structure, spatial structure (loose or tight), and internal organization (density, thickness, space covered) of green algae mats are a result of certain environmental conditions, hydrodynamic and thermal, in particular (Ozimek 1990; Pikosz and Messyasz 2015a). The coexistence of a number of taxa in the same place may cause a reaction: competition over space and nutrients. The result of this competition may be the mutual replacement or exclusion of species, as well as the secretion of chemical substances. Several "defensive" competitive strategies of filamentous algae blooms in shallow waters have been accounted for (Pikosz et al. 2017). Freshwater macroalgae may gain an advantage over their competitors thanks to these adaptation techniques: development in early spring, nutrient absorption from water, chemical makeup of thalli (presence of the stress protein – proline – in periods of high mat density), changes in morphological structure, secretion of growth and development suppression substances, and endospore creation (Ozimek 1990). They develop best in shallow, eutrophic bodies of water.

In freshwater ecosystems with high nutrient content and appropriate physicochemical parameters, varied species of filamentous green algae communities develop very intensively, forming large mats on the surface layer (free-floating forms) and the benthic layer (attached to the bottom, Figs. 3.2 and 3.3). *Cladophora glomerata* showed the highest morphological variety among all of the analyzed green algae. In standing waters, thalli are bushy and relatively short (up to 20 cm), while in running waters, the number of filaments is limited and the thalli are long (Fig. 3.3). Significant cell changes (length and width of cells, thickness of cell wall, number of pyrenoids and nuclei) have been observed in filaments of *C. glomerata* collected from rivers (Pikosz and Messyasz 2015a, 2016). These changes can stem from hydrodynamic (water currents, waves) and thermal conditions of the given river habitat. In the case of lakes and ponds, this morphological variability is much less prominent and, if it occurs, may be connected with levels of water depth (this can result in differences in shapes, allowing macroalgae to absorb light).

The development of stable, seasonal mats is possible due to a variety of growth models (taking thermal and light requirements into account) employed by filamentous green algae. Reactions of chosen taxa to certain environmental gradients, e.g., to the presence of orthophosphates in water, show that *Cladophora glomerata*, *C. globulina*, *Rhizoclonium* sp., *Stigeoclonium nanum*, and *Vaucheria* sp. inhabit strongly eutrophic waters, while algae taxa representing Zygnemataceae, *Cladophora rivularis* (L.) Hoek, *Oedogonium capillare* Kütz. ex Hirn, and *Cladophora fracta* (O.F. Müller ex Vahl) Kütz. have their optimum of development in much less fertile waters (Pikosz and Messyasz 2016). For filamentous green algae, the strategy of *outpacing the competition* is very characteristic. The development in early spring (low thermal and light requirements) of *Cladophora glomerata*, as well as of the species of the genera *Tribonema*, *Ulothrix*, and *Rhizoclonium*, allows them to create stable seasonal mats. These are taxa employing the strategy of early spring growth, making it possible to create stable mats very early. *Oedogonium capillare*, *Cladophora rivularis*, and *C. globulina* belong to thermophilic species, the growth optimum of which occurs in warm waters (Pikosz and Messyasz 2016; Pikosz et al. 2017).

The main structural compound of algal mats is usually *Cladophora glomerata*. Other species of filamentous green alga, such as *Oedogonium*, *Bulbochaete*, *Rhizoclonium*, or *Stigeoclonium*, are only detached fragments entangled in *Cladophora* threads (Table 3.1).

3.3 Architectural Forms of Patches

Intensive growth of macroscopic algae causes greater visibility of this particular group of species in bodies of water in the form of a properly defined mat (of symmetrical or asymmetrical nature) that takes specific shapes. Some macroalgae species occur dispersed in water, and they can be defined as non-defined mats, even when their occurrence is massive.

The architectural complexity of a properly defined mat consists of different structural forms: patches, mats, corridors, nets, or spider webs (Adhikary and Sahu 1992; Pascelli et al. 2013; Yniguez et al. 2015; Weiss et al. 2017). A patch of a given species or an agglomeration of macroalgae species can be defined as a relatively homogenous area. The term "mat" is connected to filamentous macroscopic algae floating freely on the water surface or attached to the bottom (Saunders et al. 2012). In Polish ecosystems in which the occurrence of macroscopic algae was observed, taxa from

Table 3.1 Characteristics of the various *Cladophora* patches

Dominant taxon	Accompanying filamentous algae	Forms	Sites
Cladophora glomerata	*Rhizoclonium* sp.	Long filaments,	Small lowland rivers
	Stigeoclonium nanum	Attached to the bottom and stones	
	Oedogonium spp.		
	Vaucheria sp.	Free-floating	
Cladophora glomerata	*Oedogonium* sp.	Free-floating	Shallow lake
	Tribonema vulgare	Dense mat	
Cladophora rivularis	*Cladophora globulina*	Dense mat like watt	Field pond
	Oedogonium capillare		
	Ulothrix variabilis		
	Tribonema aequale		
	Tribonema vulgare		
	Spirogyra spp., *Mougeotia* sp.		
Cladophora fracta	*Cladophora glomerata*	Dense mat in littoral zone (4 m from the shore), soft to the touch	Dam reservoir
	Spirogyra spp.		
	Mougeotia sp.		
Cladophora fracta	*Spirogyra* spp.	Mat mainly slippery to the touch	Urban pond
	Mougeotia spp.		
	Zygnema spp.		
	Sirogonium sp.		
	Cladophora sp.		

Chlorophyta genera were dominant – mainly *Cladophora* taxa (Pikosz and Messyasz 2016). Filamentous species from the class Xantophyceae, Zygnemataceae, and blue green algae were also noted (Messyasz et al. 2015a). According to the species' composition, the following types were distinguished:

- Green algae mats (up to seven species) with a dominance of *Cladophora* taxa – characteristic mainly for rivers
- Cyanobacteria – green algae mats (12–14 species) – characteristic for ponds and shallow lakes

The type of form is related to the size of an object or area. Moreover, forms can be discussed in terms of individual macroalgae growth habitats or the patch arrangement in water bodies. This is largely due to the morphological structure of macroalgal thalli, which include crusty thallus (*Hildenbrandia*), netlike thallus (*Hydrodictyon*), filamentous thallus (*Cladophora*), or tubular thallus (*Ulva*). Macroalgae are known for their morphological plasticity and wide variety of growth form, from encrusting to heavily branched. Branching growth forms are typically fast-growing species (e.g., *Ulva*, *Cladophora*) that extend above the benthos, allowing them to avoid interactions with neighboring organisms. In contrast, slow-growing forms (e.g., *Hildenbrandia*) are less expansive in the water body.

Extensive, compact, and dense patches of macroalgae in freshwater ecosystems occur in the character of mats, overgrowing the surface of the water floor and water table with different levels of intensity. Thalli of those species that grow into very big patches construct algal mats of unimaginable shapes, which are difficult to name. Macroalgae mat formations can be divided into:

- Free-floating mats – taking shapes of flocs, mats or felts, tufts, clouds, netlike structures
- Attached to the bottom – upright growth forms (aligned), bush-like structures, solitary thalloid growth forms
- Overgrowing the entire water column – includes both free-floating and attached forms, as well as the forms that overgrow each element that can serve as a pillar, e.g., submerged water plants

Diversification of the mats' shape can also occur because of the hydrodynamic conditions of the given water ecosystem, even concerning the same macroalgae species. An example of such diversification can be seen in *Cladophora glomerata*, a filamentous green algae. In ponds and lakes, it creates filamentous forms attached to the surface or, in its mature state, floating on the surface of the water in the form of dense mats that overgrow the entire water column (Figs. 3.2 and 3.3). In rivers, on the other hand, *Cladophora* attaches itself to the bottom and has very long threads that rise along with the current and show similarity to braids.

Figure 3.4 shows the most common examples of big patches of macroalgae in stagnant waters. According to their size, patches of all macroalgae species can be categorized into five size classes: 0–1 m^2, 1–5 m^2, 5–10 m^2, 10–50 m^2, >50 m^2. The most abundant are the communities that form mats in the littoral zone and reach up to 10 m^2. Among them, the following shapes can be distinguished: flocs (*Zygnema*, Fig. 3.4b), clouds (*Spirogyra*, *Mougeotia*, *Cladophora*, Fig. 3.4e), netlike structures (*Hydrodictyon*, Fig. 3.4g), spider webs (*Spirogyra*, *Cladophora*, Fig. 3.4h), and jelly structures (*Spirogyra*, *Ulothrix*). The second type of community consists of big (10–50 m^2), single-species patches attached to the bottom. This group is represented by stonewort meadows (*Chara*, Fig. 3.4d), bush-like structures (*Chaemisiphon*, *Cladophora*, Fig. 3.4f), cushion-like colonies (*Vaucheria*), and filamentous patches loosely attached to the bottom (*Batrachospermum*, *Draparnaldia*, *Mougeotia*, *Microspora*, *Ulothrix*, *Tribonema*).

The biggest patches of macroalgae, often covering entire surfaces of bodies of water, frequently take the shape of mats

Fig. 3.2 Images of different types of architectural macroalgal mat in small water bodies: (**a**) Zygnemataceae in the shallow pond in Poznań; (**b**) *Cladophora* and *Ulothrix* attached to the bottom; (**c**) fragments of reeds as substrate for filamentous green algae; (**d**) *Cladophora glomerata* in a loosely overgrown water column in the shallow Lake Oporzyńskie; (**e**) *Cladophora glomerata* forming cloud-shaped mats; (**f**) *Cladophora glomerata* forming dense mats in a shallow lake

and felts, which can be very dense and carpet-like (thickness up to 30 cm, *Cladophora*, Fig. 3.4a) or loose and carpet-like (thalloid algae, *Ulva*, Fig. 3.4c). Shallow pond-like lakes often have their entire volume of water covered by *Cladophora glomerata*, forming loose patches of tangled threads (Fig. 3.4j) or compact communities created by patches that are pounded into one by waves (Fig. 3.4i).

Fig. 3.3 Images of different types of architectural macroalgal mat in ponds and rivers: (**a** and **b**) *Ulva flexuosa* in the shallow Lake Laskownickie; (**c**) C_1, C_2 – *Cladophora fracta* in the artificial pond in Poznań; (**d**) benthic *Cladophora glomerata*, attached to the bottom in a lowland river; (**e**) benthic *Cladophora glomerata*, attached to the bottom in the shallow Lake Oporzyńskie; (**f**) tubular thalli of *Ulva flexuosa* within *Cladophora glomerata* mats in the small Nielba River; (**g**) *Cladophora glomerata* forming long filaments as braids in the Nielba River; (**h**) free-floating form of *Cladophora glomerata* in a pond; (**i**) *Ulva* and *Vaucheria* attached to a stone substratum

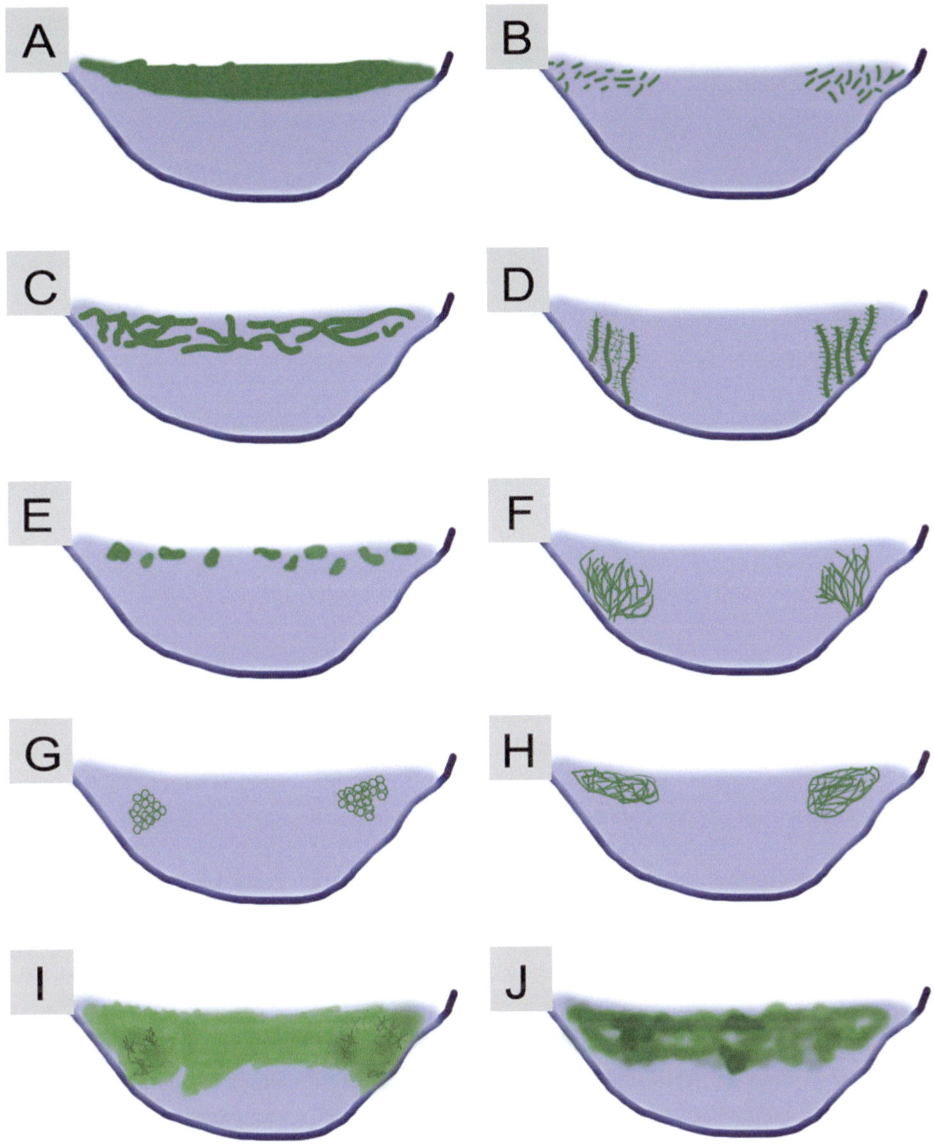

Fig. 3.4 Models of patches (mats) created by macroalgae in freshwater bodies. Types of structure: carpet-dense (**a**), flocs (**b**), carpet-loose (**c**), stonewort meadow (**d**), clouds (**e**), bushes (**f**), nets (**g**), spider webs (**h**), compact (**i**) and tangled (**j**)

3.4 The Effect of Dense Freshwater *Cladophora glomerata* Mats on Competitive Outcomes

Shallow lakes can have two alternative stable states over a wide range of nutrient concentrations: a submerged macrophyte-dominated in clear-water state and a phytoplankton- dominated in turbid-water state (Scheffer 2001). In some cases, filamentous algae may replace macrophytes (Pieczyńska 2008). According to Irfanullah and Moss (2005), mats of filamentous green algae can act as a buffer against a shift to phytoplankton dominance, thus maintaining a clear-water state; these account for up to 90% of total primary production in the ecosystem. Concentration of chlorophyll-a in mats is significantly higher than in places without filaments, which indicates that the mat is not strong enough of a buffer to protect against phytoplankton growth.

Therefore, our research revealed only lower concentrations of PO_4^{3-} in sites with mats, which indicated that nutrient supply is incorporated into the biomass (Pikosz and Messyasz 2016). It also confirmed that phosphorus is the main growth-limiting factor for *C. glomerata* (Freeman 1986; Higgins et al. 2008). Variables in nitrate and ammonium were not significant, both due to the inhibited process of assimilation that affects light and temperature conditions (Salisbury and Ross 1978) and decaying algae and biogenic compounds that may be recirculated into the water (Pieczyńska and Tarmanowska 1996). The presence of macroalgal mats in shallow water ecosystems may be a cause of reduced nutrient uptake by macroalgae, and therefore cause a significant decrease in nutrient

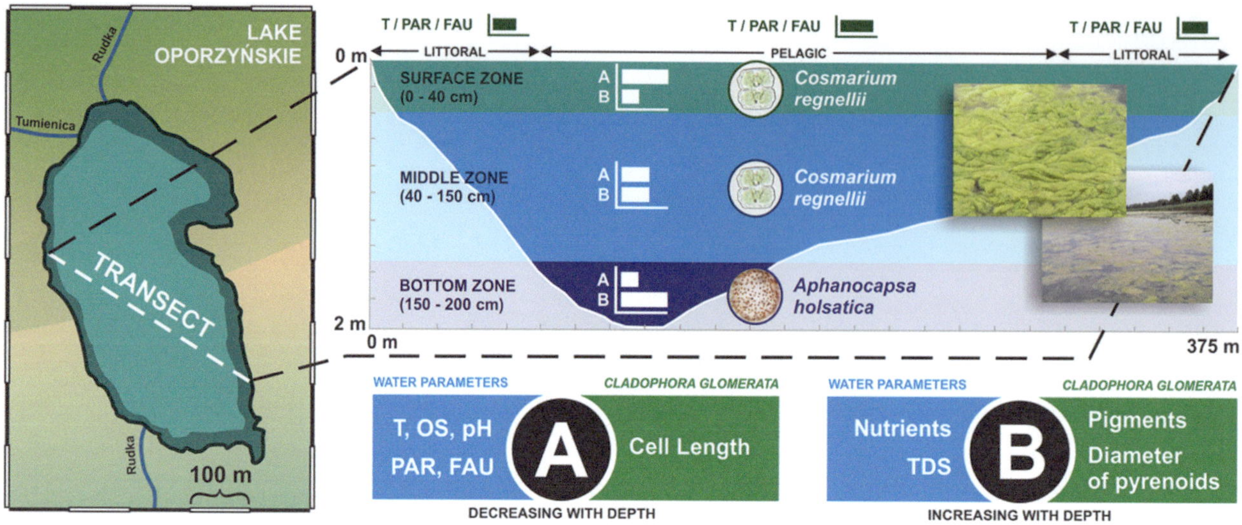

Fig. 3.5 Detailed algal mat characteristics, physical and chemical parameters of water, dominant taxa of phytoplankton, observed vertically (surface, middle, bottom) and horizontally (littoral, pelagic) in the stationary mat phase during maximum algal growth and biomass

fluxes from sediment to the overlying water column (Higgins et al. 2008). In contrast to data published by Khanum (1982), who argued that the massive development of *Cladophora* sp. has a significant impact on the total dissolved substance, the analysis of the impact that mats have on shallow Polish lakes did not show any statistical relation (Pikosz et al. 2017). The higher value of oxygen saturation in mats is due to the process of photosynthesis.

Understanding the role of dense mats in water ecosystems and environmental gradients is important, not only from the fundamental and practical perspectives but also in the aspect of biological conservation. Water temperature and light have an impact on both the vertical and horizontal distributions of *Cladophora glomerata* in dense mats in shallow lakes. Depending on the depth of the water column, the amount of light available differs considerably, which could potentially influence the growth rate of microalgae and macroalgae. The data clearly demonstrates that *C. glomerata*, which creates dense single-species mats, indicates heterogeneity in its vertical distribution and horizontal homogeneity (Fig. 3.5). This heterogeneity is attributable to the impact of water temperature, PAR, oxygen saturation (OS), pH, nutrient gradients (concentration of N-NO_3^-, NH_4^+, PO_4^{3-}), and chlorophyll-a generated by the depth of the water column. Temperature, dissolved oxygen, and PAR irradiation significantly declined with increasing depth in the mat. Pigment content in the cell of *C. glomerata*, cell length, and the diameter of pyrenoids changed from the top to the bottom of the mat. Vertical changes were also related to the phytoplankton structure (Pikosz et al. 2017).

In the horizontal profile of a dense *Cladophora* mat, phytoplankton was homogeneously distributed through the littoral and pelagic layers, and the concentrations progressively reached about 6000 cells per liter. Likewise, the differences in concentrations of chlorophyll-a between these two layers were small and not statistically significant. The situation was very different in the vertical profile, where, in the surface layer, chlorophyll-a concentration was lower than in the deeper part of the mat in the lake. The changes were statistically significant in contrast to the littoral and pelagic zones (Table 3.2). The total number of algae taxa identified in all samples equaled 117. When analyzing the phytoplankton's spatial changeability, it was found that diatoms decreased from the littoral to the pelagic, while green algae (*Chlorococcales*) reached significantly higher densities in the pelagic zone. Irrespective of the sampling site, the quantitative share of dominant *Cosmarium regnellii* Wille was similar, oscillating between 19% and 29% (Table 3.2). The richness of the phytoplankton species was clearly low (17–23; average 19) in the middle and bottom zones compared to the average number of 22–38 algae taxa per sample in the surface layer. In the vertical profile, among phytoplankton taxa, *Cosmarium regnellii* Wille was clearly dominant (24% on average) in the surface and middle parts, while the bottom zone was dominated by cyanobacterium *Aphanocapsa holsatica* (Lemm.) Cronberg et Komárek (31%). This variability also concerns the accompanying species (Table 3.2), which change along with depth: from cryptophytes (*Cryptomonas*) and chrysophytes (*Chrysochromulina*) near the surface through cyanobacteria (*Chroococcus*, *Planktolyngbya*) and the small green algae (*Desmodesmus*) at the bottom.

Desmid *Cosmarium regnellii* Wille, dominant in this system, is a small species, and its occurrence optimum has a broad pH range – from 6 to above 8 (Lenzenweger 1999). Its prevalence in the relatively high abundance of *Cosmarium regnellii* within sites of the littoral and pelagic zones of Lake

Table 3.2 The phytoplankton chlorophyll-a composition of water in Lake Oporzynskie along the vertical and spatial profiles (mean ± SD)

Sampling depth and profile	Chl a (μgL^{-1})	Dominant taxon	Accompanying taxa
Vertical profile			
Surface 0–40 cm (n = 5)	9.44 ± 6.53	*Cosmarium regnellii* Wille	*Rhodomonas lacustris* var. *nannoplanctonica* (Skuja) Javornicky
			Chrysochromulina parva Lackey
			Cryptomonas marssonii Skuja
Middle 50–150 cm (n = 11)	32.91 ± 17.42	*Cosmarium regnellii* Wille	*Chroococcus turgidus* (Kütz.) Nägeli
			Planktolyngbya limnetica (Lemm.) Komárková-Legnerová and Cronberg
			Tetraëdron minimum (A.Braun) Hansgirg
Bottom 160–200 cm (n = 5)	36.84 ± 7.48	*Aphanocapsa holsatica* (Lemm.) G.Cronberg et Komárek	*Desmodesmus communis* (E.Hegewald) E.Hegewald
			Desmodesmus opoliensis (P.G.Richter) E.Hegewald
			Crucigeniella rectangularis (Nägeli) Komárek
Friedman's ANOVA χ^2	24.39		
Kendall's W	0.37		
P	**0.00001**		
Spatial profile			
Littoral (n = 10)	18.70 ± 9.33	*Cosmarium regnellii* Wille	*Rhodomonas lacustris* var. *nanoplanctonica* (Skuja) Javornicky
			Pediastrum simplex Meyen
			Cocconeis placentula Ehrenberg
Pelagic (n = 15)	25.33 ± 14.40	*Cosmarium regnellii* Wille	*Oocystis lacustris* Chodat
			Crucigeniella rectangularis (Nägeli) Komárek
			Coelastrum astroideum De Notaris
t-statistic	−1.72		
P	0.10829		

Bold represents statistical significance, $p < 0.05$

Oporzynskie is a result of its preference for fertile waters. This species is often observed in eutrophic waters of lakes, both in phytoplankton (Gołdyn et al. 2013) and in periphyton (Messyasz and Kuczyńska-Kippen 2006). However, some minor variations were related to the accompanying species, depending on the location of the sampling site in the lake. In the littoral zone, a greater share of diatoms was found, especially *Cocconeis placentula* Ehr., which is a periphytic species. Analysis of epiphytic communities of freshwater macroalgae showed that *Cocconeis placentula* is a pioneer species in the colonization of the substrate, i.e., thalli of green algae, and can very quickly cover the macroalgal surface in a carpeting manner (Messyasz et al. 2012). In the case of high density of *Cladophora glomerata*, filaments forming dense mats, in the littoral zone of waving water, can tear off epiphytic diatoms (including *Cocconeis placentula*) from their surface, but also from reed stems, thereby enriching the water in diatoms. Confirmation of this phenomenon of periphyton's direct impact on phytoplankton structure is also seen in literature (Kitner and Poulíčková 2003; Poulíčková et al. 2008; John et al. 2009). This finding suggests that *Cladophora glomerata* is subjected to the impact of waving in the surface layer of water. In contrast, in the pelagic zone, where water temperature was clearly higher than in the littoral zone, an intensive development of small chlorococcal green algae was also observed. Dense planktonic green algae communities are usually highly abundant in eutrophicated water ecosystems, as well as in ecosystems with a strong gradient of salinity, i.e., coastal, shallow, and size-differentiated lakes (Burchardt et al. 2003).

On the vertical scale, the contrast between the dominant phytoplankton and the accompanying species composition was found in the surface and bottom zones. The classification system of phytoplankton strategies (C, S, R, and intermediate strategies C-S, C-R, R-S) developed by Reynolds (1988, 2006) is based on individual taxons' evolutionary responses to disturbance and stress and their use of limited resources. According to this system, a majority of phytoplankton species in Lake Oporzynskie belong to C-strategists (which dominate in low disturbance intensity and stress) and C-R strategists (which exhibit varying tolerance to turbidity and nutrient enrichment). Dominating in the bottom zone of the water column, cyanobacterium *Aphanocapsa holsatica* (Lemm.) G.Cronberg et Komárek is an example of an intermediate C-S strategy with low-temperature tolerance, simultaneously increasing the ability to exploit nutrients. This finding suggests that differences among the three layers (surface, middle, bottom) for phytoplankton in the vertical profile may indicate that a habitat of a dense *Cladophora glomerata* mat controls nutrients, the underwater light climate, and the temperature, and thus also the variability of strategies of microalgal species.

3.5 Biomass Production and Utilization of Macroalgae

Primary production is dependent on many environmental (including climatic) factors, as well as factors connected to habitat that are constant within a given water ecosystem (Whitton 1970; Reynolds 2006). Under normal conditions, primary production is diversified with regard to the amount of biomass produced, which is dependent on the type of ecosystem (river, pond, shallow/deep lake) and vegetative season (temperature and insolation, warm summer vs. cold summer). Eutrophic and hypertrophic waters are considered to be bodies of high macroalgae biomass production (Messyasz et al. 2015b). However, any possible water contamination may lead to damage invisible to the naked eye, which, if prolonged, can cause many long-term effects on basic processes: growth of macroalgal thalli and an increase in their biomass. Such damage can be observed in the incorrect forming of chloroplasts and changes in physiological processes. It decreases not only the biomass production itself but also the quality of the biomass in terms of biocenosis or economy. Many methods can be used to evaluate the production of biomass, but the most common one is simply the calculation of macroalgae biomass produced in a given area in a given time period. Results are given per dry mass of algae, and in case of experiments, it is also compared with the control (Khuantrairong and Traichaiyaporn 2009; Lenzi et al. 2011; Messyasz et al. 2015c). Macroalgae biomass production can be converted to g/m^2/day or to the amount of energy in the biomass. The diversification of biomass can be presented in accordance with the surface of the body of water, the place of the mat's growth, and its architectural structure. From the economic point of view (amount of biomass and costs of gathering), the growth of filamentous green algae, which form dense carpet-like mats that flow along the water's surface, is the most cost-effective.

The problem of excessive amounts of algae can be solved by using mats with large surfaces for economic purposes (Fig. 3.6). The most frequently used method of removing mats is gathering the green algae at the shore and letting it decompose. Macroalgae show the ability to accumulate trace elements, and therefore the usage of algae biomass can be successfully exploited: (i) in biological water treatment (to eliminate nutrients such as nitrogen and phosphorous), (ii) as an effective measure to prevent cyanobacteria blooms (natural algaecide), (iii) as a cheap absorbent in wastewater treatment plants (e.g., biosorption of heavy metals), and (iv) as an ingredient in fertilizers used to enrich soil with elements, because algae are a rich source of amino acids, peptides, proteins, and carbohydrates (Michalak and Chojnacka 2009; Michalak et al. 2015, 2016, 2017; Pankiewicz et al. 2016). They also consist of high amounts of essential unsaturated fatty acids (Messyasz et al. 2015b).

The thalli of the green alga *Cladophora*, when applied as a substrate performing the role of a biofilter, are more effective than aquatic plants in taking nutrient elements from the deposits in which they are rooted (Pieczyńska 1988). Thalli of freshwater forms of *Cladophora* or *Ulva* taxa are tychoplanktonic, rarely attached to the substratum, which allows for them to absorb nutrient elements and heavy metals through the entirety of their surface (Whitton 1984; Chmielewská and Medved 2001; Rybak et al. 2012). Using the thalli of these macro-green algae as biofilters seems to be a very promising appliqué method for the improvement of water quality and, at the same time, makes it possible for the algae to reach high biomass. The assumption that the thalli of

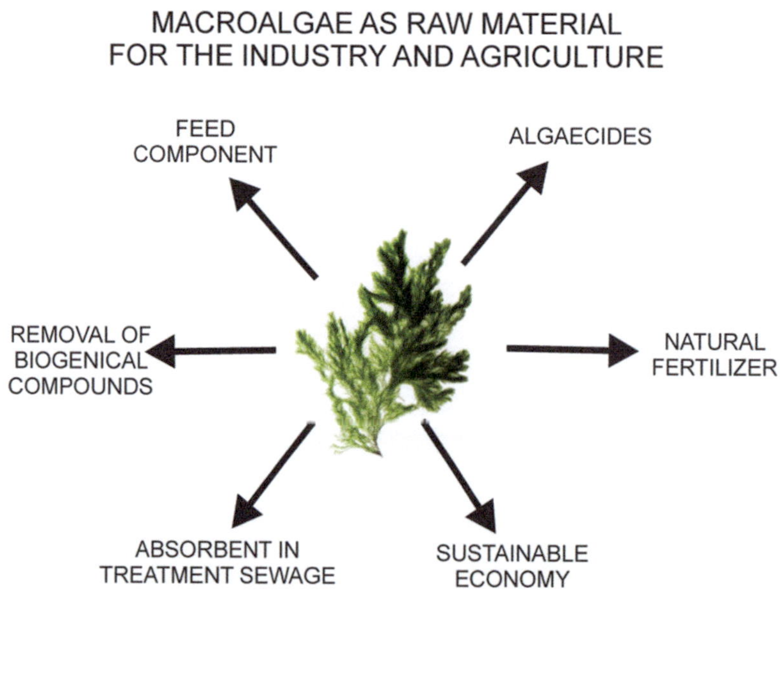

Fig. 3.6 Scheme showing the economic capacity of freshwater macroalgae biomass utilization

freshwater forms of macro-green algae are characterized by a great dual ability to absorb nitric and phosphoric compounds, as well as microelements, from water and accumulate these elements in cells results both from the fast rate of biomass increase and promising research findings regarding the sea forms of these algae (Miyata et al. 1977; Markager and Sand-Jensen 1996; Wosnitza and Barrantes 2005; Michalak and Chojnacka 2009; Alcaraz et al. 2013). The effectiveness of macroalgae thalli as a substrate performing the role of a biofilter is closely connected to the determination of their physiological abilities to absorb and their accumulation of nutrient compounds from water. On that score, the examination of the qualitative and quantitative composition of nutrient elements accumulated in macroalgae thalli will concern developing populations, under both experimental and farm conditions. Comparative breeding should demonstrate the optimal conditions for the growth of thalli and their accumulation of nutrient elements and mineral substances. At the same time, it will determine the minimal level at which desired changes in water quality are occurring, and the level of success and effectiveness of the method itself. Application of the phenomenon of N and P accumulation by macroalgae *Cladophora* or *Ulva* as a biofilter or algaecides is opening up new prospects for aquatic economy and their practical usage. According to the authors of this chapter, interest in this biological method will be significant, on account of the efficient elimination of nutrient elements from wastewater and reduced cyanobacteria blooms.

References

Adhikary SP, Sahu JK (1992) Distribution and seasonal abundance of algal forms in Chilka Lake, East Coast of India. Jpn J Limnol (Rikusuigaku Zasshi) 53(3):197–205

Alcaraz JLM, Canales LM, Sanjurjo MA (2013) Morphological description and ecology of some rare macroalgae in south-central Spanish rivers (Castilla-La Mancha Region). An Jard Bot Madrid 70(1):81–90

Bellinger EG, Sigee DC (2010) Freshwater algae: identification and use as bioindicators. Wiley-Blackwell, Oxford

Berner T, Dubinsky Z, Wyman K, Falkowski PG (1989) Photoadaptation and the "package effect" in *Dunaliella tertiolecta* (Chlorophyceae). J Phycol 25(1):70–78

Bischof K, Peralta G, Kräbs G, van de Poll WH, Pérez-Lloréns JL, Breeman AM (2002) Effects of solar UV-B radiation on canopy structure of Ulva communities from southern Spain. J Exp Bot 53(379):2411–2421

Blomster J, Hoey EM, Maggs CA, Stanhope MJ (2000) Species-specific oligonucleotide probes for macroalgae: molecular discrimination of two marine fouling species of *Enteromorpha* (*Ulvophyceae*). Mol Ecol 2:177

Borja A, Bricker SB, Dauer DM, Demetriades NT, Ferreira JG, Forbes AT, Hutchings P, Jia X, Kenchington R, Carlos Marques J, Zhu C (2008) Overview of integrative tools and methods in assessing ecological integrity in estuarine and coastal systems worldwide. Mar Pollut Bull 56:1519–1537

Bricker SB, Longstaff B, Dennison W, Jones A, Boicourt K, Wicks C, Woerner J (2008) Effects of nutrient enrichment in the nation's estuaries: a decade of change. Harmful Algae 8:21–32

Burchardt L, Messyasz B, Owsianny PM, Pełechata A, Stefaniak K (2003) Chlorococcales algae from four lakes in the Slowinski National Park (northern Poland). Biologia (Bratisl) 58(4):467–474

Chemello R, Milazzo AM (2002) Effect of algal architecture on associated fauna: some evidence from phytal molluscs. Mar Biol 140:981–990

Chmielewská E, Medved J (2001) Bioaccumulation of heavy metals by green algae *Cladophora glomerata* in a refinery sewage lagoon. Croat Chem Acta 74(1):135–145

Chudyba H (1965) *Cladophora glomerata* and accompanying algae in the Skawa River. Acta Hydrobiol 7(1):93–126

Dąmbska I, Karpiński J (1954) Ramienice, klucz do oznaczania gatunków krajowych. PWN, Warszawa

Das S, Deshmukhe G, Dwivedi A (2014) Grazing of selected genera of green, red and brown macroalgae. Appl Ecol Environ Res 12(3):717–725

Dere S, Günes T, Sivaci R (1998) Spectrophotometric determination of chlorophyll-a, b and total carotenoid contents of some algae species using different solvents. Turk J Bot 22:13–17

Dodds WK, Gudder DA (1992) The ecology of Cladophora. J Phycol 28:415–427

Duarte CM (1995) Submerged aquatic vegetation in relation to different nutrient regimes. Ophelia 41:87–112

Ensminger I, Hagen C, Braune W (2000) Strategies providing success in a variable habitat. II. Ecophysiology of photosynthesis of *Cladophora glomerata*. Plant Cell Environ 23:1129–1136

Ensminger I, Xyländer M, Hagen C, Braune W (2001) Strategies providing success in a variable habitat: III. Dynamic control of photosynthesis in *Cladophora glomerata*. Plant Cell Environ 24(8):769–779

Ensminger I, Schmidt L, Tittmann S, Lloyd J (2005) Will photosynthetic gain of boreal evergreen conifers increase in response to a potentially longer growing season? In: Carpentier R, Bruce D, van der Est A (eds) Photosynthesis: fundamental aspects to global perspectives, vol 2. Allen Press, Lawrence, pp 976–978

Farina J, Castilla J, Ojeda F (2003) The idiosyncratic effect of a sentinel species on contaminated rocky intertidal communities. Ecol Appl 6:1533–1552

Fletcher RL (1996) The "Green Tide" problem. In: Schramm W, Nienhuis P (eds) Ecological studies, marine benthic vegetation. EEC, Brussels, pp 29–33

Freeman MC (1986) The role of nitrogen and phosphorus in the development of *Cladophora glomerata* (L.) Kutzing in the Manawatu River, New Zealand. Hydrobiologia 131:23–30

Gąbka M (2007) Distribution of *Chara tenuispina* A. Braun 1835 (Characeae) in Poland. Oceanol Hydrobiol Stud 36(1):241–248

Gąbka M (2009) Charophytes of the Wielkopolska region (NW Poland): distribution, taxonomy and autecology. Bogucki Wydawnictwo Naukowe Poznań, p 109

Garcia ME, Aboal M (2014) Environmental gradients and macroalgae in Mediterranean marshes: the case of Pego-Oliva marsh (East Iberian Peninsula). Sci Total Environ 475:216–224

Gołdyn R, Messyasz B, Domek P (2013) The response of Lake Durowskie ecosystem to restoration measures. Carpathian J Earth Environ Sci 8(3):43–48

Green L, Fong P (2015) The good, the bad and the Ulva: the density dependent role of macroalgal subsidies in influencing diversity and trophic structure of an estuarine community. Oikos. https://doi.org/10.1111/oik.02860

Guiry MD, Guiry GM (2017) AlgaeBase. World-wide electronic publication. National University of Ireland, Galway. http://www.algaebase.org. Searched on 05–26.05.2017

Hainz R, Wöber C, Schagerl M (2009) The relationship between *Spirogyra* (*Zygnematophyceae*, *Streptophyta*) filament type groups and environmental conditions in Central Europe. Aquat Bot 91(3):173–180

Hard DD (1992) Community organization in streams: the importance of species interactions, physical factors, and chance. Oecologia 91:220–228

Herbst RP (1969) Ecological factors and the distribution of *Cladophora glomerata* in the Great Lakes. Am Midl Nat 82:90–98

Higgins SN, Malkin SY, Howell TE (2008) An ecological review of *Cladophora glomerata* (Chlorophyta) in the Laurentian Great Lakes. J Phycol 44(4):839–854

Hoffmann JP, Graham LE (1984) Effects of selected physicochemical factors on growth and zoosporogenesis of *Cladophora glomerata* (Chlorophyta). J Phycol 20:1–7

Hutchinson GE (1957) Concluding remarks. Cold Spring Harb Symp Quant Biol 22(2):415–427

Irfanullah HM, Moss B (2005) A filamentous green algae-dominated temperate shallow lake: variations on the theme of clear-water stable states? Arch Hydrobiol 163(1):25–47

John R, Kusber WH, Romero OE (2009) *Cocconeis pediculus* Ehrenberg and *C. placentula* Ehrenberg *var. placentula* (Bacillariophyta): typification and taxonomy. Fottea 9(2):275–288

Khanum A (1982) An ecological study of freshwater algal mats. Bot Bull Acad Sin 23:89–104

Khuantrairong T, Traichaiyaporn S (2009) Production of biomass, carotenoid and nutritional values of *Cladophora* sp. (Kai) by cultivation in mass culture. Phycologia 48(4 Supplement):60

Kitner M, Poulićková A (2003) Littoral diatoms as indicators for the eutrophication of shallow lakes. Hydrobiologia 506–509:519–524

Krause-Jensen D, McGlathery K, Rysgaard S, Christensen PB (1996) Production within dense mats of the filamentous macroalga *Chaetomorpha linum* in relation to light and nutrient availability. Mar Ecol Prog Ser 134:207–216

Krause-Jensen D, Christensen PB, Rysgaard S (1999) Oxygen and nutrient dynamics within mats of the filamentous macroalga *Chaetomorpha linum*. Estuaries 22:31–38

Leliaert F, Verbruggen H, Wysor B, De Clerck O (2000) DNA taxonomy in morphologically plastic taxa: algorithmic species delimitation in the *Boodlea* complex (Chlorophyta: Cladophorales). Mol Phylogenet Evol 53:122–133

Leliaert F, Smith DR, Moreau H, Herron M, Verbruggen H, Delwiche CF, De Clerck O (2012) Phylogeny and molecular evolution of the green algae. Crit Rev Plant Sci 31:1–46

Lenzenweger R (1999) Desmidiaceenflora von Österreich, Teil 3. Bibliotheca Phycologica Band, vol 104. J. Cramer in der Gebrüder Borntraeger Verlagsbuchhandlung, Berlin, pp 1–218

Lenzi M, Renzi M, Nesti U, Gennaro P, Persia E, Porello S (2011) Vegetation cyclic shift in eutrophic lagoon. Assessment of dystrophic risk indices based on standing crop evaluation. Estuar Coast Shelf Sci 132:99–107

Lenzi M, Gennaro P, Mercatali I, Persia E, Solari D, Porello S (2013) Physicochemical and nutrient variable stratifications in the water column and in macroalgal thalli as a result of high biomass mats in a non-tidal shallow-water lagoon. Mar Pollut Bull 75(1–2):98–104

Lester WW, Adams MS, Farmer AM (1988) Effects of light and temper ature on photosynthesis of the nuisance alga *Cladophora glomerata* (L.) Kütz from Green Bay, Lake Michigan. New Phycol 109:53–58

Lichtenthaler HK, Wellburn AR (1983) Determination of total carotenoids and chlorophyll a and b of leaf extract in different solvents. Biochem Soc Trans 11:591–592

Malta E, Draisma SGA, Kamermans P (1999) Free-floating *Ulva* in the southwest Netherlands: species or morphotypes? A morphological, molecular and ecological comparison. Eur J Phycol 34:443–454

Markager S, Sand-Jensen K (1996) Implications of thallus thickness for growth-irradiance relationships of marine macroalgae. Eur J Phycol 31:79–87

McGlathery KJ, Krause-Jensen D, Rysgaard S, Christensen PB (1997) Patterns of ammonium uptake within dense mats of the filamentous macroalga *Chaetomorpha linum*. Aquat Bot 59:99–115

Messyasz B, Kuczyńska-Kippen N (2006) Periphytic algal communities: a comparison of Typha angustifolia L. and *Chara tomentosa* L. beds in three shallow lakes (West Poland). Pol J Ecol 54(1):15–27

Messyasz B, Rybak A (2009) The distribution of green algae species from the *Enteromorpha* genera [syn. *Ulva*; Chlorophyta] in Polish inland waters. Oceanol Hydrobiol Stud 38(1):121–138

Messyasz B, Rybak A (2011) Abiotic factors affecting the development of *Ulva* sp. (Ulvophyceae; Chlorophyta) in freshwater ecosystems. Aquat Ecol 45(1):75–87

Messyasz B, Pikosz M, Rybak A, Łepkowska K (2012) Epiphytic diatom community and calcium carbonate crystals characteristics on the surface of freshwater *Ulva* thalli. Teka Kom Ochr Kszt Środ Przyr – OL PAN 9:96–106

Messyasz B, Łęska B, Fabrowska J, Pikosz M, Rój E, Cieślak A, Schroeder G (2015a) Biomass of freshwater Cladophora as a raw material for agriculture and the cosmetic industry. Open Chem 13:1108–1118. https://doi.org/10.1515/chem-2015-0124

Messyasz B, Łęska B, Fabrowska J, Pikosz M, Cieślak A, Schroeder G (2015b) Effects of organic compounds on the macroalgae culture of freshwater *Aegagropila*. Open Chem 13:1040–1044

Messyasz B, Pikosz M, Schroeder G, Łęska B, Fabrowska J (2015c) Chapter 2: Identification and ecology of macroalgae species existing in Poland. In: Kim SK, Chojnacka K (eds) Marine algae extracts: processes, products and applications, 1st edn, pp 17–39

Michalak I, Chojnacka K (2009) Edible macroalga *Ulva prolifera* as microelemental feed supplement for livestock: the fundamental assumptions of the production method. World J Microbiol Biotechnol 25:997–1005

Michalak I, Dmytryk A, Wieczorek PP, Rój E, Łęska B, Górka B, Messyasz B, Lipok J, Mikulewicz M, Wilk R, Schroeder G, Chojnacka K (2015) Supercritical algal extracts: a source of biologically active compounds from nature. J Chem – Hindawi 2015:597140. https://doi.org/10.1155/2015/597140

Michalak I, Górka B, Wieczorek PP, Rój E, Lipok J, Łęska B, Messyasz B, Wilk R, Schroeder G, Dobrzyńska-Inger A, Chojnacka K (2016) Supercritical fluid extraction of algae enhances levels of biologically active compounds promoting plant growth. Eur J Phycol 51:243–252. https://doi.org/10.1080/09670262.2015.1134813

Michalak I, Chojnacka K, Saeid A (2017) Plant growth biostimulants, dietary feed, supplements and cosmetics formulated with supercritical CO_2 algal extracts. Molecules 22(66):1–17. https://doi.org/10.3390/molecules22010066

Miyata M, Okazaki M, Furuya K (1977) Site and nature of calcium carbonate deposits on calcareous Brown alga *Padina japonica* – (Studies on the calcium carbonate deposition of alga). Bull Jpn Soc Phycol 25:1–6

Montoya-Moreno Y, Aguirre-Ramírez N (2013) Knowledge to ecological preferences in a tropical epiphytic algae to use with eutrophication indicators. J Environ Protect 4:27–35

Ozimek T (1990) Aspects of the ecology of a filamentous alga in a eutrophicated lake. Hydrobiologia 191:23–27

Painter DS, Kamaitis G (1987) Reduction of Cladophora biomass and tissue phosphorus in lake Ontario, 1972–83. Can J Fish Aquat Sci 44(12):2212–2215

Pankiewicz R, Łęska B, Messyasz B, Fabrowska J, Sołoducha M, Pikosz M (2016) First isolation of polisacharidic ulvans from the cell wall of freshwater algae. Algal Res 19:348–354. https://doi.org/10.1016/j.algal.2016.02.025

Pascelli C, Riul P, Riosmena-Rodriguez R, Scherner F, De Castro Nunes JM, Hall-Spencer JM, de Oliveira EC, Horta PA (2013) Seasonal and depth-driven changes in rhodolith bed structure and associated macroalgae off Arvoredo island (southeastern Brazil). Aquat Bot 111:62–65

Pełechaty M (2005) Does spatially varied phytolittoral vegetation with significant contribution of charophytes cause spatial and temporal heterogeneity of physical-chemical properties of the pelagic waters of a tachymictic lake? Polish J Environ Stud 14(5):63–73

Pełechaty M, Gąbka M (2003) Two approaches to lake-naturalness determination – a case study from four mid-forest Polish lakes. Polish J Environ Stud 12(4):440–445

Philips DJH (1990) Use of macroalgae and invertebrates as monitor of heavy metals in estuaries and coastal waters. In: Furness RW, Rainbow PS (eds) Heavy metals in marine environment. CRC Press, Boca Raton, pp 81–99
Pieczyńska E (1988) Rola makrolitów w kształtowaniu trofii jezior. Wiad Ekol 34(4):375–404
Pieczyńska E (2008) Eutrofizcja płytkich jezior – znaczenie makrofitów. Wiad Ekol 54(1):3–28
Pieczyńska E, Tarmanowska A (1996) Effect of decomposing filamentous algae on the growth of *Elodea canadensis* Michx. (a laboratory experiment). Aquat Bot 54:313–319
Pikosz M, Messyasz B (2015a) Composition and seasonal changes in filamentous algae in floating mats. Oceanol Hydrobiol Stud 44(2):273–281
Pikosz M, Messyasz B (2015b) New data to distribution, morphology and ecology of *Oedogonium capillare* Kützing ex Hirn (Oedogoniales, Chlorophyta) in Central Europe (Poland). Biodivers Res Conserv 40:31–36
Pikosz M, Messyasz B (2016) Characteristic of Cladophora and coexisting filamentous algae against a background of environmental factors in freshwater ecosystems in Poland. Oceanol Hydrobiol Stud 45(2):202–2015
Pikosz M, Messyasz B, Gąbka M (2017) Functional structure of algal mat (Cladophora glomerata) in a freshwater in western Poland. Ecol Indic 74:1–9. https://doi.org/10.1016/j.ecolind.2016.09.041
Poulíčková A, Hašler P, Lysaková M, Spears B (2008) The ecology of freshwater epipelic algae: an update. Phycologia 47(5):437–450
Reynolds CS (1988) Functional morphology and the adaptive strategies of freshwater phytoplankton. In: Sandgren CD (ed) Growth and reproductive strategies of freshwater phytoplankton. Cambridge University Press, Cambridge, pp 388–433
Reynolds CS (2006) Ecology of phytoplankton. Cambridge University Press, Cambridge, p 535
Rybak A, Messyasz B, Łęska B (2012) Freshwater *Ulva* (Chlorophyta) as a bioaccumulator of selected heavy metals (Cd, Ni and Pb) and alkaline earth metals (Ca and Mg). Chemosphere 89:1066–1076
Sakayama H (2008) Taxonomy of *Nitella* (Charales, Charophyceae) based on comparative morphology of oospores and multiple DNA marker phylogeny using cultured material. Phycol Res 56:202–215
Sakayama H, Kasai F, Nozaki H, Watanabe MM, Kawachi M, Shigyo M, Nishihiro J, Washitani I, Krienitz L, Ito M (2009) Taxonomic reexamination of *Chara globularis* (Charales, Charophyceae) from Japan based on oospore morphology and *rbc*L gene sequences, and the description of *C. leptospora* sp. nov. J Phycol 45:917–927
Salisbury FB, Ross CW (1978) Plant physiology, 2nd edn. Wadsworth Publ. Co., Inc, Belmont
Saunders LL, Kilham SS, Fairchild GW, Verb R (2012) Effects of small-scale environmental variation on metaphyton condition and community composition. Freshw Biol 57:1884–1895. https://doi.org/10.1111/j.1365-2427.2012.02851.x
Scheffer M (2001) Alternative attractors of shallow lakes. Sci World 1:254–263
Simons J, van Beem AP (1990) Spirogyra species and accompanying algae from pools and ditches in The Netherlands. Aquat Bot 37(3):247–269
Stewart HL, Carpenter RC (2003) The effects of morphology and water flow on photosynthesis of marine macroalgae. Ecology 84(11):2999–3012
Thybo-Christesen M, Blackburn TH (1993) Internal N-cycling, measured by $^{15}NH_4^+$ dilution, in *Cladophora sericea* in a shallow Danish bay. Mar Ecol Prog Ser 100:283–296
Van den Hoek C, Mann DG, Jahns HM (1995) Algae. An introduction to phycology, 1st edn. Cambridge University Press, Cambridge, p 623
Vergara JJ, Pérez-Lloréns JL, Peralta G, Hernández I, Niell FX (1997) Seasonal variation of photosynthetic performance and light attenuation in Ulva canopies from Palmones river estuary. J Phycol 33:773–779
Weiss A, Costa R, Wichard T (2017) Morphogenesis of Ulva mutabilis (Chlorophyta) induced by Maribacter species (Bacteroidetes, Flavobacteriaceae). Bot Mar 60(2):197–206
Whitton BA (1970) Biology of freshwater *Cladophora*. Water Res 4:457–476
Whitton BA (1984) Algae as monitors of heavy metals in freshwaters. In: Shubert LE (ed) Algae as ecological indicators. Academic, London, pp 257–280
Wosnitza MAT, Barrantes JG (2005) Utilization of seaweed *Ulva* sp. in Paracas Bay (Peru): experimenting with compost. J Appl Phycol 18:27–31
Yniguez AT, McManus JW, Collado-Vides L (2015) Consequences of morphological plasticity and fragmentation on space occupation of coral reef macroalgae. Ecol Model 309–310:128–142
Yoshii Y, Hanyuda T, Wakana K, Miyaji K, Arai S, Ueda K, Inouye I (2004) Carotenoid compositions of Cladophora balls (*Aegagropila linnaei*) and some members of the Cladophorales (Ulvophyceae, Chlorophyta): their taxonomic and evolutionary implication. J Phycol 40:1170–1177

Algae in Biotechnological Processes

4

Emilia Niemczyk, Beata Żyszka-Haberecht, Damian Drzyzga, Monika Lenartowicz, and Jacek Lipok

Abstract

As photoautotrophic organisms, algae possess all of the valuable features that determine their role as the primary producers in the biosphere. A wide range of tolerance based on their extremely efficient adaptation to biochemical processes, as well as the specific cellular structure of these organisms, when correlated with the ecological plasticity of microalgae in particular, predispose these biota to growing and developing under either laboratory or industrial conditions. Hence, the natural features of algae have opened wide the door for the multidirectional biotechnological use of these organisms, with a dynamically growing number of such applications fully supporting this thesis. Among the variety of examples, however, there are two main areas of activity that involve algae in biotechnological processes. The first has arisen historically out of the long tradition of the use of biomass of algae or algal isolates as a source of substances with qualities of interest. The second area is based on the impressive biochemical machinery of algae that are able to produce, de novo, a huge number of organic compounds, as well as transform all of them. This approach allows for the use of algae as effective biocatalysts. This chapter is composed of four short stories, two of which illustrate algae as a source of select valuable chemicals (phycobiliproteins, polyphenols) and the other two of which are dedicated to the biocatalytical abilities of those organisms to protect ecosystems against organic pollutants and transition metal ions.

E. Niemczyk · B. Żyszka-Haberecht · D. Drzyzga · M. Lenartowicz
Faculty of Chemistry, University of Opole, Opole, Poland

J. Lipok (✉)
Faculty of Chemistry, Department of Analytical and Ecological Chemistry, Opole University, Opole, Poland
e-mail: jalip@uni.opole.pl

Keywords

Microalgae · Cyanobacteria · Natural products · Biocatalysis · Phycobiliproteins · Polyphenols · Organic pollutants · Metal ions

4.1 Introduction

Algae seem to be a group of organisms especially predisposed to various biotechnological purposes. As photoautotrophic organisms, these biota possess all sorts of valuable biochemical features that determine their role as the primary producers in the biosphere. The ability to produce the majority of known organic compounds due to an extremely efficient combination of primary and secondary metabolisms places these organisms at the center of interest for biotechnologists. It is not insignificant that the nutritional requirements of algae are relatively poor, especially if we are comparing the chemical nature of nutrients – simple inorganic salts – with their products, structurally complex organic compounds.

All of these metabolic features, coupled with maintenance in water as a medium and the possibility of using both sunlight and artificial light as sources of necessary energy, have determined that algae nowadays, especially microalgae, play an increasing role as a biotechnological source of pharmaceuticals and cosmetics (bioactive molecules of novel modes of action or of new chemical scaffolds) and nutraceuticals (functional foods, probiotics, antioxidants), as well as substances of industrial significance, i.e., biofuels and enzymes. It is worth adding that because of historical (old taxonomy) and functional (photosynthesis) issues, some strains of cyanobacteria, such as blue-green algae, are often included in the microalgae group. Due to the specific metabolism of these photosynthetic bacteria, the biotechnological meaning of the entire group of microalgae (algae) is significantly enriched, especially in regard to the matter of bio-

© Springer International Publishing AG, part of Springer Nature 2018
K. Chojnacka et al. (eds.), *Algae Biomass: Characteristics and Applications*, Developments in Applied Phycology 8,
https://doi.org/10.1007/978-3-319-74703-3_4

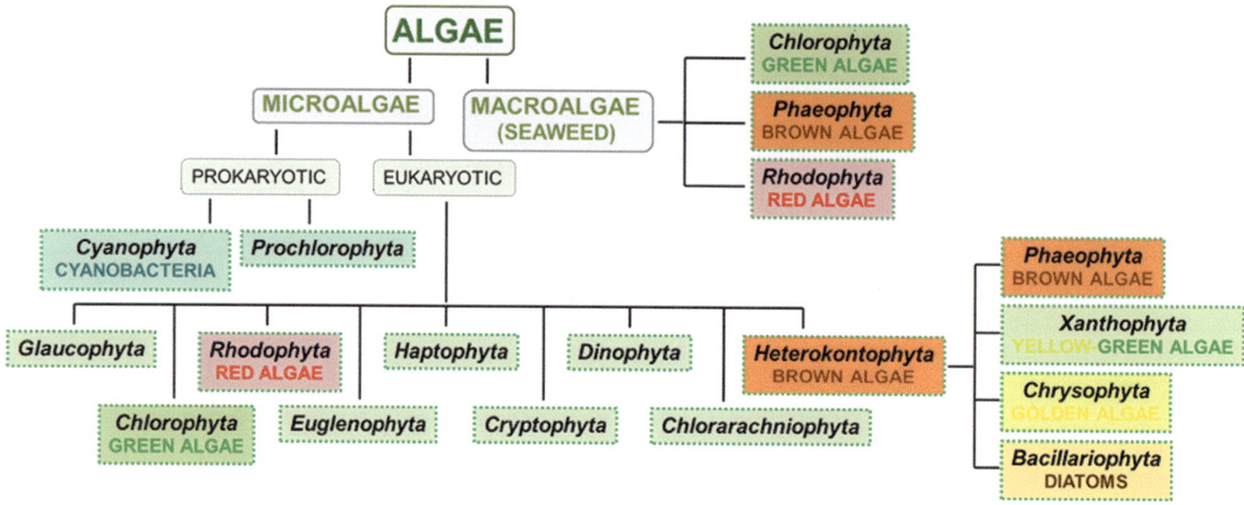

Fig. 4.1 Classification scheme of the different algal groups

transformation processes, including biodegradation, in the context of environmental biotechnology (Fig. 4.1).

However, the benefits that are obtained due to the controlled usage of algae, or their functional parts in biotechnology, depend significantly on a few constraints, among which product extraction and purification, strain stability, productivity, and downstream processing, as well as engineering problems, seem to be the most important. Thus, the advantages and disadvantages that are expressed nowadays regarding the use of algae in biotechnological processes are presented in respect to algae as a source of valuable chemicals, as well as in regard to the biocatalytic capacities of these biota.

The ability to produce a variety of organic compounds due to the extremely efficient combination of primary and secondary metabolisms possesses special meaning in the context of the formation of chemicals of highly desirable biological activity.

4.2 Photosynthetic Constituents of Microalgae and Their Biotechnological Usefulness: Phycobilisomes and Phycobiliproteins

The key element of the photosynthetic apparatus of microalgae and macroalgae is constituted by phycobiliproteins, which aggregate each other into phycobilisomes. Phycobilisomes are giant subcellular complexes, composed of different phycobiliproteins and linker proteins in a manner that increases absorption of the energy of light and its transfer through the photosystem apparatus. Phycobilisomes consist mostly of phycobiliproteins (80%) – a protein backbone to which linear tetrapyrrole prosthetic groups (called phycobilins) are covalently attached (Sekar and Chandramohan 2007). Phycobiliproteins are arranged into phycobilisomes and are stabilized by colorless link proteins (Tandeau de Marsac and Cohen-Bazire 1977). Aside from the main functions of phycobilisomes, they serve as an element of a photoprotective system under conditions of intense light (Kirilovsky and Kerfeld 2013). Furthermore, when access to sources of nutrients becomes limited, under the conditions of nitrogen and/or phosphorus starvation, the degradation of phycobiliproteins occurs so as to ensure the survival of the cells (Baier et al. 2004).

Phycobiliproteins belong to the family of water-soluble pigment proteins, which consists of over ten of these pigments. These proteins are found in prokaryotic cyanobacteria, eukaryotic red algae, and glaucophytes, in which they construct phycobilisomes. Similarly, cryptomonads produce phycobiliproteins, but they are not arranged into multiplexes (Glazer 1994). The four main classes of phycobiliprotein, based on their spectral features, are bluish-green allophycocyanin (APC, λ_{Amax} = 650–655 nm, λ_{Fmax} = 657–660 nm), blue phycocyanin (PC, λ_{Amax} = 610–620 nm, λ_{Fmax} = 645–653 nm), reddish-orange phycoerythrin (PE, λ_{Amax} = 540–570 nm, λ_{Fmax} = 575–590 nm), and reddish-purple phycoerythrocyanin (PEC, λ_{Amax} = 560–600, λ_{Fmax} = 610–625). The number and type of phycobilins are responsible for the color of the phycobiliproteins. Two bilins are found in allophycocyanin and three bilins each in phycocyanin and phycoerythrocyanin, with phycoerythrin carrying five or six prosthetic groups (Bryant 1982; Stadnichuk et al. 2015). Phycobilin biosynthetic pathways with 5-aminolevulinic acid, heme, or biliverdin as precursors have been suggested by various researches (Brown et al. 1984; Beale and Cornejo 1991). Phycobiliproteins (especially phycocyanin and phycoerythrin) are unique natural products that present advantageous qualities, with the wide spectrum of their properties making them appropriate for application in the food, cosmetic, and pharmaceutical industries, as well as in bioresearches. Phycocyanin, as a nontoxic and non-carcinogenic pigment, is used as a food, drug, and cosmetic colorant. This blue protein obtained from *Arthrospira platensis* is used in coloring chewing gum, desserts, milk products, ice creams, and sweets; however, in the USA, the extract obtained from the dried biomass of this species has only been permitted for use as a safe colorant since 2014 (Sarada et al. 1999; Borowitzka 2013; FDA 2014). Phycocyanin demonstrates great antioxidant activity, which, in the future, may lead to medical use of the protein, since many diseases, like Alzheimer's or Parkinson's, are connected to the formation of reactive oxygen species (Simonian and Coyle 1996; Romay et al. 2003). Moreover, many studies have reported a great number of other advantageous health and nutraceutical effects of phycobiliproteins, including neuroprotective, anti-inflammatory, anticancer, and hepatoprotective (Romay et al. 1998; Rimbau et al. 1999; Sathyasaikumar et al. 2007; Vázquez-Sánches et al. 2009). Therefore, phycobiliproteins are no longer just pigments but have become known as valuable bioactive compounds, which, in the cosmetic industry, could function as vital antiaging factors or as additives for functional food.

Because of their fluorescent properties, phycobiliproteins possess growing importance in biomedical research and immunodiagnostics (Sarada et al. 1999). Due to its bright fluorescent, high photostability, quantum yield, and large Stokes shift, phycoerythrin is considered as the highest-quality fluorophore. Glazer (1994) showed that phycoerythrin is suitable for conjugation with biological molecules, like protein A, antibodies, biotin, and others. Phycoerythrin's fluorescence opens up the doorway for its widespread application in flow cytometry, histometry, fluorescent microscopy, and immunoassay (Kronick 1988). Most common flow cytometers are equipped with detectors of orange-emitting phycoerythrin and fluorescein (Telford et al. 2012). Phycoerythrin fulfills specific criteria and works well as a fluorescent label for commercial confocal microscopes with krypton or argon lasers (Wells and Johnson 1994). The increasing range of application of phycobiliproteins entails demand for these compounds with various purity ratios.

4.2.1 Cultivation Strategies Leading to Obtainment of Phycobiliproteins

Biotechnological production of phycobiliproteins is a multistep process, involving numerous factors and conditions, since, in order to obtain high efficiency and the required purity, both the biomass rate and the pigment content in the cells must be adjusted. Conditions of microalgal growth, such as the composition and pH of the medium, light intensity and regime, temperature, and concentration of CO_2, are often modified to enhance the biosynthesis of phycobiliproteins by algal cells (Borowitzka 2016). These protein pigments can be formed with strategies that use different trophic conditions, such as photoautotrophic, mixotrophic, or heterotrophic (Borowitzka and Moheimani 2013).

Currently, there are two approaches to photoautotrophic production of phycobiliproteins under sunlight: cultivation of red algae or cyanobacteria in open pond systems and cultivation in closed bioreactors. Open pond culture systems are considered the main type for large-scale cultivation of algae at relatively low cost. However, their use is limited by the poor and uneven availability of light, lack of control over cultivation conditions, and vulnerability to contamination from other bacteria or algae (Jorquera et al. 2010; Usher et al. 2014). Therefore, the production of phycobiliproteins through outdoor methods is restricted to a small number of algal species. Today, the species most often used in this type of culture is *Arthrospira platensis* (Milledge 2011). Because of the inability to use many species and the possibility of contamination, closed photo-bioreactors seem to be preferred for production of phycobiliproteins over open ponds. Such systems feature equalized light delivery, increased productivity, and better CO_2 distribution. Moreover, these methods of cultivation allow for control of the temperature and other culture criteria, i.e., light intensity, composition, and pH of medium or mixing parameters. However, despite all of these advantages, closed cultivation systems have still generated extremely high costs that might only be acceptable for the production of refined, natural chemicals.

Mixotrophic methods of production of these proteins are carried out with the addition of a carbon source, mostly glucose, in enclosed photo-bioreactors. The main issue related to this system is the possibility of contamination of heterotrophs. The presence of glucose in the nutrient medium results in a higher growth rate and higher phycocyanin production compared to photautotrophic cultivation (Marquez et al. 1993; Chen et al. 1996). Yu et al. (2011) found that mixotrophic conditions changed an arrangement of photosynthetic pigments in *Anabeana* sp. In this case, the content of phycobiliproteins relative to chlorophyll *a* was decreased. Red algae, *Galdieria sulphuraria*, a potential candidate for production of blue phycocyanin, showed increased production of this pigment when grown mixotrophically on glycerol (Sloth et al. 2006). Bachchhav et al. (2016) have observed that phycocyanin content produced by *Arthrospira* cultivated with LEDs under mixotrophic conditions was six times higher than under autotrophic conditions.

Although it is considered that photosynthetic pigments are produced specifically under photoautotrophic conditions, some algae are able to biosynthesize them heterotrophically in the dark. In the fully heterotrophic processes, algae are maintained in the dark and use an organic carbon source as a substrate for growth. The most important advantages of heterotrophic systems seem to be independence of light intensities and increased production of certain valuable metabolites. Nevertheless, this kind of cultivation has several limitations, such as risk of contamination by other organisms, higher costs and energy consumption (necessary heating), unsuitability for pigment production, and a small number of species with the capability for heterotrophic metabolism (Chen and Zhang 1997). Cyanobacterial and red algal species, which grow in extreme habitats, show the ability to benefit from a variety of surrounding carbon sources. Gross and Schnarrenberger (1995) have reported that an extremophilic *G. sulphuraria*, strain 074G, can produce phycocyanin under heterotrophic conditions. Sloth et al. (2006) studied carbon-limited but nitrogen-sufficient heterotrophic batch cultures of *G. sulphuraria*, in which the amount of phycocyanin increased repeatedly Another research demonstrated that *G. sulphuraria*, in heterotrophic growth, synthesizes higher amounts of phycocyanin than *A. platensis* in outdoor cultivation (Graverholt and Eriksen 2007). Moreover, other studies show that *Arthrospira* strains are able to grow heterotrophically on different carbon sources, in the absence of light. However, very low accumulation of pigments and small growth rates make this an unlikely option for producing phycobiliproteins through *A. platensis* (Mühling et al. 2005; Chojnacka and Zielińska 2012).

Due to the fact that phycobilisomes are part of a photosynthetic apparatus, in *Cyanobacteria* and Rhodophyta, light plays an important role in their formation. Generally, microalgae with phycobilisomes grow more efficiently under lower light intensity; in fact, overly intensive illumination may induce bleaching, poor growth, or death of the culture (Lorenz et al. 2005). The impact of bright light on the presence of phycobilisomes was also investigated in the case of red macroalgae. López-Figueroa (1992) have summarized that the amount of phycobiliproteins per cell was higher when reduced light was applied. Light intensity and color can also influence the composition of phycobiliproteins, for example, Talaraico and Cortese (1993) indicated that white, blue, and green illumination, used in *Audouinella saviana* cultures, caused various effects in the ratios of pigments and thus in the conjugation of phycobilisomes.

Temperature is another very important factor that may have an impact on the composition of pigments in algal cells. Experiments with *Oscillatoria* strains showed that temperature might be an effective agent for target cultivation. Strain N9DM, which was grown at 30 ± 2 °C, produced mostly phycoerythrin, while at a higher temperature (55 ± 2 °C), the same strain produced only phycocyanin (Singh et al. 2012). The optimal temperatures for phycobiliprotein production range widely from 20 to 37 °C (Chaloub et al. 2015).

Phycobiliproteins, arranged in phycobilisomes, serve as a nutrient supply under unfavorable conditions, and therefore defining the impact of nitrogen on the production of phycobiliproteins is a very reasonable course. Peter et al. (2010) reported that nitrogen starvation causes modifications in the composition of phycobilisomes in *Arthrospira platensis*. Simeunović et al. (2013) studied the effect of $NaNO_3$, as a nitrogen source in culture media, on the production of phycobiliproteins by *Anabaena* and *Nostoc* genera. They found that this form of nitrogen affects the proportions of phycobiliproteins in cells but does not change their total concentration. While recent studies on this matter have mainly focused on microalgae, the response of macroalgae to nitrogen stress conditions is poorly understood. Yu and Yang (2008) presented biochemical and physiological characteristics, including phycoerythrin production, of macroalgae *Gracilaria lemaneiformis* under high nitrogen and phosphorus concentrations. The results of their study indicated that overly high concentrations of N/P induce the inhibition of phycoerythrin synthesis.

4.2.2 Isolation and Purification of Phycobiliproteins

The extraction and downstream processing steps should be optimized each time in order to obtain the proper purity ratio for the desired applications. The criterion that is commonly used to determine the purity of received phycobiliproteins is the ratio of specific visible light maximum absorbance to the absorbance at 280 nm, which defines the presence of contaminating proteins. The phycocyanin purity ratio is defined by the A_{620}/A_{280} ratio. Phycocyanin's application as a food pigment requires an absorbance ratio of ≥0.7. The purity ratio of 3.9 refers to the reagent grade, while the ratio greater than 4 reflects the analytical grade of this compound (Borowitzka 2013). Phycoerythrin extract purity is determined as the absorbance ratio of A_{565}/A_{280} and must be higher than 4 for its use in clinical research or molecular biology (Benavides and Rito-Palomares 2004). The general purification procedure includes protein extraction, recovery, and purification by chromatographic methods. Phycobiliproteins may be extracted from wet or dry algae biomass using a variety of cell disruption methods (like repeated freezing and thawing cycles, osmotic shocks, sonication, lysozyme treatment, or mechanical homogenization). The right cell disruption technique is the key to achieving maximized product recovery (Günerken et al. 2015). The next stage of the phycobiliprotein procurement process is usually carried out through selective precipitation with ammonium sulfate. The fractionation is featured as fast and relatively inexpensive. The purification is a crucial step toward obtaining ultrapure phycobiliproteins, especially phycoerythrin. Different column chromatography methods, such as ion exchange chromatography, gel filtration, adsorption chromatography, or exchange bed chromatography, allow for achieving the high-quality proteins. Many studies have been conducted on the purification of phycobiliproteins, especially phycoerythrin, with maximum purity and yield. Bermejo Román et al. (2002) isolated and purified phycoerythrin from *Porphyridium cruentum* using a fast two-step chromatographic method. In their method, expanded bed adsorption chromatography was followed by an anionic chromatography (DEAE cellulose), and it allowed them to reach a final phycoerythrin recovery of 32%. Soni et al. (2008) focused on purification of phycocyanin from *Porphyridium fragile* by using ammonium sulfate precipitation combined with hydrophobic interaction chromatography. The authors noted that the purity of the phycocyanin was 4.52. Sørensen et al. (2013) studied different methods of extraction and purification of phycocyanin from microalgae *Galdieria sulphuraria*. According to those authors, ammonium sulfate precipitation combined with anion exchange chromatography is required to enhance a high purity ratio (3.5, 4.5). Fractionation only ensured phycocyanin sediment with a purity of 0.7. Sun et al. (2009) extracted and purified phycoerythrin from macroalgae *Heterosiphonia japonica*. Proteins were extracted using an ultrasonic cell disruptor in 50 mM phosphate buffer and was precipitated by salting out. Isolation of the protein was carried out in two gel filtration steps, with Sepharose CL-4B and Sephadex G-200. Ion exchange chromatography on DEAE-Sepharose was used to purify the phycoerythrin. This protocol yielded a final ratio of A_{565}–A_{280} of 4.89. Patil and Raghavarao (2007) separated phycocyanin and allophycocyanin using an aqueous biphasic system. This procedure increased the purity ratio of the phycocyanin from an initial 1.18 to 3.52. In studies by Denis et al. (2009), ultrafiltration was described as the method of concentration and pre-purification of the phycoerythrin solution from macroalgae *Grateloupia turuturu*. The results indicated that this process allowed them to retain 100% of the phycoerythrin in its non-denatured form. Recently, Munier et al. (2015) demonstrated a one-step chromatographic technique of purification of phycoerythrin from *G. turuturu*. The purity ratio and recovery yield achieved after using DEAE-Sepharose Fast Flow chromatography were 2.89% and 27%, respectively. Different methods of protein extraction, including phycoerythrin, from macroalgae *Palmaria palmata* were compared by Harnedy and FitzGerald (2013). The result showed that high shear force is a more efficient method of cell disruption for the isolation of water-soluble proteins than osmotic shock.

In recent years, studies on biotechnological production of phycobiliproteins and their recovery from micro- and macroalgal cells have undergone an important expansion. Novel, advanced purification techniques have been developed to meet the standards of the different fields, especially for molecular biology and pharmaceutical researches. Similarly, improved, more extensive studies on the upstream process, including pigment accumulation, have been implemented.

4.3 Algal Phenolic Compounds of Biotechnological Soundness

Algae are an excellent source of many novel and biologically functional substances. Apart from high-quality proteins, containing essential amino acids, protein dyes, dietary fiber, essential fatty acids, minerals, and vitamins, algae could also be a good source of phenolic compounds (Mišurcová 2011; Ambrozova et al. 2014).

Phenolic compounds are one of the most widely occurring groups of phytochemicals known for their antioxidant activity, synthesized from phenylpropanoids, primarily through the pentose phosphate, shikimate, and phenylpropanoid pathways. Synthesis of phenolic compounds is triggered by external factors such as microbial infections, photooxidation stress, reactive oxygen species, harm, UV light, diseases, and herbivores (Ross and Kasum 2002). Chemically, the structures of natural polyphenols can be divided into several classes, including phenolic acids (hydroxybenzoic acids, hydroxycinnamic acids), flavonoids (flavones, flavonols, flavanones, flavanonols, flavanols, anthocyanins, chalcones), isoflavonoids (isoflavones, coumestans), stilbenes, lignans, lignins, and phenolic polymers (proanthocyanidins – condensed tannins and hydrolysable tannins) (Manach et al. 2004) (Fig. 4.2).

This large and diverse group of secondary metabolites can be found in both terrestrial and aquatic environments. Polyphenols occur mainly in fruits and beverages, such as tea, wine, and coffee, and also in vegetables, leguminous plants, and cereals. Their concentrations in foods differ according to many factors (genetic, environmental, technologic, etc.); generally, phenolic acids account for one-third of the total intake and flavonoids for the remaining two-thirds, with the most abundant flavonoids in the diet being flavanols, anthocyanins, and their oxidation products (Manach et al. 2004). Bioavailability within polyphenols differs considerably. As far as some compounds are concerned, it also depends on their form in their respective dietary sources (Manach et al. 2004). Generally, their primary function in plants is as protection against ultraviolet radiation and pathogens (Manach et al. 2004). Other roles include the pigmentation, reproduction, and growth of plants (Zern and Fernandez 2005). Recently, polyphenolic compounds have become very common constituents of the human diet and have received increasing interest from consumers, as well as food manufacturers, for many reasons, the health benefits mentioned above being the most significant (Manach et al. 2004).

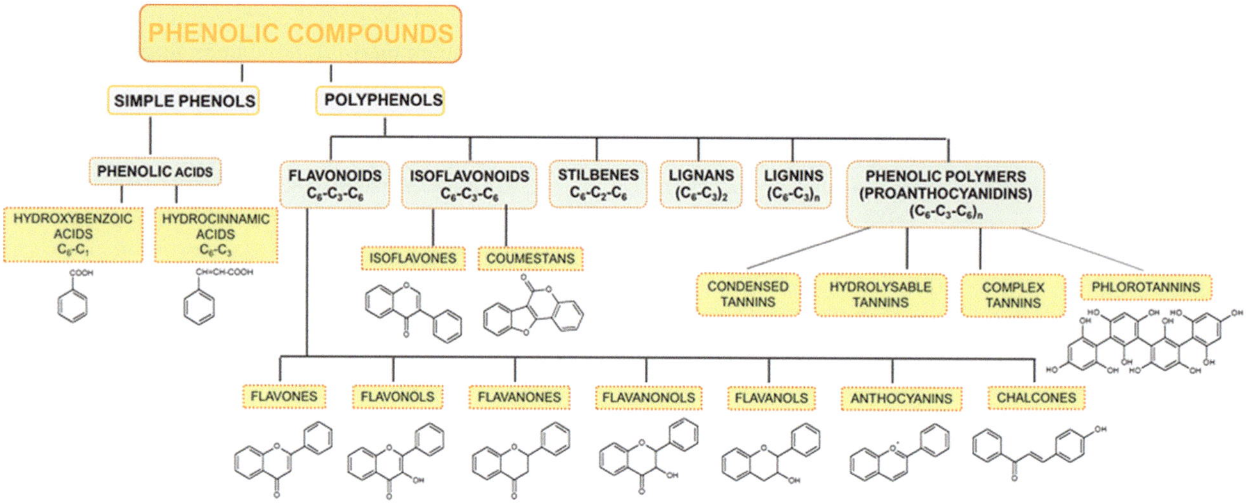

Fig. 4.2 Classification of naturally occurring phenolic compounds

It is commonly believed that possession of phenolic compounds is a unique feature of terrestrial plants that evolved in response to UV-light exposure during their transition from aquatic to terrestrial environments. The results of current studies unequivocally indicate that algae living in aquatic ecosystems also contain a wide range of phenolic compounds. Among these substances are also potential precursors, key intermediates, and end products of the flavonoid biosynthetic pathway, which means that algae must possess the enzyme pool required for their biosynthesis. The biological function(s) of phenolic compounds in algae remains to be clarified, especially since low levels of these substances in the cells of the microorganisms in question limit their role as antioxidants. Interestingly, the least complex phenolic composition contained the cells of prokaryotic microalgae, namely, cyanobacteria (Goiris et al. 2014). In response to variations in abiotic and biotic stress conditions, algae may produce and accumulate large quantities of secondary metabolites that function as defense compounds during their development (Mannino et al. 2014). Algae possess antioxidant systems to counteract environmental stresses and produce bioactive compounds, including polyphenols (Maharana et al. 2015). Besides their strong antioxidant properties, the naturally occurring algal polyphenols are known to have numerous biological properties. The possible use of these active compounds as functional ingredients in many industrial applications is therefore very clear. Phenolic compounds are present as major chemical components in most algal cells. They are secondary metabolites and are thus not directly involved in processes of primary algal metabolism, such as photosynthesis, cell division, and reproduction. Commonly characterized as stress compounds, phenolics are involved in protective chemical mechanisms against biotic factors, such as grazing, settlement of bacteria, or other fouling organisms, and abiotic stressors, such as UV radiation and metal contaminants. Phenolic compounds found in algae also include the phlorotannins found in brown algae and in some red algae in lower amounts. Phlorotannins, a subgroup of tannins, can be found in brown algae in relatively high percentages (up to 15% of dry weight, depending on the species) and are formed as polymers of phloroglucinol (1,3,5-trihydroxybenzene); however, their chemical composition is greatly variable, considering that there are different types of phlorotannin with very different degrees of polymerization (Koivikko et al. 2007). It is widely accepted that phlorotannins are found within the algal cells that form complexes with different components of the cell walls, such as alginic acid (Kim et al. 2013), and therefore the protocols for obtaining extracts enriched in these components should be optimized to improve their extractability. Phlorotannins exhibit primary functions, for example, in the growth and development of cell walls in Fucales. They are integral structural components of cell walls, but they have also been studied because of their other secondary ecological functions, as well as their therapeutic properties (anticancer, antioxidative, antibacterial, anti-allergic, antidiabetic, antiaging, anti-inflammatory, and anti-HIV activities) (Li et al. 2011; Thomas and Kim 2011). A survey of the currently available data suggests that phlorotannins are the main classes of anti-inflammatory and antioxidant phenolic compound found in marine algae. They demonstrate promising activity in in vitro systems, showing a potential for further development as therapeutic agents.

Phenolic compounds are considered to be dominant contributors to antioxidant activity. Different algal products provide diverse total phenolic contents due to many influencing factors, such as algal species, geographical origin or area of cultivation, and seasonal, physiological, or environmental variations (Marinho-Soriano et al. 2006). Total phenolic contents are generally affected by different abiotic factors, such as temperature, irradiance levels, nutrient availability, salinity and herbivory, or biotic factors, such as grazing pressure and the reproductive state of the algae (Connan et al. 2004). Algae past their peak of maturity and fertility show higher phenolic content (Steinberg 1989; Stiger et al. 2004). All of these factors could act on the spatiotemporal regulation of the phenolic metabolic expressions, inducing marked qualitative and quantitative variations among individuals at a very small scale, together with intraindividual variations (Amsler and Fairhead 2006; Connan et al. 2004). In the case of photoautotrophic organisms, the quality and quantity of light are two of the most important parameters. Although sunlight is the most cost-effective source of energy for the production of microalgae, artificial light may become commercially viable for the production of some important chemical compounds, including phenolic compounds. For example, cyanobacteria are mainly able to use red, yellow, and green light but are much less amenable to blue light. It seems that the light sources used for algae cultures should contain far-red emitters (Kula et al. 2016). It has been shown, inter alia, that UV light stimulates the synthesis of phenolic compounds (Goiris et al. 2012). Hence, the manipulation of cyanobacteria growth conditions and subsequent processing of their biomass might be useful for maximizing the amount of chemicals in biotechnological production, even on an industrial scale. As an example referring to the processing of biomass, it was found that higher medicinal effect was obtained from dry algae samples than from fresh samples (Padmini Sreenivasa Rao and Karmarkar 1986). Moreover, it is worth emphasizing that polyphenols from algae are more potent than their analogs from terrestrial plants (Nagayama et al. 2002). Antioxidant activity, an important property of bioactive algal compounds, has been ascribed to their reactive oxygen species' scavenging ability, quenching singlet oxygen, reducing power, and chelating ability (Andrade et al. 2013; Maharana et al. 2015).

The development of new analytical procedures able to provide a more systematic chemical characterization of the compounds found in algae and/or in the extracts obtained from these biota has been performed intensively over recent years. It is known that chemical components obtained from algal cells (including phenolic compounds) and their activity also depend on the solvents being used for the extraction, as well as on the type of alga (Manilal et al. 2009). The polarity of the extracting solvent, as well as the technique of extraction, is crucial in regard to the yield of obtained polyphenols (Hayouni et al. 2007). Solvent polarity plays a key role in total phenolic recovery, with total amounts typically increasing with increasing polarity. Nowadays, there is a need to develop novel extraction techniques in order to isolate natural, biologically active compounds that can be utilized in the cosmetic, pharmaceutical, and food industries. Different extraction techniques, including traditional ones such as solvent extraction and novel ones such as supercritical fluid extraction (SFE), microwave-assisted extraction (MAE), ultrasound-assisted extraction (UAE), enzyme-assisted extraction (EAE) and pressurized liquid extraction (PLE), and their combinations are widely used to extract biologically active compounds from algae (Michalak and Chojnacka 2014). Among these techniques, the least amount of scientific attention has been paid to the enzymatic extraction of algae. The accessibility of active compounds from algae can be limited due to the composition of the algal cell wall. The high content of various cell wall polysaccharides, especially cellulose, can limit the isolation of active compounds. Therefore, the enzymatic degradation of cell wall polymers is becoming an attractive alternative to chemical and mechanical processes, because it helps to release bioactive components that are linked to cell wall components, thus increasing the efficiency of their extraction (Wijesinghe and Jeon 2012; Adalbjörnsson and Jónsdóttir 2015). Enzyme-assisted extraction is also considered a green technique, due to its efficiency and mild reactive conditions, and can be an alternative to conventional extraction techniques that use organic solvents. Moreover, this type of extraction is considered safe for labile active compounds such as phenolics (Boulila et al. 2015). μ-SPE plate methodology used in hyphenated techniques with liquid chromatography-electrospray ionization-tandem mass spectrometry (LC-ESI–MS) assays provides information about specific phenols in sea algae with a sensitivity and precision that has not yet been experienced. HPTLC (high-performance thin-layer chromatography) combined with post-derivatization DPPH• assay and ferric chloride assay is also used to successfully screen algae samples for antioxidant activity and polyphenolic content. These assays are based on their ability: (a) to scavenge non-biological stable free radicals (DPPH•) or (b) to chelate Fe^{3+} ions. Strong, positive, and significant correlations between total phenolic content and DPPH• radical scavenging activity have shown that phenolic compounds, including flavonoids, are the main contributors of antioxidant activity in algae (Agatonovic-Kustrin et al. 2016).

Various algal species, variable solvents, and different methods of extraction, together with the presence of phenols as the major chemical constituents responsible for the high antioxidant activity of algae, make them interesting candidates for potent antimicrobial agents that could replace synthetic chemical products, representing promising new horizons for the pharmaceutical industry. Consequently, the use of natural antioxidants of algal origin, in the food, cosmetic, and therapeutic industries, is a promising alternative to synthetic antioxidants with respect to low cost, high compatibility with dietary intake, and no harmful effects inside the human body

(Lobo et al. 2010). Moreover, natural products are generally safer than synthetic chemicals due to the absence of chemical contamination and are therefore better accepted by the society and food industries than synthetic chemicals (Patra et al. 2008).

4.4 Algae in Environmental Biotechnology: The Guards of Ecosystem Homeostasis

4.4.1 Biotransformations of Organophosphonate Xenobiotics

The fate of any xenobiotics in the environment depends on the physicochemical properties of the compound, the qualities of the biotope, and meteorological factors (Salmon-Monviola et al. 2011). However, coexistence of these substances with microorganisms in ecosystems inclines us to investigate their mutual impact. Despite the fact that even more rigorous legal restrictions have been put on the management of chemicals, especially in the field of their disposal, the presence of many xenobiotics is still detected in ecosystems. One of the groups of xenobiotics of especially bad fame is the organophosphonates, which are highly water-soluble and stable in aqueous solutions. Moreover, specific to these substances, the direct carbon-to-phosphorus (C-P) bond is resistant to chemical hydrolysis, thermal decomposition, and photolysis, thus raising concern about their potential accumulation in aquatic ecosystems. What is more, methods for the effective removal of phosphonates from the environment have yet to be developed (Drzyzga et al. 2017). This concern is made all the bigger by the fact that aminophosphonates (into the structure of which the atoms of phosphorus and nitrogen are placed) and the products of their decomposition could actively stimulate growth of certain water organisms, particularly the group of microalgae known as cyanobacteria (Lipok et al. 2007; Forlani et al. 2008, 2013). The reason may be that the total content of P and N is enriched, which usually leads to eutrophication (Sharpley and Wang 2014). According to a statement by the US Environmental Protection Agency, that phenomenon was listed as one of the main problems impairing water quality (US_Environmental Protection Agency 2011), and it quite frequently results in blooms of cyanobacteria and harmful algae. It has been suggested that the occurrence of these organisms is connected to algal abilities to transform phosphonate chemicals, even if their features make them resistant to most of the known methods of chemical and/or enzymatic hydrolysis (Studnik et al. 2015). Considering that glyphosate (*N*-phosphonomethyl glycine) is the primary aminophosphonate herbicide, applied worldwide on an unprecedented scale, and keeping in mind cyanobacterial abilities to decrease its concentration, the role of these microalgae in the biocatalytic transformation of this substance is particularly important. Glyphosate is the active ingredient in the popular, nonselective herbicide Roundup® and is one of the most studied, in fact, most modeled, compounds, when considering the potential of a given microbial strain for biodegradation (Kafarski et al. 2000). It has been suggested that glyphosate is an environmentally friendly herbicide, the use of which does not have a negative impact, with a few exceptions (Williams et al. 2000). Recent studies have proven its toxic effects on some aquatic organisms, within algae, diatoms, and other freshwater phytoplankton strains (Pesce et al. 2009; Vendrell et al. 2009; Vera et al. 2010). However, the presence of glyphosate could also stimulate the growth of halophilic and freshwater cyanobacteria, which decompose this substance and use it as a source of phosphorus (Ravi and Balakumar 1998; Forlani et al. 2013). Therefore, cyanobacteria could be concerned as a key organism in cycling processes of bioelements, leading to a decrease in amounts of phosphonate in surface waters (Benitez-Nelson 2015; Heimann and Cires 2015). Currently, it is thought that up to 25% of all organic phosphorus in aquatic environments exists in the form of phosphonates; however, the value depends on the studies, and it is more likely that this value is approximately 10% (Van Mooy et al. 2015; Lin et al. 2016). Investigations into cyanobacterial metabolism proved that many strains could also transform phosphonates other than glyphosate. Two great examples of such abilities are the proven decomposition of hexamethylenediamine-*N*,*N*,*N'*,*N'*-tetrakis (methylphosphonic acid), also known as Dequest® 2054, by *Arthrospira platensis* (Forlani et al. 2011, 2013) and diethylenetriamine penta (methylenephosphonic) acid (DTPMP), known as Dequest® 2060S, by the cell extract of *Anabaena variabilis* (Drzyzga et al. 2017). Those studies reveal algae, and particularly cyanobacteria, as being real, effective biocatalysts. Their outstanding enzymatic potential is expressed in the transformation of different compounds that are considered dangerous for the environment and human beings. Decomposition of many of these xenobiotics, carried out actively by cyanobacteria, eliminates them from the soil or water and results in the same improved quality of ecosystems (Haritash and Kaushik 2009; Balcerzak et al. 2014).

Besides the organophosphonates, microalgae are able to control the level of certain metal ions, such as arsenic (As) and mercury (Hg), harmful to health and life. Some strains of microalgae can accumulate substantial amounts of these metals and, by methylation, diminish their toxicity (Lefebvre et al. 2007; Yin et al. 2011). The most important contribution is that of algal and cyanobacterial consortia in the utilization of polycyclic aromatic hydrocarbons (PAHs). These aromatic hydrocarbons with two or more fused benzene rings are the very first environmental carcinogens, whose difficult degradation under natural conditions leads to bioaccumulation of such substances (Haritash and Kaushik 2009). Algal strains, i.e., *Chlorella vulgaris*, *Scenedesmus platydiscus*, *Scenedesmus quadricauda*, and *Selenastrum capricornutum*, involved in PAHs' degradation appear to be extremely useful for reduction of the concentration of xenobiotics in their habitats. However, the efficiency of those processes is dependent on the species and the applied toxicant, as in case of 2,4,6-trinitrotoluene (TNT), which is the subject of reductive transformation carried out by *Anabaena* sp. The concentration of TNT in groundwaters has been estimated as being high and potentially hazardous, which, together with the proven mutagenic and toxic effect of this substance, has intensified efforts to find effective methods for its removal. One such method, batch cyanobacterial cultures, seems to be a convenient method for TNT bioremediation, since the efficiency of removal was about 96% (Pavlostathis and Jackson 2002). On the other hand, algae *Merismopedia quadruplicata* (formerly *Agmenellum quadruplicatum*) possess the ability to utilize aniline and its amine derivatives, which could be found in the effluent from some factories (Cerniglia et al. 1981). As it turns out, strains belonging to the genera *Synechococcus* sp., *Leptolyngbya* sp., and *Phormidium* sp. could be involved in bioremediation processes of dyes present in wastewater from the textile industry with 50% efficiency (Silva-Stenico et al. 2012). The biotransformation potential of algae has been revealed in the reduced level of methyl parathion (*O*,*O*-dimethyl *O*-*p*-nitrophenyl phosphorothioate), an insecticide extensively used in rice cultivation (Megharaj et al. 1994). Oxidation and hydrolyzing as detoxification routes of another organophosphorous pesticide, fenamiphos (*O*-ethyl-*O*[3-methyl-4-methylthiophenyl]-isopropylamidophosphate), were documented for five species of green alga and cyanobacteria, among which were *Nostoc muscorum*, *Nostoc* sp., and *Anabaena* sp. (Caceres et al. 2008).

These few examples illustrate how rich and potent the enzymatic machinery of algae is in respect to the neutralization of onerous xenobiotics. Therefore, the ability to transform various organic compounds makes algae, including cyanobacteria, an important component of the system that guards environmental sustainability. What is more, these features of the described biota make them practically ready to use as a key component of biotechnological systems for removing environmental toxins.

4.4.2 Bioremediation of Metal Ions by Algae

Nowadays, it is not only organophosphorous, or, more widely, organic compounds in general, that may cause serious problems for water environments. Metal ions, which occur naturally in ecosystems, may also be as toxic as other xenobiotics under some conditions. Most of them are important, essential micronutrients that are necessary for the proper growth of all organisms and especially microorganisms. The presence of metal ions is crucial during processes such as respiration, photosynthesis, the folding of proteins, and enzyme activation. But the transport of metal cations into the interior of the cells has to be strictly controlled in order to maintain the proper homeostasis and prevent the accumulation of toxic concentrations of these elements (Waldron and Robinson 2009; Huertas et al. 2014). Currently, dynamically developing sectors of industry and agriculture are the main sources of excessive amounts of metal ions released into related ecosystems, which consequently leads to pollution of the environment (O'Connell et al. 2008; Tsekova et al. 2010; Zeraatkar et al. 2016). Exploring innovative methods for the effective treatment of wastewater can protect freshwater resources and aquatic ecosystems. For the last few decades, the research on algae-based wastewater treatment has indicated that they can play a valuable role in the remediation of industrial pollutions (Hoffmann 1998; Oswald 2003; Zeraatkar et al. 2016).

Bioremediation technology may be divided into two groups: in situ and ex situ. The in situ methods are carried out by microorganisms right in the polluted area itself and are considered cheaper and easier to use than other methods. They are described as effective in the presence of significant surface contamination but in relatively low concentrations. This type of remediation can be used, e.g., by nuclear power plants to remove radionuclides or to bio-leach rare metals. The ex situ techniques are initiated outside of the polluted area, where the chemical methods are first implemented, after which biotechnological decontamination is used (Boopathy 2000; Vidali 2001). Bioremediation processes may be classified according to different criteria. At first, because of the metabolism of microorganisms, bioremediation can be divided into that dependent on metabolism (bioaccumulation) and that independent of metabolism (biosorption). Secondly, because of the localization of metal after the finished process, it can be further distinguished: intracellular accumulation, extracellular accumulation/precipitation, and cell surface adsorption/precipitation (Veglio and Beolchini 1997). Biosorption concerning simple, metabolically passive, physicochemical reactions, involving, e.g., complexation, adsorption, ion exchange, or precipitation, is the process in which metal ions are quickly, and often reversibly, bound to functional groups of compounds that build cells (Pirszel et al. 1995; De Philippis et al. 2011). Due to the chemical composition of algal cells, especially the cell walls and their secretions, the chemical character of algal biomass is like that of ion exchange resins or activated charcoal (Veglio and Beolchini 1997). The use of metabolically inactive cells allows us to carry out biosorption within a wide range of pH (3–9) and temperature (4–90 °C) (Pirszel et al.

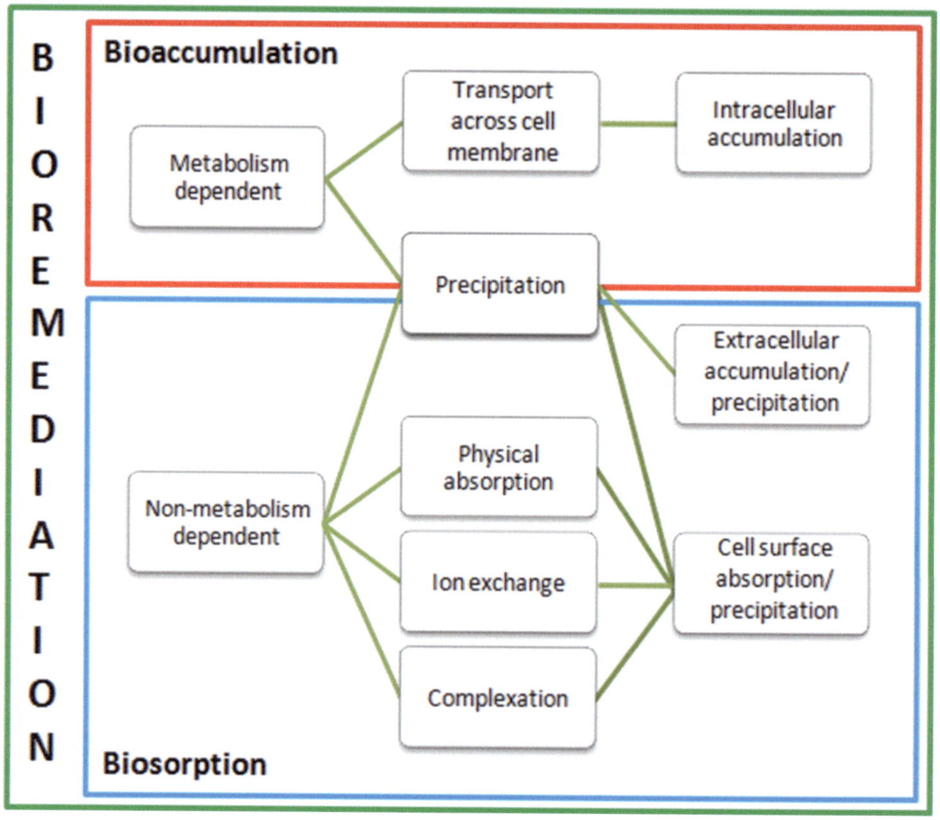

Fig. 4.3 Classification of bioremediation processes dependent on the metabolism of the cells and on the location of the place where this process is carried out

1995). After the finished process, the biomass with bounded metal ions can be regenerated and used in the next cycle of biosorption. The number of cycles depends on the physical and mechanical properties of the biosorbent (Das et al. 2008; Pal and Paul 2008). Bioaccumulation is a more complex process and requires metabolically active cells. The first stage of this process is the same as in biosorption: it is fast and includes adsorption of metal ions to the cell surface. Next, the persistent chemical bond between the metal and functional groups may be formed, or the ions may be transported into the cytoplasm of cells. Bioaccumulation processes are often the result of defense mechanisms activated by microorganisms against excessive concentrations of metal ions present in their environment (Veglio and Beolchini 1997; Das et al. 2008; Chojnacka 2010; Zeraatkar et al. 2016) (Fig. 4.3).

The most common metals occurring in industrial waste are copper, zinc, nickel, cobalt, lead, chromium, and cadmium (Kaushik et al. 1999; Holm et al. 2002). At the same time, the widespread presence of algae, including cyanobacteria, in aquatic environments and their known ability to accumulate high concentrations of metal ions and other pollutants make the bioremediation process much more advantageous compared to conventionally used methods (Prasanna et al. 2008). The research on algae clearly indicates that this group of organism has great potential to be used as a biosorbent for removing metal ions from wastewater (Zeraatkar et al. 2016). The level of toxicity of metal ions can be strain specific and, importantly, determines the potential remediation capacity of using a specific algal strain. To avoid numerous disturbances and the time-consuming research needed to determine these levels for certain strains, the researchers started to use nonliving cells in such experiments and processes (Monteiro et al. 2010; Zeraatkar et al. 2016). It was found that dried algal biomass exhibited a greater capacity for metal ion biosorption compared to living cells (Zeraatkar et al. 2016), and it can be regarded as an assemblage of polymers, e.g., sugars, cellulose, pectins, or glycoproteins, which may be used as adsorbents with the potential for cost-effective wastewater treatment (Volesky 2007). The first barrier against the biosorption of metal cations is the cell walls reached in chemically variable binding groups: hydroxyl, phosphoryl, carboxyl, sulfur, amine, imidazole, sulfate, phosphate, carbohydrate, etc. (Schiewer and Volesky 2000; Kaplan 2013). Metal ion biosorption capacity has been attributed to the presence of the groups mentioned and depends on factors such as the number of functional groups in the algal cells, the coordination number of the sorbed cation, the accessibility of binding groups for metal ions, the formation constants of those complexes, and the chemical state of these sites (Mehta and Gaur 2005).

The most investigated species among microalgae is the halophilic strain *Arthrospira* (*Spirulina*) *platensis*, whose sorption capacity was tested in regard to cations such as copper (Greene et al. 1987; Al-Homaidan et al. 2014), zinc, nickel (Balaji et al. 2013), cadmium (Rangsayatorn et al. 2002), lead (Seker et al. 2008), and chromium (Gokhale et al. 2007; Lodi et al. 2008). Most of the results of those researches confirmed information obtained much earlier (Greene et al. 1987) that the observed biosorption mainly has an ion exchange character. The other research carried out with the freshwater species *Nostoc calcicola* has shown that, typically, the copper(II) ions had been taken in two stages. During the first 10 minutes, fast absorption at the surface of the cells was observed, and the transport of ions into the interior of cells, which is dependent on metabolism, was noticed next. Moreover, in this case, it was indicated that the energy necessary for transport came mainly from the reactions carried out in photosystem II (Verma and Singh 1990). Data in the literature show that the biosorption process may be very fast, which was confirmed by the research carried out on the *Chroococcus paris* strain. This species of cyanobacteria was able to uptake as much as 90% of metal ions in 1 minute (Les and Walker 1984).

Besides the metals, which are considered an environmental contamination, only a few data indicate that the cations of precious metal may also be bioremediated. Among the organisms able to sorb silver and gold ions are filamentous fungi (Pethkar and Paknikar 1998; Pethkar et al. 2001; Eisler 2003), yeast (Savvaidis 1998; Simmons and Singleton 1996), algae (Cordery et al. 1994; Niu and Volesky 1999; Romero-Gonzalez et al. 2003; Mata et al. 2009), and cyanobacteria (Colica et al. 2012). Besides binding silver and gold cations (Savvaidis 1998; De Corte et al. 2010; Deschatre et al. 2015), bacteria are also able to sorb and remediate palladium and platinum ions (De Vargas et al. 2004). An interesting aspect of bioremediation is the possibility of immobilization of the cyanobacterial biomass at both natural and synthetic polymers. The immobilization procedure has some advantages, e.g., easier processes of regeneration of the biosorbent and its reuse, easier separation of biomass from waste and prevention of loss of biomass during the biosorption cycle, and increased sorption capacity as well (Pradhan and Rai 2000). Moreover, immobilized cells enhance photosynthetic capacity (Bailliez et al. 1986) and reduce the toxicity of some substances (Bozeman et al. 1989). Examples of such a type of bioremediation are the cells of *Microcystis* sp. immobilized in a natural polymer alginate, able to remove copper ions (Pradhan and Rai 2000). The same procedure was used for the species *Anacystis nidulans*, which uptakes nickel, zinc, and cadmium (Awasthi and Rai 2004). Furthermore, the immobilization of powdered cells of *Chlorella vulgaris* results in a twofold increase in nickel removal in comparison to free cells (Al-Rub et al. 2004).

The high cost of conventional technologies used to remove metal ions from wastewaters and, on the other hand, the good sorption capacity presented by algae make this group of organisms an interesting subject for research concerning genetic modification in order to obtain new, cost-effective, "green" biosorbents (Zeraatkar et al. 2016). Both the eukaryotic and prokaryotic organisms have many genes that are responsible for creation of the systems involved in metal uptake, detoxification, and tolerance (Rosen 2002). Numerous researches have been dedicated to genetic modifications of *S. cerevisiae* (Kuroda et al. 2002) and *E. coli* (Sousa et al. 1998; Krishnaswamy and Wilson 2000; Bae et al. 2003; Kiyono and Pan-Hou 2006), due to the enhanced potential of sorption. However, in the case of algae, to date, relatively little attention has been paid to investigating the recombinant strains for metal ion biosorption. Thus, there is great potential in the study of both the development of genetically modified algae and subsequent testing of the safety of their use on a wider scale.

4.5 Biotechnological Benefits of the Use of Algae

Four short stories about certain aspects of the use of algae in biotechnology do not thoroughly exhaust the problem, but, of course, this was not the aim of this paper. Our intention was to show two different, but fully complementary, approaches: (i) the beneficial treatment of algae as a source of natural, valuable substances of desired characteristics and activity, and (ii) the dynamically growing role of algae as natural biocatalysts, able to conduct even the most sophisticated processes related to the safety of human existence in the biosphere. These approaches seem to be the most common in a natural way, due to the role of algae in ecosystems. As primary producers, these biota are responsible for providing other organisms with an energy stored in the form of certain substances and specific constituents that promote the growth and safe development of biocenoses. On the other hand, being pioneer autotrophic organisms, algae, especially microalgae, are not only able to adapt themselves to the conditions of a given habitat, but more importantly, these biota are able to change their living environment in the manner of biocatalytic transformations. Impressive enzymatic machinery in particular predisposes them to such activity. Those qualities of algae, when combined with relatively small nutritional requirements and a wide range of environmental tolerance, fully justify the dynamic growth of interest in the application of these biota in any possible way. Increasing richness of the human population and higher awareness of and demand for goods produced in the most natural and environmentally friendly ways possible favor the use of algae, just as scientific and environmental considerations do. In regard to biotechnological purposes, it may be concluded that the observed renaissance of the use of algae is not a transitional fashion, but rather a well-thought-out choice that is only beginning to reap benefits. Hence, enhancing the range and efficiency of the use of photoautotrophic primary producers in biotechnology appears to offer a chance to resolve many economic and ecological concerns.

References

Adalbjörnsson BV, Jónsdóttir R (2015) Enzyme-enhanced extraction of antioxidant ingredients from algae. Methods Mol Biol 1308:145–150

Agatonovic-Kustrin S, Morton DW, Ristivojević P (2016) Assessment of antioxidant activity in Victorian marine algal extracts using high performance thin-layer chromatography and multivariate analysis. J Chromatogr A 1468:228–235

Al-Homaidan AA, Al-Houri HJ, Al-Hazzani AA, Elgaaly G, Moubayed NMS (2014) Biosorption of copper ions from aqueous solutions by *Spirulina platensis* biomass. Arab J Chem 7:57–62

Al-Rub FA, El-Naas M, Benyahia F, Ashour I (2004) Biosorption of nickel on blank alginate beads, free and immobilized algal cells. Process Biochem 39:1767–1773

Ambrozova JV, Misurcova L, Vicha R, Machu L, Samek D, Baron M, Mlcek J, Sochor J, Jurikova T (2014) Influence of extractive solvents on lipid and fatty acids content of edible freshwater algal and seaweed products, the green microalga *Chlorella kessleri* and the cyanobacterium *Spirulina platensis*. Molecules 19:2344–2360

Amsler CD, Fairhead VA (2006) Defensive and sensory chemical ecology of brown algae. Adv Bot Res 43:1–91

Andrade PB, Barbosa M, Matos RP, Lopes G, Vinholes J, Mouga T, Valentão P (2013) Valuable compounds in macroalgae extracts. Food Chem 138:1819–1828

Awasthi M, Rai LC (2004) Adsorption of nickel, zinc and cadmium by immobilized green algae and cyanobacteria: a comparative study. Ann Microbiol 54:257–267

Bachchhav MB, Kulkarni MV, Ingale AG (2016) Enhanced phycocyanin production from *Spirulina platensis* using light emitting diode. Dig J Inst Eng India Ser E. https://doi.org/10.1007/s40034-016-0090-8

Bae W, Wu CH, Kostal J, Mulchandani A, Chen W (2003) Enhanced mercury biosorption by bacterial cells with surface-displayed MerR. Appl Environ Microbiol 69:3176–3180

Baier K, Lehmann H, Stephan DP, Lockau W (2004) NblA is essential for phycobilisome degradation in *Anabaena* sp. strain PCC 7120 but not for development of functional heterocysts. Microbiology 150:2739–2749. https://doi.org/10.1099/mic.0.27153-0

Bailliez C, Largeau C, Berkaloff C, Casadevall E (1986) Immobilization of *Botryococcus braunii* in alginate: influence on chlorophyll content, photosynthetic activity and degeneration during batch cultures. Appl Microbiol Biotechnol 23:361–366

Balaji S, Kalaivani T, Rajasekaran C (2013) Biosorption of zinc and nickel and its effect on growth of different *Spirulina* strains. Clean Soil Air Water 42:507–512

Balcerzak L, Lipok J, Strub D, Lochynski S (2014) Biotransformations of monoterpenes by photoautotrophic micro-organisms. J Appl Microbiol 117:1523–1536

Beale SI, Cornejo J (1991) Biosynthesis of Phycobilins. 15,16-Dihydrobiliverdin IXα is a partially reduced intermediate in the formation of phycobilins from biliverdin IXα. J Biol Chem 266:22341–22345

Benavides J, Rito-Palomares M (2004) Bioprocess intensification: a potential aqueous two-phase process for the primary recovery of B-phycoerythrin from *Porphyridium cruentum*. J Chromatogr B 807:33–38. https://doi.org/10.1016/j.jchromb.2004.01.028

Benitez-Nelson C (2015) Ocean chemistry. The missing link in oceanic phosphorus cycling? Science 348:759–760

Bermejo Román R, Alvárez-Pez JM, Acién Fernández FG, Molina Grima E (2002) Recovery of pure B-phycoerythrin from the microalga *Porphyridium cruentum*. J Biotechnol 93:73–85

Boopathy R (2000) Factors limiting bioremediation technologies. Bioresour Technol 74:63–67

Borowitzka MA (2013) High-value products from microalgae – their development and commercialisation. J Appl Phycol 25:743–756. https://doi.org/10.1007/s10811-013-9983-9

Borowitzka MA (2016) Algal physiology and large-scale outdoor cultures of microalgae. In: Borowitzka MA, Beardall J, Raven JA (eds) The physiology of microalgae. Springer, Dordrecht, pp 601–652

Borowitzka MA, Moheimani NR (2013) Open pond culture systems. In: Borowitzka MA, Moheimani NR (eds) Algae for biofuels and energy. Springer, Dordrecht, pp 133–152

Boulila A, Hassen I, Haouari L, Mejri F, Ben Amor I, Casabianca H, Hosni K (2015) Enzyme-assisted extraction of bioactive compounds from bay leaves (*Laurus nobilis* L.). Industrial Crops and Products 74:485–493

Bozeman J, Koopman B, Bitton G (1989) Toxicity testing using immobilized algae. Aquat Toxicol 14:345–352

Brown SB, Holroyd JA, Vernon DI (1984) Biosynthesis of phycobiliproteins. Incorporation of biliverdin into phycocyanin of the red alga *Cyanidium caldarium*. Biochem J 219:905–909

Bryant DA (1982) Phycoerythrocyanin and phycoerythrin: properties and occurrence in cyanobacteria. J Gen Microbiol 128:835–844

Caceres TP, Megharaj M, Naidu R (2008) Biodegradation of the pesticide fenamiphos by ten different species of green algae and cyanobacteria. Curr Microbiol 57:643–646

Cerniglia CE, Freeman JP, Van Baalen C (1981) Biotransformation and toxicity of aniline and aniline derivatives in cyanobacteria. Arch Microbiol 130:272–275

Chaloub RM, Motta NMS, de Araujo SP, de Aguiar PF, da Silva AF (2015) Combined effects of irradiance, temperature and nitrate concentration on phycoerythrin content in the microalga *Rhodomonas* sp. (Cryptophyceae). Algal Res 8:89–94. https://doi.org/10.1016/j.algal.2015.01.008

Chen F, Zhang Y (1997) High cell density mixotrophic culture of *Spirulina platensis* on glucose for phycocyanin production using a fed-batch system. Enzym Microb Technol 20:221–224. https://doi.org/10.1016/S0141-0229(96)00116-0

Chen F, Zhang Y, Guo S (1996) Growth and phycocyanin formation of *Spirulina platensis* photoheterotrophic culture. Bitechnol Lett 18:603–308. https://doi.org/10.1007/BF00140211

Chojnacka K (2010) Biosorption and bioaccumulation – the prospects for practical applications. Environ Int 36:299–307

Chojnacka K, Zielińska A (2012) Evaluation of growth yield of *Spirulina* (*Arthrospira*) sp. in photoautotrophic, heterotrophic and mixotrophic cultures. Word J Microbiol Biotechnol 28:437–445. https://doi.org/10.1007/s11274-011-0833-0

Colica G, Caparrotta S, Bertini G, De Philippis R (2012) Gold biosorption by exopolysaccharide producing cyanobacteria and purple nonsulphur bacteria. J Appl Microbiol 113:1380–1388

Connan S, Goulard F, Stiger V, Deslandes E, Ar Gall E (2004) Interspecific and temporal variation in phlorotannin levels in an assemblage of brown algae. Bot Mar 47:410–416

Cordery J, Will AJ, Atkinson K, Wills BA (1994) Extraction and recovery of silver from low grade liquors using microalgae. Miner Eng 7:1003–1015

Das N, Vimala R, Karthika P (2008) Biosorption of heavy metals – an overview. Indian J Biotechnol 7:159–169

De Corte S, Hennebel T, Verschuere S, Cuvelier C, Verstraete W, Boon N (2010) Gold nanoparticle formation using *Shewanella oneiden-*

sis: a fast biosorption and slow reduction process. J Chem Technol Biotechnol 86:547–553

De Philippis R, Colica G, Micheletti E (2011) Exopolysaccharide-producing cyanobacteria in heavy metal removal from water: molecular basis and practical applicability of the biosorption process. Appl Microbiol Biotechnol 92:697–708

De Vargas I, Macaskie LE, Guibal E (2004) Biosorption of palladium and platinum by sulfate-reducing bacteria. J Chem Technol Biotechnol 79:49–56

Denis C, Massé A, Fleurence J, Jaouen P (2009) Concentration and pre-purification with ultrafiltration of a R-phycoerythrin solution extracted from macro-algae *Grateloupia turuturu*: process definition and up-scaling. Sep Purif Technol 69:37–42. https://doi.org/10.1016/j.seppur.2009.06.017

Deschatre M, Ghillebaert F, Guezennec J, Simon-Colin C (2015) Study of biosorption of copper and silver by marine bacterial exopolysaccharides. WIT Trans Ecol Environ 196:549–559

Drzyzga D, Forlani G, Vermander J, Kafarski P, Lipok J (2017) Biodegradation of the aminopolyphosphonate DTPMP by the cyanobacterium *Anabaena variabilis* proceeds via a C-P lyase-independent pathway. Environ Microbiol 19:1065–1076

Eisler R (2003) Biorecovery of gold. Indian J Exp Biol 41:967–971

FDA (2014) Federal register volume 79, number 70: listing of color additives exempt from certification; *Spirulina* extract. https://www.gpo.gov/fdsys/pkg/FR-2014-04-11s/html/2014-08099.htm. Accessed 26 Oct 2017

Forlani G, Pavan M, Gramek M, Kafarski P, Lipok J (2008) Biochemical bases for a widespread tolerance of cyanobacteria to the phosphonate herbicide glyphosate. Plant Cell Physiol 49:443–456

Forlani G, Prearo V, Wieczorek D, Kafarski P, Lipok J (2011) Phosphonate degradation by *Spirulina* strains: cyanobacterial biofilters for the removal of anticorrosive polyphosphonates from wastewater. Enzym Microb Technol 48:299–305

Forlani G, Bertazzini M, Giberti S, Wieczorek D, Kafarski P, Lipok J (2013) Sublethal detergent concentrations increase metabolization of recalcitrant polyphosphonates by the cyanobacterium *Spirulina platensis*. Environ Sci Pollut Res Int 20:3263–3270

Glazer A (1994) Phycobiliproteins – a family of valuable, widely used flourophores. J Appl Phycol 6:105–112. https://doi.org/10.1007/BF02186064

Goiris K, Muylaert K, Fraeye I, Foubert I, De Brabanter J, De Cooman L (2012) Antioxidant potential of microalgae in relation to their phenolic and carotenoid content. J Appl Phycol 24:1477–1486

Goiris K, Muylaert K, Voorspoels S, Noten B, De Paepe D, E Baart GJ, De Cooman L (2014) Detection of flavonoids in microalgae from different evolutionary lineages. J Phycol 50:483–492

Gokhale SV, Jyoti KK, Lele SS (2007) Kinetic and equilibrium modeling of chromium (VI) biosorption on fresh and spent *Spirulina platensis*/*Chlorella vulgaris* biomass. Bioresour Technol 99:3600–3608

Graverholt OS, Eriksen NT (2007) Heterotrophic high-cell-density fed-batch and continuous-flow cultures of *Galdieria sulphuraria* and production of phycocyanin. Appl Microbiol Biotechnol 77:69 75. https://doi.org/10.1007/s00253-007-1150-2

Greene B, McPherson R, Darnall D (1987) Algal sorbents for selective metal ion recovery. In: Patterson JW, Passion R (eds) Metal speciation, separation and recovery. Lewis Publishers, Chelsea., 1987, pp 315–338

Gross W, Schnarrenberger C (1995) Heterotrophic growth of two strains of the acido-thermophilic red alga *Galdieria sulphuraria*. Plant Cell Physiol 36:633–638

Günerken E, D'Hondt E, Eppink MHM, Garcia-Gonzalez L, Elst K, Wijffels RH (2015) Cell disruption for microalgae biorefineries. Biotechnol Adv 33:243–260. https://doi.org/10.1016/j.biotechadv.2015.01.008

Haritash AK, Kaushik CP (2009) Biodegradation aspects of polycyclic aromatic hydrocarbons (PAHs): a review. J Hazard Mater 169:1–15

Harnedy PA, Fitz Gerald RJ (2013) Extraction of protein from the macroalga *Palmaria palmata*. LWT – Food Sci Technol 51:375–382. https://doi.org/10.1016/j.lwt.2012.09.023

Hayouni EA, Abedrabba M, Bouix M, Hamdi M (2007) The effects of solvents and extraction method on the phenolic contents and biological activities in vitro of Tunisian *Quercus coccifera* L. and *Juniperus phoenicea* L. fruit extracts. Food Chem 105:1126–1134

Heimann K, Cires S (2015) N_2 -Fixing cyanobacteria: ecology and biotechnological applications. In: Se-Kwon K (ed) Handbook of microalgae: biotechnology advances. Academic, London, pp 501–515

Hoffmann JP (1998) Wastewater treatment with suspended and nonsuspended algae. J Phycol 34:757–763

Holm O, Hansen E, Lassen C, Stuer-Lauridsen F, Kjolholt J (2002) Heavy metals in waste. European Commission DG ENV. E3, Project ENV.E.3/ETU/2000/0058. Final report

Huertas MJ, Lopez-Maury L, Giner-Lamia J, Sanchez-Riego AM, Florencio FJ (2014) Metals in Cyanobacteria: analysis of the copper, nickel, cobalt and arsenic homeostasis mechanisms. Life 4:865–886

Jorquera O, Kiperstok A, Sales EA, Embiruçu M, Ghirardi ML (2010) Comparative energy life-cycle analyses of microalgal biomass production in open ponds and photobioreactors. Bioresour Technol 101:1406–1413. https://doi.org/10.1016/j.biortech.2009.09.038

Kafarski P, Lejczak B, Forlani G (2000) Biodegradation of pesticides containing carbon to phosphorus bond. In: Hall JC, Hoagland RE, Zablotowicz RM (eds) Pesticide biotransformation in plants and microorganisms, Similarities and Divergencies. ACS, Washington, DC, pp 145–163

Kaplan D (2013) Absorption and adsorption of heavy metals by microalgae. In: Richmond A, Hu Q (eds) Handbook of microalgal culture: applied phycology and biotechnology. Blackwell Publishing, Ames, pp 439–447

Kaushik S, Sahu BK, Lawania RK, Tiwari RK (1999) Occurrence of heavy metals in lentic water of Gwalior region. Pollut Res 18:137–140

Kim SM, Kang SW, Jeon JS, Jung YJ, Kim WR, Kim CY, Um BH (2013) Determination of major phlorotannins in *Eisenia bicyclis* using hydrophilic interaction chromatography: seasonal variation and extraction characteristics. Food Chem 138:2399–2406

Kirilovsky D, Kerfeld CA (2013) The orange carotenoid protein: a blue-green light photoactive protein. Photochem Photobiol Sci 12:1135–1143. https://doi.org/10.1039/c3pp25406b

Kiyono M, Pan-Hou H (2006) Genetic engineering of bacteria for environmental remediation of mercury. J Health Sci 52:199–204

Koivikko R, Loponen J, Pihlaja K, Jormalainen V (2007) High-performance liquid chromatographic analysis of phlorotannins from the brown alga *Fucus vesiculosus*. Phytochem Anal 18:326–332

Krishnaswamy R, Wilson DB (2000) Construction and characterization of an *Escherichia coli* strain genetically engineered for Ni (II) bioaccumulation. Appl Environ Microbiol 66:5383–5386

Kronick MN (1988) Phycobiliproteins as labels in immunoassay. In: Ngo TT (ed) Nonisotopic immunoassay, 1st edn. Springer, New York, pp 163–185

Kula M, Rys M, Saja D, Tys J, Skoczowski A (2016) Far-red dependent changes in the chemical composition of *Spirulina platensis*. Eng Life Sci 16:777–785

Kuroda K, Ueda M, Shibasaki S, Tanaka A (2002) Cell surface-engineered yeast with ability to bind, and self-aggregate in response to, copper ion. Appl Microbiol Biotechnol 59:259–264

Lefebvre DD, Kelly D, Budd K (2007) Biotransformation of Hg(II) by cyanobacteria. Appl Environ Microbiol 73:243–249

Les A, Walker RW (1984) Toxicity and binding of copper, zinc and cadmium by the blue- green alga *Chroococcus paris*. Water Air Soil Pollut 23:129–139

Li YX, Wijesekara I, Li Y, Kim SK (2011) Phlorotannins as bioactive agents from brown algae. Process Biochem 46:2219–2224

Lin S, Litaker RW, Sunda WG (2016) Phosphorus physiological ecology and molecular mechanisms in marine phytoplankton. J Phycol 52:10–36

Lipok J, Owsiak T, Młynarz P, Forlani G, Kafarski P (2007) Phosphorus NMR as a tool to study mineralization of organophosphonates—the ability of *Spirulina* spp. to degrade glyphosate. Enzyme Microb Technol 41:286–291

Lobo V, Patil A, Phatak A, Chandra N (2010) Free radicals, antioxidants and functional foods: impact on human health. Pharmacogn Rev 4:118–126

Lodi A, Soletto D, Solisio C, Converti A (2008) Chromium(III) removal by *Spirulina platensis* biomass. Chem Eng J 136:151–155

López-Figueroa F (1992) Diurnal variation in pigment content in *Porphyra laciniata* and *Chondrus crispus* and its relation to the diurnal changes of underwater light quality and quantity. PSZNI: Mar Ecol 13:285–305

Lorenz M, Friedl T, Day JG (2005) Perpetual maintenance of actively metabolizing microalgal cultures. In: Andersen RA (ed) Algal culturing techniques, 1st edn. Elsevier, Amsterdam, pp 145–156

Maharana D, Das PB, Verlecar XN, Pise NM, Gauns MU (2015) Oxidative stress tolerance in intertidal red seaweed *Hypnea musciformis* (Wulfen) in relation to environmental components. Environ Sci Pollut Res Int 22:18741–18749

Manach C, Scalbert A, Morand C, Rémésy C, Jiménez L (2004) Polyphenols: food sources and bioavailability. Am J Clin Nutr 79:727–747

Manilal A, Sujith S, Kiran GS, Selvin J, Shakir C, Gandhimathi R, Lipton AP (2009) Antimicrobial potential and seasonality of red algae collected from the Southwest coast of India tested against shrimp, human and phytopathogens. Ann Microbiol 59:207–219

Mannino AM, Vaglica V, Oddo E (2014) Seasonal variation in total phenolic content of *Dictyopteris polypodioides* (Dictyotaceae) and *Cystoseira amentacea* (Sargassaceae) from the Sicilian coast. Fl Medit 24:39–50

Marinho-Soriano E, Fonseca PC, Carneiro MAA, Moreira WS (2006) Seasonal variation in the chemical composition of two tropical seaweeds. Bioresour Technol 97:2402–2406

Marquez FJ, Sasaki K, Kakizono T, Nishio N, Nagai S (1993) Growth characterization of *Spirulina platensis* in mixotrophic and heterotrophic conditions. J Ferment Bioeng 76:408–410. https://doi.org/10.1016/0922-338X(93)90034-6

Mata YN, Torres E, Blázquez ML, Ballester A, González F, Munoz JA (2009) Gold (III) biosorption and bioreduction with the brown alga *Fucus vesiculosus*. J Hazard Mater 166:612–618

Megharaj M, Madhavi DR, Sreenivasulu C, Umamaheswari A, Venkateswarlu K (1994) Biodegradation of methyl parathion by soil isolates of microalgae and cyanobacteria. Bull Environ Contam Toxicol 53:292–297

Mehta SK, Gaur JP (2005) Use of algae for removing heavy metal ions from wastewater: progress and prospects. Crit Rev Biotechnol 25:113–152

Michalak I, Chojnacka K (2014) Algal extracts: technology and advances. Eng Life Sci 14:581–591

Milledge JJ (2011) Commercial application of microalgae other than as biofuels: a brief review. Rev Environ Sci Biotechnol 10:31–41. https://doi.org/10.1007/s11157-010-9214-7

Mišurcová L (2011) Chemical composition of seaweeds. In: Kim SK (ed) Handbook of marine macroalgae: biotechnology and applied phycology. Wiley, Chichester, pp 173–192

Monteiro C, Castro PL, Malcata FX (2010) Cadmium removal by two strains of *Desmodesmus pleiomorphus* cells. Water Air Soil Pollut 208:17–27

Mühling M, Belay A, Whitton BA (2005) Screening *Arthrospira* (*Spirulina*) strains for heterotrophy. J Appl Phycol 17:129–135. https://doi.org/10.1007/s10811-005-7214-8

Munier M, Morançais M, Dumay J, Jaouen P (2015) One-step purification of R-phycoerythrin from the red edible seaweed *Grateloupia turuturu*. J Chromatogr B 992:23–29. https://doi.org/10.1016/j.jchromb.2015.04.012

Nagayama K, Iwamura Y, Shibata T, Hirayama I, Nakamura T (2002) Bactericidal activity of phlorotannins from the brown alga *Ecklonia kurome*. J Antimicrob Chemoth 50:889–893

Niu H, Volesky B (1999) Characteristics of gold biosorption from cyanide solution. J Chem Technol Biotechnol 74:778–784

O'Connell DW, Birkinshaw C, O'Dwyer TF (2008) Heavy metal adsorbents prepared from the modification of cellulose: a review. Bioresour Technol 99:6709–6724

Oswald W (2003) My sixty years in applied algology. J Appl Phycol 15:99–106

Padmini Sreenivasa Rao PP, Karmarkar SM (1986) Antibacterial substances from brown algae II. Efficiency of solvents in the evaluation of antibacterial substances from *Sargassum johnstonii* Setchell et Gardner. Bot Mar 29:503–507

Pal A, Paul AK (2008) Microbial extracellular polymeric substances: central elements in heavy metal bioremediation. Indian J Microbiol 48:49–64

Patil G, Raghavarao KSMS (2007) Aqueous two phase extraction for purification of C-phycocyanin. Biochem Eng J 34:156–164. https://doi.org/10.1016/j.bej.2006.11.026

Patra JK, Rath SK, Jena KB, Rathod VK, Thatoi H (2008) Evaluation of antioxidant and antimicrobial activity of seaweed (*Sargassum* sp.) extract: a study on inhibition of glutathione-S-transferase activity. Turk J Biol 32:119–125

Pavlostathis SG, Jackson GH (2002) Biotransformation of 2,4,6-trinitrotoluene in a continuous-flow *Anabaena* sp. system. Water Res 36:1699–1706

Pesce S, Batisson I, Bardot C, Fajon C, Portelli C, Montuelle B, Bohatier J (2009) Response of spring and summer riverine microbial communities following glyphosate exposure. Ecotoxicol Environ Saf 72:1905–1912

Peter P, Phaninatha Sarma A, Azeem ul Hasan MD, Murthy SDS (2010) Studies on the impact of nitrogen starvation on the photosynthetic pigments through spectral properties of the cyanobacterium, *Spirulina platensis*: identification of target phycobiliprotein under nitrogen chlorosis. Bot Res Int 3:30–34

Pethkar AV, Paknikar KM (1998) Recovery of gold from solutions using *Cladosporium cladosporioides* biomass beads. J Biotechnol 63:121–136

Pethkar AV, Kulkarni SK, Paknikar KM (2001) Comparative studies on metal biosorption by two strains of *Cladosporium cladosporioides*. Bioresour Technol 80:211–215

Pirszel J, Pawlik B, Skowroński T (1995) Cation exchange capacity of algae and cyanobacteria: a parameter of their sorption abilities. J Ind Microbiol 14:319–322

Pradhan S, Rai LC (2000) Optimization of flow rate, initial metal ion concentration and biomass density for maximum removal of Cu^{2+} by immobilized Microcystis. World J Microbiol Biotechnol 16:579–584

Prasanna R, Jaiswal P, Kaushik BD (2008) Cyanobacteria as potential options for environment al sustainability – promises and challenges. Indian J Microbiol 48:89–94

Rangsayatorn N, Upatham ES, Kruatrachue M, Pokethitiyook P, Lanza GR (2002) Phytoremediation potential of *Spirulina* (*Arthrospira*) *platensis*: biosorption and toxicity studies of cadmium. Environ Pollut 119:45–53

Ravi V, Balakumar H (1998) Biodegradation of the C-P bond in glyphosate by the cyanobacterium *Anabaena variabilis* L. J Sci Ind Res 57:790–794

Rimbau V, Caminis A, Romay C, González PM (1999) Protective effects of C-phycocyanin against kainic acid-induced neuronal damage in rat hippocampus. Neurosci Lett 276:75–78

Romay C, Armesto J, Remirez D, Goznález R, Ledon N, Garcia I (1998) Antioxidant and anti-inflammatory properties of C-phycocyanin from blue-green algae. Inflamm Res 47:36–41. https://doi.org/10.1007/s000110050256

Romay C, Goznález R, Ledón N, Remirez D, Rimbau V (2003) C-phycocyanin: a biliprotein with antioxidant, anti-inflammatory and neuroprotective effects. Curr Protein Pept Sci 4:207–216

Romero-Gonzalez ME, Williams CJ, Gardiner PHE, Gurman SJ, Habesh S (2003) Spectroscopic studies of the biosorption of gold(III) by dealginated seaweed waste. Environ Sci Technol 37:4163–4169

Rosen BP (2002) Transport and detoxification systems for transition metals, heavy metals and metalloids in eukaryotic and prokaryotic microbes. Comp Biochem Physiol A Mol Integr Physiol 133:689–693

Ross JA, Kasum CM (2002) Dietary flavonoids: bioavailability, metabolic effects, and safety. Ann Rev Nutr 22:19–34

Salmon-Monviola J, Gascuel-Odoux C, Garcia F, Tortrat F, Cordier M-O, Masson V, Trépos R (2011) Simulating the effect of technical and environmental constraints on the spatio-temporal distribution of herbicide applications and stream losses. Agric Ecosyst Environ 140:382–394

Sarada R, Pillai M, Ravishankar G (1999) Phycocyanin from *Spirulina* sp: influence of processing of biomass on phycocyanin yield, analysis of efficacy of extraction methods and stability studies on phycocyanin. Process Biochem 34:795–801

Sathyasaikumar KV, Swapna I, Reddy PVB, Murthy CRK, Roy KR, Dutta Gupta A, Senthilkumaran B, Reddanna P (2007) Co-administration of C-Phycocyanin ameliorates thioacetamide-induced hepatic encephalopathy in Wistar rats. J Neurol S 252:64–75. https://doi.org/10.1016/j.jns.2006.10.014

Savvaidis I (1998) Recovery of gold from thiourea solutions using microorganisms. Bio Metal 11:145–151

Schiewer S, Volesky B (2000) Biosorption by marine algae. In: Valdes J (ed) Bioremediation. Springer, Netherlands, pp 139–169

Sekar S, Chandramohan M (2007) Phycobiliproteins as a commodity: trends in applied research, patents and commercialization. J Appl Phycol 20:113–136. https://doi.org/10.1007/s10811-007-9188-1

Seker A, Shahwan T, Eroglu AE, Yılmaz S, Demirel Z, Dalay MC (2008) Equilibrium, thermodynamic and kinetic studies for the biosorption of aqueous lead(II), cadmium(II) and nickel(II) ions on *Spirulina platensis*. J Hazard Mater 154:973–980

Sharpley A, Wang X (2014) Managing agricultural phosphorus for water quality: lessons from the USA and China. J Environ Sci (China) 26:1770–1782

Silva-Stenico ME, Vieira FDP, Genuário DB, Silva CSP, Moraes LAB, Fiore MF (2012) Decolorization of textile dyes by cyanobacteria. J Braz Chem Soc 23:1863–1870

Simeunović J, Bešlin K, Svirčev Z, Kovač D, Babić O (2013) Impact of nitrogen and drought on phycobiliprotein content in terrestrial cyanobacterial strains. J Appl Phycol 25:597–607. https://doi.org/10.1007/s10811-012-9894-1

Simmons P, Singleton I (1996) A method to increase silver biosorption by an industrial strain of *Saccharomyces cerevisiae*. Appl Microbiol Biotechnol 45:278–285

Simonian NA, Coyle JT (1996) Oxidative stress in neurodegenerative diseases. Annu Rev Pharmacol Toxicol 36:83–106. https://doi.org/10.1146/annurev.pa.36.040196.000503

Singh NK, Parmar A, Sonani RR (2012) Madamwar D Isolation, identification and characterization of novel thermotolerant *Oscillatoria* sp. N9DM: change in pigmentation profile in response to temperature. Process Biochem 47:2472–2479. https://doi.org/10.1016/j.procbio.2012.10.009

Sloth JK, Wiebe MG, Eriksen NT (2006) Accumulation of phycocyanin in heterotrophic and mixotrophic cultures of the acidophilic red alga *Galdieria sulphuraria*. Enzyme Microb Tech 38:168–175. oi:10.1016/j.enzmictec.2005.05.010

Soni B, Trivedi U, Madamwar D (2008) A novel method of single step hydrophobic interaction chromatography for the purification of phycocyanin from *Phormidium fragile* and its characterization for antioxidant property. Bioresour Techhnol 99:188–194. https://doi.org/10.1016/j.biortech.2006.11.010

Sørensen L, Hantke A, Eriksen NT (2013) Purification of the photosynthetic pigment C-phycocyanin from heterotrophic *Galdieria sulphuraria*. J Sci Food Agric 93:2933–2938. https://doi.org/10.1002/jsfa.6116

Sousa C, Kotrba P, Ruml T, Cebolla A, De Lorenzo V (1998) Metalloadsorption by *Escherichia coli* cells displaying yeast and mammalian metallothioneins anchored to the outer membrane protein LamB. J Bacteriol 180(9):2280–2284

Stadnichuk IN, Krasilnikov PM, Zlenko DV (2015) Cyanobacterial phycobilisomes and phycobiliproteins. Microbiology 84:101–111. https://doi.org/10.1134/S0026261715020150

Steinberg PD (1989) Biogeographical variation in brown algal polyphenolics and other secondary metabolites: comparison between temperate Australasia and North America. Oecologia 78:373–382

Stiger V, Deslandes E, Payri CE (2004) Phenolic contents of two brown algae, *Turbinaria ornata* and *Sargassum mangarevense* on Tahiti (French Polynesia): interspecific, ontogenic and spatio-temporal variations. Bot Mar 47:402–409

Studnik H, Liebsch S, Forlani G, Wieczorek D, Kafarski P, Lipok J (2015) Amino polyphosphonates – chemical features and practical uses, environmental durability and biodegradation. New Biotechnol 32:1–6

Sun L, Wang S, Gong X, Zhao M, Fu X, Wang L (2009) Isolation, purification and characteristics of R-phycoerythrin from a marine macroalga *Heterosiphonia japonica*. Protein Expr Purif 64:146–154. https://doi.org/10.1016/j.pep.2008.09.013

Talaraico L, Cortese A (1993) Response of *Audouinella saviana* (Meneghini) Woelkerling (Nemaliales, Rhodophyta) cultures to monochromatic light. Hydrobiologia 260:477–484. https://doi.org/10.1007/BF00049059

Tandeau de Marsac N, Cohen-Bazire N (1977) Molecular composition of cyanobacterial phycobilisomes. Proc Natl Acad Sci U S A 74:1635–1639

Telford WG, Hawley T, Subach F, Verkhusha V, Hawley RG (2012) Flow cytometry of fluorescent proteins. Methods 57(3):318–330

Thomas NV, Kim SK (2011) Potential pharmacological applications of polyphenolic derivates from marine brown algae. Environ Toxicol Pharmacol 32:325–335

Tsekova K, Todorova D, Ganeva S (2010) Removal of heavy metals from industrial wastewater by free and immobilized cells of *Aspergillus niger*. Int Biodeterior Biodegr 64:447–451

US_Environmental Protection Agency (2011) Working in partnership with states to address phosphorus and nitrogen pollution through use of a framework for state nutrient reductions

Usher PK, Ross AB, Camargo-Valero MA, Tomlin AS, Gale WF (2014) An overview of the potential environmental impacts of large-scale microalgae cultivation. Biofuels 5:331–349. https://doi.org/10.1080/17597269.2014.913925

Van Mooy BA, Krupke A, Dyhrman ST, Fredricks HF, Frischkorn KR, Ossolinski JE, Repeta DJ, Rouco M, Seewald JD, Sylva SP (2015) Phosphorus cycling. Major role of planktonic phosphate reduction in the marine phosphorus redox cycle. Science 348:783–785

Vázquez-Sánches J, Ramón-Gallegos E, Mojica-Villegas A, Madrigal-Bujaidar E, Pérez-Pastén-Borja, Chamorro-Cevallos G (2009) *Spirulina maxima* and its protein extract protect against hydroxyurea-teratogenic insult in mice. Food Chem Toxicol 47:2785–2789. https://doi.org/10.1016/j.fct.2009.08.013

Veglio F, Beolchini F (1997) Removal of metals by biosorption: a review. Hydrometallurgy 44:301–316

Vendrell E, Ferraz DG, Sabater C, Carrasco JM (2009) Effect of glyphosate on growth of four freshwater species of phytoplankton: a microplate bioassay. Bull Environ Contam Toxicol 82:538–542

Vera MS, Lagomarsino L, Sylvester M, Perez GL, Rodriguez P, Mugni H, Sinistro R, Ferraro M, Bonetto C, Zagarese H, Pizarro H (2010) New evidences of Roundup (glyphosate formulation) impact on the periphyton community and the water quality of freshwater ecosystems. Ecotoxicology 19:710–721

Verma SK, Singh SP (1990) Factors regulating copper uptake in cyanobacterium. Curr Microbiol 21:33–37

Vidali M (2001) Bioremediation. An overview. Pure Appl Chem 73:1163–1172

Volesky B (2007) Biosorption and me. Water Res 41:4017–4029

Waldron KJ, Robinson NJ (2009) How do bacterial cells ensure that metalloproteins get the correct metal? Nat Rev Microbiol 6:25–35

Wells S, Johnson I (1994) Fluorescent Labels for Confocal Microscopy. In: Buetow DE, Stevens JK, Cameron IT, Mills LR, Padilla GM, Trogadis JE Zimmerman AM (eds) Three-Dimensional Confocal Microscopy: Volume Investigation of Biological Specimens. Elsevier, San Diego, pp 101–129

Wijesinghe WAJP, Jeon YJ (2012) Enzyme-assistant extraction (EAE) of bioactive components: a useful approach for recovery of industrially important metabolites from seaweeds: a review. Fitoterapia 83:6–12

Williams GM, Kroes R, Munro IC (2000) Safety evaluation and risk assessment of the herbicide roundup and its active ingredient, glyphosate, for humans. Regul Toxicol Pharmacol 31:117–165

Yin XX, Chen J, Qin J, Sun GX, Rosen BP, Zhu YG (2011) Biotransformation and volatilization of arsenic by three photosynthetic cyanobacteria. Plant Physiol 156:1631–1638

Yu J, Yang YF (2008) Physiological and biochemical response of seaweed *Gracilaria lemaneiformis* to concentration changes of N and P. J Exp Mar Biol Ecol 367:142–148. https://doi.org/10.1016/j.jembe.2008.09.009

Yu G, Shi D, Cai Z, Cong W, Ouyang F (2011) Growth and physiological features of cyanobacterium *Anabaena* sp. strain PCC 7120 in glucose-mixotrophic culture. Chin J Chem Eng 19:108–115

Zeraatkar AK, Ahmadzadeh H, Talebi AF, Moheimani NR, McHenry MP (2016) Potential use of algae for heavy metal bioremediation, a critical review. J Environ Manag 181:817–831

Zern TL, Fernandez ML (2005) Cardioprotective effects of dietary polyphenols. J Nutr 135:2291–2294

The Methods of Algal Biomass Extraction: Toward the Application

5

Agnieszka Dmytryk, Katarzyna Chojnacka, and Edward Rój

Abstract

Increasing customer need for natural products of both high and well-defined activity has enhanced the use of biological materials, such as algal biomass. At the same time, algae transport and storage issues, concerning fresh seaweeds in particular, have led to the development of biomolecule isolation, including extraction. To the best of our knowledge, various approaches have been successfully applied in extracting biologically active compounds from algal biomass, among which solvent and temperature treatment are the most common. Considering novel methods, processing under high pressure (pressurized liquid and supercritical fluid extraction) and ultrasound-, microwave-, and enzyme-assisted extraction have been reported. The approaches differ in their efficacy and selectivity and extract purity, determining the usability of the final product in either bulk manufacturing or as a high-value material. Application of algae-based constituents in food and beverage products, dietary and feed supplements, cosmetics, and pharmaceuticals is being widely discussed. Recently, the usability of algae oil in the technology of biofuels has been extensively examined. In the current work, preparation of algal extracts and formulations for potential industrial use are discussed.

Keywords

Phycocolloids · Astaxanthin · β-carotene · Algae oil · Eicosapentaenoic acid · Docosahexaenoic acid

A. Dmytryk (✉) · K. Chojnacka
Faculty of Chemistry, Department of Advanced Material Technologies, Wrocław University of Science and Technology, Wrocław, Poland
e-mail: agnieszka.dmytryk@pwr.edu.pl

E. Rój
Supercritical Extraction Department, New Chemical Syntheses Institute, Puławy, Poland

5.1 Introduction

The extraction of natural materials has been practiced since the Paleolithic Period (Herrero et al. 2010) and was developed by various civilizations in a quite advanced manner (Chemat et al. 2012). In coastal and island regions, there is also a several-thousand-year tradition of using algal biomass for daily-life products – including food, feed and fertilizers, as well as specialized formulas related to health, body care, and beauty (Newton 1951; Hallsson 1964; Hoppe 1979; Chapman and Chapman 1980; Tseng 1981; Bradford and Bradford 1996; Caliceti et al. 2002; Dillehay et al. 2008; Craigie 2011; Perosa et al. 2015). Contrarily, the commercial potential of algal biomass and products thereof was largely untapped in continental Europe until the bulk production of polysaccharide colloids (phycocolloids) as foodstuff stabilizers and thickeners started in the interwar period (Naylor 1976). A breakthrough in public awareness about the health benefits of algae derivatives came after World War II – when a series of research activities on microalgal proteins led to the unexpected discovery of compounds showing antibiotic activity (Borowitzka 1995; Cornet 1998; Becker 2004). In parallel, the first report about algae extraction appeared (Milton 1952), describing the technology of liquid fertilizer further adapted for industrial-scale production. The development of liquefaction of algal biomass, performed since then, enabled easier transport, and hence wider application, of the derived products (Craigie 2011).

The global algae market is increasing, following customer demand for high-value food products, which has recently exceeded consumption of algae-based food. Such a trend has entailed the elaboration of several extraction methods that are both safe and efficient. In this chapter, recent trends in the processing of algal biomass into deliverables are reviewed.

© Springer International Publishing AG, part of Springer Nature 2018
K. Chojnacka et al. (eds.), *Algae Biomass: Characteristics and Applications*, Developments in Applied Phycology 8,
https://doi.org/10.1007/978-3-319-74703-3_5

5.2 Various Approaches to Algae Extraction for Obtaining Valuable Products

Conventional (traditional) extraction methods include inorganic and organic solvent processes, usually combined with heating, among which the Soxhlet technique, hydrodistillation, and alcohol-based maceration might be indicated (Wang and Weller 2006; Michalak and Chojnacka 2014). Such an approach is, however, both time- and energy-consuming and might influence the structure, and hence the properties, of the isolated compounds (Puri et al. 2012). The separation of the final product from the extractant also constitutes an environmental issue, as large amounts of production-related wastes containing chemicals are released (EPA 2014).

Over time, the large-scale use of organic solvents led to concern over environmental damage, which resulted in an international agreement – the Montreal Protocol (1989) – aimed at restricting the industrial application of ozone-depleting substances. As the protocol was gaining more signatories, the interest in deployment of new technologies increased (Ramsey et al. 2009). In parallel – since the early 1990s – green chemistry has been developed to limit the risk of chemical exposure for living organisms by reducing or eliminating the use and/or generation of hazardous substances (Ibañez et al. 2012; Chemat et al. 2012). To achieve this goal, principles of natural product extraction have been established to facilitate green-labelled processes:

I. The implementation of innovation through variety selection and the involvement of plant-based renewables
II. The use of alternative solvents, including, in particular, water or agro-solvents
III. The reduction of energy consumption through energy recovery and the application of innovative technologies
IV. The production of side streams and residues, instead of waste, capable of being valorized within the bio- and agro-refining industry
V. The reduction of unit operations and favoring of safety, robustness, and control in processes
VI. Aiming for non-denatured, contaminant-free, and biodegradable extracts

Following these guidelines enables to introduce solutions that are efficient in terms of raw materials, solvents, and energy consumption, including within industrial-scale activities (Chemat et al. 2012).

Traditional methods, reported in the late 1980s and 1990s, are still used for extraction of phycocolloids, including treatment of algal biomass with a hot alkali solution (alkali hydrolysis) and precipitation of fibers in the presence of inorganic salts and/or alcohol afterward (Armisen and Galatas 1987; McHugh 1987, 2003; Porse 1998). The process also enables to obtain soluble proteins and peptides (McHugh 2003; Kadam et al. 2013). At the same time, the large-scale use of organic solvents involves hexane, chloroform, and both methanolic and ethanolic alcohol to provide fractions of carotenoids, fatty acids, and phenolic compounds, respectively (Li et al. 2002; Borowitzka 2013a; Kadam et al. 2013). However, there are a number of researches concerning novel processes, such as pressurized liquid extraction (PLE), supercritical fluid extraction (SFE), and enzyme-, microwave-, and ultrasound-assisted extraction (EAE, MAE, UAE, respectively), which would meet the green extraction principles (Ibañez et al. 2012; Wijesinghe and Jeon 2012). Since the costs of novel methods are higher compared to conventional operations, they are dedicated mostly to high-value products, as follows:

- PLE – phenolic compounds, fatty acids and xanthophylls
- SFE, particularly with supercritical CO_2 (SC-CO_2) as the extractant – fatty acids and carotenoids
- MAE – phenolic compounds and carbohydrates
- UAE – phenolic compounds and minerals
- EAE – carbohydrates, antioxidants

Moreover, both SFE and MAE provide efficient extraction of algae oil. Yet, there are still some obstacles to overcome in implementing the above mentioned approaches into industrial practice. EAE substrate specificity and operational conditions represent its main limitations, while PLE and MAE are not suitable for obtaining thermolabile compounds. In the near future, SFE and UAE are the most likely to be widely adapted into the large-scale production of algae derivatives (Kadam et al. 2013; Esquivel-Hernández et al. 2017 and relevant references therein).

5.3 Global Algae Market

Half of the net primary production of biomass is assumed to be aquatic (Bowles 2007). Indeed, global algae production is about $28.5 \cdot 10^6$ tonnes of fresh weight per year – besides catches from midwater outside the official statistics (FAO 2016), expected to accomplish a 5.3% increase in compound annual growth rate (CAGR) by 2024 (Transparency Market Research 2016). According to the available data, the production of microalgae is much lower compared to seaweeds because of different target applications, as well as cultivation issues and poor reflection in the worldwide reporting system (Borowitzka 1999; FAO 2016).

In commercial algal farming, Asian countries, particularly China and Indonesia, are predominant, as they represent more than 80% of global production. Such a market share is rooted in the traditional use of algae, with direct consumption ($9 \cdot 10^6$ tonnes of fresh weight) and further processing for food and/or food additives remaining a great utilization of aquaculture (FAO 2014, 2016).

In 2012, the global algae market was worth US\$ $6.4 \cdot 10^9$ (FAO 2014). At the same time, the revenue is expected to reach over US\$ $1.1 \cdot 10^9$ at 7.4% CAGR by 2024 (Transparency Market Research 2016). There is steady growth in the market for algal metabolites, comprising, besides polysaccharides and derivatives, pigments, mainly carotenoids, and fatty acids (Esquivel-Hernández et al. 2017). The latter is also involved in the algae oil market, which constitutes an object of even greater interest (Grand View Research 2017a). The variety of biologically active compounds extractable from algal biomass enables to obtain products of different customer value. The overview of the current global algae market is shown in Fig. 5.1.

Among seaweed metabolites, phycocolloids, including alginates, agar, and carrageenan, are still of the greatest interest, as their market volumes are estimated at $2.5 \cdot 10^4$ (considering propylene glycol alginate of food and pharma grade only), $1.5 \cdot 10^4$, and $5.8 \cdot 10^4$ tonnes, respectively. Although sales of carrageenan are the highest, market development for that phycocolloid is currently quite slow, at 2% annual average growth rate (AAGR). The fastest-developing market is for agar, as it provides AAGR at a level of 7%. Contrarily, AAGR for alginates is negative, while their price is still high – US\$ 14 per kg, as compared to US\$ 9 for carrageenan and US\$17 for agar (Porse and Rudolph 2017).

The volume of the phycocolloid market is reflected in its value, which reaches US\$ $3.8 \cdot 10^8$ (Markets and Markets 2015), US\$ $2.5 \cdot 10^8$ (Grand View Research 2017b; Mordor Intelligence 2017a), and US\$ $7.6 \cdot 10^8$ (Micromarket Monitor 2017; Mordor Intelligence 2017b) for alginates, agar, and carrageenan, respectively. A review of the commercial application of phycocolloids is shown in Table 5.1.

When considering microalgae derivatives, carotenoids comprise a billion-dollar market, with astaxanthin in the leading position, US\$ $4.0 \cdot 10^8$ (Algae Industry Magazine 2015; BCC Research 2015), and β-carotene in second place – US\$ $1.7 \cdot 10^8$ (Grand View Research 2016; Borowitzka 2013a). Algae are an important source for industrial-scale production of carotenoids – providing 26 and 38% of the total contribution to the market revenue of astaxanthin and β-carotene, respectively. Approaches with different biologically originated materials, as well as fermentation and chemical synthesis, have also been implemented. The latter is, however, considered cost consuming and disadvantageous, particularly in regard to products for human consumption (Grand View Research 2016, 2017c). The commercial extraction of β-carotene from microalgae *Dunaliella salina* exceeds the processing of other natural sources, such as plants (including carrots, sweet potatoes, and pumpkins),

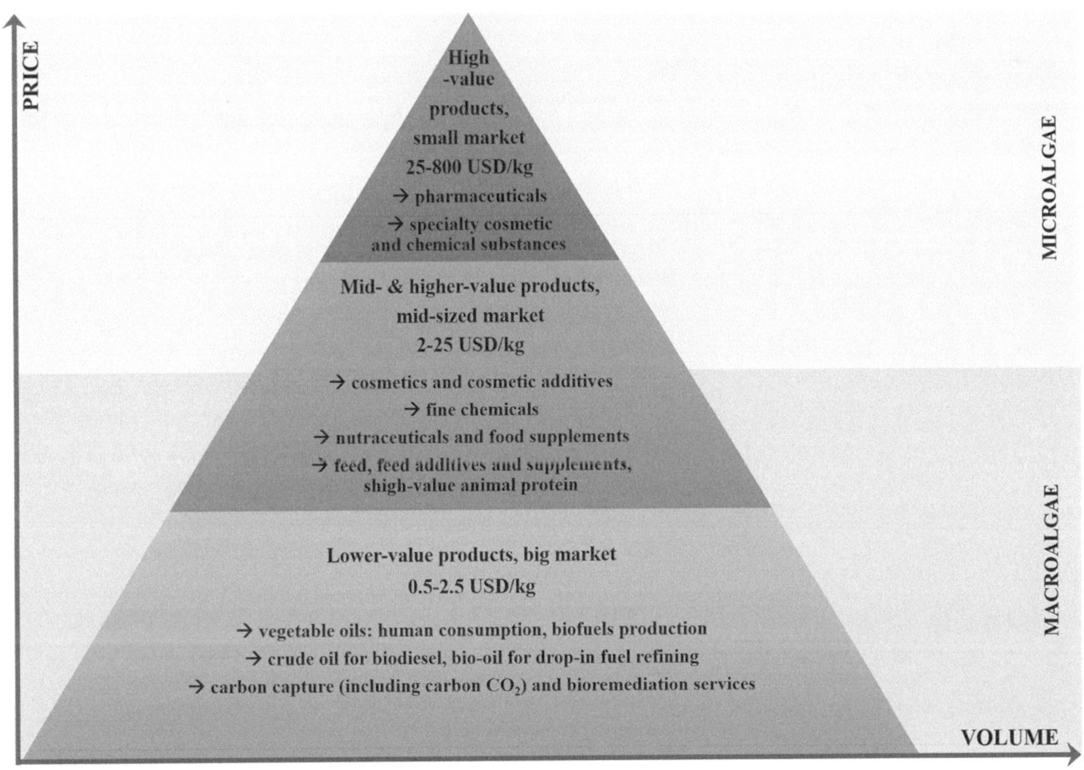

Fig. 5.1 Global value market for algae products (Thurmond 2009; Hennig and Jain 2017)

Table 5.1 Commercial use of algal colloids (polysaccharides and derivatives) (McHugh 2003, Commission Regulation (EU) No 1129/2011 of 11 November 2011)

Compound	Algae source	Application
Acl. alginic acid, AA; sodium alginate, SA; calcium alginate, CA; propylene glycol alginate, PGA)	Phaeophyceae: species of *Ascophyllum*, *Durvillaea*, *Ecklonia*, *Saccharina*, *Laminaria*, *Lessonia*, *Macrocystis*, *Saccharina*, *Sargassum*	Food (E 400–E 404, PGA – E 406):
		Thickening agent for the following products:
		Fillings and icings for baked goods (pastries and cakes)
		Sauces, syrups, and ice cream toppings
		Mayonnaise and salad dressings
		Dairy products, such as yogurt (PGA), chocolate milk, whipped cream
		Fruit drinks with fruit pulp added (PGA)
		Stabilizing agent for the following products:
		Ice cream
		Beer foam
		Gelling agent for the following products:
		Artificial fruits (e.g., cherries) (CA)
		Jellies (SA)
		Paste to stuff olives with
		Binding agent to restructure and reform products, such as meat, fish, onion rings, and shrimp substitutes (SA, CA)
		Gel-coating layer to preserve frozen fish (CA)
		Fining agent to clarify wine
		Textiles:
		Thickening agent for dye-containing paste in textile printing
		Biotechnology
		Carrier to immobilize biocatalysts – enzymes or cells – by entrapping them with beads (CA)
		Pharmaceuticals and healthcare sector:
		Component of polymer fibers for wound dressings (CA)
		Tablet disintegrant (AA)
		Component of dietary foods and tablets for heartburn/indigestion to swell in the stomach and reduce the sense of hunger and the likelihood of reflux-derived irritation, respectively (AA)
		Component of controlled-release systems for medicinal drugs
		Paper:
		Surface sizing agent
		Film former to improve ink holdout and printability
		Component of starch adhesives
		Paper coating
		Other
		Component of coatings for welding rods or electrodes
		Binder in fish feed
		Mold release agents for plaster molds and fiberglass
		Plastics
		Coating for anti-tack paper – a release agent for synthetic
		Resin decorative boards (SA)
Agar	Rhodophyta: species of *Gelidium*, *Gracilaria*, and *Gelidiella*	Food (90% of global agar production, E 406):
		Stabilizing and thickening agent for the following products:
		Fillings and icings for baked goods (pastries and cakes), meringues
		Confections, such as jellies and candies
		Ices and sherbets
		Dairy products, such as cheese, cream, and yogurt
		Gelled meat and fish
		Fining agent to clarify wine (mainly plum wine)
		Biotechnology:
		Culture media, mainly bacteriological
		Ornamental plant cultivation:
		Growth substrate for obtaining plant clones or copies in orchid nurseries
		Pharmaceuticals:
		Smooth laxative

(continued)

Table 5.1 (continued)

Compound	Algae source	Application
Carrageenan	Rhodophyta: *Kappaphycus alvarezii*, *Eucheuma denticulatum*, *Betaphycus gelatinum*, *Chondrus crispus*, *Gigartina skottsbergii*, *Sarcothalia crispata*, *Mazzaella laminarioides*, *Chondracanthus canaliculatus*	Food (E 407, processed *Eucheuma* seaweed E 407a):
		Stabilizing and thickening agent for the following products:
		Dairy products, such as cottage cheese, chocolate milk, UHT sterilized milk, cream
		Ice cream and sherbets
		Liquid coffee whiteners
		Dry instant chocolate mixes
		Low-oil or no-oil salad dressings (also thickened by carrageenan)
		Reduced-oil mayonnaise (also thickened by carrageenan)
		Gelling and thickening agent for the following products:
		Jellies
		Mousse desserts
		Fruit drink mixes
		Ham (retainment of soluble protein during cooking of ham in brine solutions)
		Precooked poultry products
		Low-fat meat and poultry products (e.g., frankfurters)
		Fining agent to clarify beer
		Pet food:
		Gelling and thickening agent for canned food
		Household chemistry:
		Component of air freshener gels
		Cosmetics:
		Component of toothpaste
		Biotechnology:
		Carrier to immobilize biocatalysts – enzymes or cells – by entrapping them with beads

fungi, palm oil, and microbial production (Borowitzka 2013b; Grand View Research 2016; Oilgae 2016). At the same time, the most known source of astaxanthin, microalgae *Haematococcus pluvialis*, is increasingly competing with krill, shrimp by-products, and yeast (*Phaffia* species) (Grand View Research 2017c; Transparency Market Research 2017), while remaining predominant toward Pacific sockeye salmon and *Paracoccus* bacteria (Transparency Market Research 2017).

As natural pigments, both astaxanthin and β-carotene are primarily used to improve the color of food products (Algae Industry Magazine 2016); however, various trends in their application have been developed (Borowitzka 2013a). The biggest share in the astaxanthin market (about 40%) belongs to the sector of feed additive enhancing pigmentation in aquaculture animals (fish, e.g., salmon and trout, and crustaceans, e.g., shrimp), which, however, mostly involves synthetic and microbial-derived compounds (Research and Markets 2015; Transparency Market Research 2017). Astaxanthin of natural origin, algae in particular, shows great antioxidant activity (determined by its oxygen radical absorbance capacity) and thus has become more profitable, being launched for food and beverages, nutritional and dietary supplements, and nutraceuticals (Enzing et al. 2014, Algae Industry Magazine 2015, BBC Research 2015, Research and Markets 2015; Grand View Research 2017c). The latter holds the second-biggest market share (about 32%), and the new concept of nutraceutical delivery with soft gels is projected to reach the fastest growth rate (Grand View Research 2017c).

Besides preventing oxidative stress, the compound also has capability for healing UV-induced damage, which has led to its successful use in the treatment of cardiovascular disease, neurodegenerative disease (Alzheimer's, Parkinson's), and both ophthalmic and orthopedic disorders (Enzing et al. 2014; Grand View Research 2017c). For the same reason, astaxanthin is applied as an antiaging agent in cosmetics, cosmeceuticals, and personal care products, forming the third-largest sector in terms of market share (about 18%) (Research and Markets 2015; Grand View Research 2017c; Transparency Market Research 2017).

Despite lower market value than astaxanthin, β-carotene was the first high-value compound extracted commercially from microalgae (Borowitzka 2013a) and is generally the most prominent carotenoid (BBC Research 2015). Being a precursor to vitamin A and both an antioxidative and anti-inflammatory agent, β-carotene is suited to a wide range of applications (Enzing et al. 2014; Grand View Research 2016), yet its use as a color additive (E 160a) seems to remain the most prominent (Commission Regulation (EU) No

1129/2011 of 11 November 2011). In fact, the food and beverage sector holds the leading market share, since it contributes one-third of the total revenue and shows a further upward trend. A slightly smaller market share, still 30%, belongs to dietary supplements, which are followed by cosmetics and personal care products – almost 22% of total revenue. The third-largest applicant sector involves a great variety of formulations, starting from bath and shower preparations, moving through face cleansers, aftershave lotions, shampoos, and hair conditioners, and ending at makeup and sun-care products (Grand View Research 2016). The beneficial health properties of β-carotene encourage its application in animal feed as well (Enzing et al. 2014).

According to the conventional approach, neither eicosapentaenoic nor docosahexaenoic acid (EPA and DHA, respectively) is an essential fatty acid. Yet, they are customarily classified as such due to functionality criteria (Hassam et al. 1977a, b; NIH 2016) and the limited efficiency of natural synthesis (Harris 2010). Since diets are usually low in EPA and DHA, and thus these fatty acids are commonly deficient, there is a great opportunity for the sectors of dietary and nutritional supplements to be advanced (Joint WHO/FAO Expert Consultation 2003). EPA- and DHA-based supplements held the biggest share – over US$ $1.0 \cdot 10^9$, in the revenue of the omega-3 ingredients market, which is valued at about US$ $3.5 \cdot 10^9$ and is expected to reach US$ $4.0 \cdot 10^9$ within the next 5 years. The other profitable area of application is divided into four sectors, as follows: functional food, infant formulas, pet food, and pharmaceuticals (Global Market Insights 2016). Relating to the latter, EPA and DHA ingredients show a protective effect against cardiovascular conditions, arthritis, and cancer (Joint WHO/FAO Expert Consultation 2003; Global Market Insights 2016). More than 75% of feedstock for omega-3 fatty acid production is fish oil, mostly from sardines and anchovies. However, its unpleasant organoleptic properties, along with the increasing popularity of the vegetarian lifestyle among customers, enhance the position of algae oil, currently at about 8% of market share, as an alternative source material (Global Market Insights 2016).

Though economically beneficial, the use of fatty acids for human consumption – including supplements, as well as foods and beverages – is the smallest applicant sector of the US$ $1.5 \cdot 10^9$-worth algae oil market. The leading market share belongs to fuel grade products, such as biodiesel, jet fuel, and gasoline, which provide almost 40% of the total revenue. Algae oil is also successfully applied, at about 30% of total revenue, to animal feed (Grand View Research 2017a)

Besides separate fractions, algal extracts within mixtures of biologically active compounds are also gaining attention. The applicability of such products will be reviewed in subsequent chapters of this book.

5.4 Conclusion

Algal biomass is a rich source of various biologically active compounds. In industrial production, phycocolloids, carotenoids, and essential fatty acids are of particular importance. Biofuel production from algae oil has also been exhaustively verified. While phycocolloids have been successfully extracted through the use of alkali hydrolysis, for the other constituents, novel techniques – improved in terms of efficiency, solvent use, and both time and cost consumption – have been developed. Among the known approaches, ultrasound-assisted and SC-CO_2 extraction are considered promising for industrial implementation, yet the latter seems to be the most suitable to fulfill the market demand.

References

Algae Industry Magazine (2015) Boom in astaxanthin boosts carotenoid market. http://www.algaeindustrymagazine.com/boom-in-astaxanthin-boosts-carotenoid-market/. Accessed on 4 Sept 2017

Algae Industry Magazine (2016) Astaxanthin leading carotenoids market growth. http://www.algaeindustrymagazine.com/astaxanthin-leading-carotenoids-market-growth/. Accessed on 4 Sept 2017

Armisen R, Galatas F (1987) Production, properties and uses of agar In: McHugh DJ (ed) Production and utilization of products from commercial seaweeds, FAO Fisheries Technical Paper 288. FAO, Rome. ISBN 92-5-102612-. http://www.fao.org/docrep/X5822E/x5822e03.htm#chapter%201%20%20%20production,%20properties%20and%20uses%20of%20agar. Accessed on 4 Sept 2017

BCC Research (2015) The Global Market for carotenoids. https://www.reportbuyer.com/product/96628/the-global-market-for-carotenoids.html. Accessed on 4 Sept 2017

Becker W (2004) Microalgae in human and animal nutrition. In: Richmond A (ed) Handbook of microalgal culture: biotechnology and applied phycology. Blackwell Science, Oxford, pp 312–351

Borowitzka MA (1995) Microalgae as sources of pharmaceuticals and other biologically active compounds. J Appl Phycol 7(1):3–15

Borowitzka MA (1999) Economic evaluation of microalgal processes and products. In: Cohen Z (ed) Chemicals from microalgae. Taylor & Francis, London, pp 387–409

Borowitzka MA (2013a) High-value products from microalgae – their development and commercialization. J Appl Phycol 25(3):743–756

Borowitzka MA (2013b) *Dunaliella*: biology, production, and markets. In: Richmond A, Hu Q (eds) Handbook of microalgal culture. Wiley, Oxford, pp 359–368

Bowles D (2007) Micro-and macro-algae: utility for industrial applications. Outputs from the EPOBIO project. CNAP, University of York/CPL Press, Newbury

Bradford P, Bradford M (1996) Cooking with sea vegetables. Thorsons Publishing Group, Wellinborough

Caliceti M, Argese E, Sfriso A, Pavoni B (2002) Heavy metal contamination in the seaweeds of the Venice lagoon. Chemosphere 47:443–454

Chapman VJ, Chapman DJ (1980) Seaweeds and their uses. Chapman and Hall, New York

Chemat F, Vian MA, Cravotto G (2012) Green extraction of natural products: concept and principles. Int J Mol Sci 13(7):8615–8627

Commission Regulation (EU) No 1129/2011 of 11 November 2011 amending Annex II to Regulation (EC) No 1333/2008 of the European Parliament and of the Council by establishing a Union list of food additives (Text with EEA relevance) (n.d.)

Cornet JF (1998) Les photobioréacteurs. Biofutur 176:1–10

Craigie JS (2011) Seaweed extract stimuli in plant science and agriculture. J Appl Phycol 23(3):371–393

Dillehay TD, Ramirez C, Pino M, Collins MB, Rossen J, Pino-Navarro JD (2008) Monte Verde: seaweed, food, medicine, and the peopling of South America. Science 320(5877):784–786

Enzing C, Ploeg M, Barbosa M, Sijtsma L (2014) Microalgae-based products for the food and feed sector: an outlook for Europe. IPTS Institute for Prospective Technological Studies, JRC, Seville

Esquivel-Hernández DA, Ibarra-Garza IP, Rodríguez-Rodríguez J, Cuéllar-Bermúdez SP, de J Rostro-Alanis M, Rostro-Alanis M, Alemán-Nava GS, García-Pérez JS, Parra-Saldívar R (2017) Green extraction technologies for high-value metabolites from algae: a review. Biofuels Bioprod Biorefin 11:215–231

Food and Agriculture Organization of the United Nations (2014) The state of world fisheries and aquaculture. Opportunities and challenges, FAO, Rome. E-ISBN 978-92E-ISBN 978-92-5-108276-8-5-108276-8 pp 1–223. http://www.fao.org/3/a-i3720e.pdf. Accessed on 4 Sept 2017

Food and Agriculture Organization of the United Nations (2016) The state of world fisheries and aquaculture 2016. Contributing to food security and nutrition for all. FAO, Rome. ISBN 978-92-5-109185-2 pp 1–190. www.fao.org/3/a-i5555e.pdf. Accessed on 4 Sept 2017

Global Market Insights (2016) EPA/DHA (Omega 3) ingredients market size by application (dietary supplements, pharmaceuticals, functional foods, pet & animal feed, infant formulas), by source (anchovy/sardine, high concentrates, medium concentrates, low concentrates, algae oil, tuna oil, cod liver oil, salmon oil, krill oil, menhaden oil), industry analysis report, regional analysis, application potential, price trend, competitive market share & forecast, 2015–2022. https://www.gminsights.com/industry-analysis/EPA-DHA-omega-3-ingredients-market. Accessed on 4 Sept 2017

Grand View Research (2016) Beta-carotene market analysis by source (algae, fruits & vegetables, & synthetic), by application (food & beverages, dietary supplements, cosmetics, & animal feed) and segment forecasts to 2024. http://www.grandviewresearch.com/industry-analysis/beta-carotene-market. Accessed on 4 Sept 2017

Grand View Research (2017a) Algae oil market analysis, by grade (fuel, food, feed), by application (biofuel, dietary supplement, F&B, animal feed), by region (North America, Europe, Asia Pacific, South & Central America & MEA), and segment forecasts, 2014–2025. http://www.grandviewresearch.com/industry-analysis/algae-oil-market. Accessed on 4 Sept 2017

Grand View Research (2017b) Agar agar gum market analysis by product (powder, square, strips), by application (confectioneries, bakery & pastry, retail, meat, microbiological & molecular), by region, and segment forecasts, 2014–2025. http://www.grandviewresearch.com/industry-analysis/global-agar-agar-gum-market. Accessed on 4 Sept 2017

Grand View Research (2017c) Astaxanthin market analysis by source (natural [yeast, krill/shrimp, microalgae] and synthetic), by product (dried biomass/powder, oil, soft gels, liquid), by application, and segment forecasts, 2014–2025. http://www.grandviewresearch.com/industry-analysis/global-astaxanthin-market. Accessed on 4 Sept 2017

Hallsson SV (1964) The uses of seaweed in Iceland. Proc Int Seaweed Symp 4:398–405

Harris WS (2010) Omega-3 fatty acids. In: Coates PM, Betz JM, Blackman MR, Cragg GM, Levine M, Moss J, White JD (eds) Encyclopedia of dietary supplements, 2nd edn. Informa Healthcare, London, pp 577–586

Hassam AG, Rivers JPW, Crawford MA (1977a) Metabolism of gamma-linolenic acid in essential fatty acid-deficient rats. J Nutr 107(4):519–524

Hassam AG, Rivers JPW, Crawford MA (1977b) Potency of γ-linoleic acid (18: 3ω6) in curing essential fatty acid deficiency in the rat. Ann Nutr Metab 21(Suppl. 1):190–192

Hennig J, Jain M (2017) Algae products. Fish 2.0 investment insights. Available via Fish 2.0. http://www.fish20.org/9-news/440-report-algae-market-to-reach-45bn-by-2023. Accessed on 4 Sept 2017

Herrero M, Mendiola JA, Cifuentes A, Ibañez E (2010) Supercritical fluid extraction: recent advances and applications. J Chromatogr A 1217:2495–2511

Hoppe HA (1979) Marine algae and their products and constituents in pharmacy. In: Hoppe HA, Levring T, Tanaka Y (eds) Marine algae in pharmaceutical science, vol 1. Walter de Gruyter, Berlin, pp 25–119

Ibañez E, Herrero M, Mendiola JA, Castro-Puyana M (2012) Extraction and characterization of bioactive compounds with health benefits from marine resources: macro and micro algae, cyanobacteria, and invertebrates. In: Hayes M (ed) Marine bioactive compounds. Springer Science+Business Media, New York, pp 55–98

Joint WHO/FAO Expert Consultation (2003) Diet, nutrition and the prevention of chronic diseases. World Health Organ Tech Rep Ser 916(i–viii):1

Kadam SU, Tiwari BK, O'Donnell CP (2013) Application of novel extraction technologies for bioactives from marine algae. J Agric Food Chem 61(20):4667–4675

Li X, Fan X, Han L, Lou Q (2002) Fatty acids of some algae from the Bohai Sea. Phytochemistry 59:157–161

Markets and Markets (2015) Alginates & derivatives market by type (sodium alginate, calcium alginate, potassium alginate, pga, others), application (food & beverage, industrial, pharmaceuticals, others), & by region – Global Trends & Forecast to 2019. http://www.marketsandmarkets.com/Market-Reports/alginates-derivatives-market-77102420.html. Accessed on 4 Sept 2017

McHugh DJ (1987) Production, properties and uses of alginates. In: DJ MH (ed) Production and utilization of products from commercial seaweeds, FAO Fisheries Technical Paper 288. FAO, Rome. ISBN 92-5-102612-2. http://www.fao.org/docrep/x5822e/x5822e04.htm#chapter2production,propertiesandusesofalginates. Accessed on 4 Sept 2017

McHugh DJ (2003) A guide to the seaweed industry, FAO Fisheries Technical Paper 441. FAO, Rome, pp 1–105. ISSN 0429-9345, www.fao.org/3/a-y4765e.pdf. Accessed on 4 Sept 2017

Michalak I, Chojnacka K (2014) Algal extracts: technology and advances. Eng Life Sci 4(6):581–591

Micromarket Monitor (2017) Global Carrageenan Market research report. http://www.micromarketmonitor.com/market-report/carrageenan-reports-9953387913.html. Accessed on 4 Sept 2017

Milton RF (1952) Improvements in or relating to horticultural and agricultural fertilizers. British Patent 1952/664989

Montreal Protocol on substances that deplete the ozone layer (1989)

Mordor Intelligence (2017a) Global agar market – growth, trend and forecasts (2017–2022). https://www.mordorintelligence.com/industry-reports/agar-market. Accessed on 4 Sept 2017

Mordor Intelligence (2017b) Global Carrageenan Market – trends and forecasts: (2017–2022) (Segmented by types, applications, grades and geography). https://www.mordorintelligence.com/industry-reports/global-carrageenan-market-industry. Accessed on 4 Sept 2017

National Institute of Health (NIH). Office of Dietary Supplements (2016) Omega-3 fatty acids. https://ods.od.nih.gov/factsheets/Omega3FattyAcids-HealthProfessional/#en3. Accessed on 4 Sept 2017

Naylor J (1976) Production, trend and utilization of seaweeds and seaweed products, FAO Fisheries Technical Paper 159. FAO, Rome. ISBN 92-5-100091-3 pp 1–73. http://www.fao.org/docrep/005/ac860e/AC860E00.htm#TOC. Accessed on 4 Sept 2017

Newton GW (1951) Seaweed utilization. Sampson Low, London

Oilgae (2016) Business opportunities & market size of beta carotene. http://www.oilgae.com/non_fuel_products/betacarotene.html. Accessed on 4 Sept 2017

Perosa A, Bordignon G, Ravagnan G, Zinoviev S (eds) (2015) Algae as a potential source of food and energy in developing countries: sustainability, technology and selected case studies. Edizioni Ca' Foscari – Digital Publishing, Venezia

Porse H (1998) Global seaweed market trends. Plenary presentation at the 16th International seaweed symposium, Cebu City, Philippines, 12–17 April 1998

Porse H, Rudolph B (2017) The seaweed hydrocolloid industry: 2016 updates, requirements, and outlook. J Appl Phycol. https://doi.org/10.1007/s10811-017-1144-0:1-14

Puri M, Sharma D, Barrow CJ (2012) Enzyme-assisted extraction of bioactives from plants. Trends Biotechnol 30(1):37–44

Ramsey E, Sun Q, Zhang Z, Zhang C, Wei G (2009) Mini-review: green sustainable processes using supercritical fluid carbon dioxide. J Environ Sci 21:720–726

Research and Markets (2015) Global astaxanthin market – sources, technologies and application. https://www.researchandmarkets.com/reports/3129287/global-astaxanthin-market-sources-technologies. Accessed on 4 Sept 2017

Thurmond W (2009) Biofuels and algae markets, strategies and commercialization outlook. Emerging Markets Online report. https://www.slideshare.net/tonya.m.britton/wil-thurmondemerging-markets-online. Accessed on 4 Sept 2017

Transparency Market Research (2016) Algae market, by application, by cultivation technology, and geography – global industry analysis, size, share, growth, trends, and forecast – 2016–2024. http://www.transparencymarketresearch.com/pressrelease/algae-market.htm. Accessed on 4 Sept 2017

Transparency Market Research (2017) Astaxanthin Market – global industry analysis, size, share, growth, trends and forecast 2016–2024. http://www.transparencymarketresearch.com/astaxanthin-market.html. Accessed on 4 Sept 2017

Tseng CK (1981) Commercial cultivation. In: Lobban CS, Wynne MJ (eds) The biology of seaweeds. University of California Press, Berkeley, pp 680–725

United States Environmental Protection Agency (EPA) (2014) 2014 toxics release inventory national analysis complete report. https://www.epa.gov/toxics-release-inventory-tri-program/2014-tri-national-analysis-complete-report. Accessed on 4 Sept 2017

Wang L, Weller CL (2006) Recent advances in extraction of nutraceuticals from plants. Trends Food Sci Technol 17:300–312

Wijesinghe WAJP, Jeon YJ (2012) Enzyme-assistant extraction (EAE) of bioactive components: a useful approach for recovery of industrially important metabolites from seaweeds: a review. Fitoterapia 83(1):6–12

Seaweeds As a Component of the Human Diet

6

Izabela Michalak and Katarzyna Chojnacka

Abstract

Algae (microalgae and seaweeds) have been used for centuries as food, animal fodder and a source of chemicals for the pharmaceutical, cosmetic, food and chemical industries. In the last few decades, knowledge about the influence of algae in diet on health and well-being has increased. Algae still gain attention due to their unique composition. They are known to be a rich concentrate of biologically active compounds, such as carbohydrates, proteins, minerals, oils, fats, polyunsaturated fatty acids, antioxidants (polyphenols, tocopherols, vitamin C, mycosporine-like amino acids) and pigments. Besides the nutritional value of algae, bioactive compounds play a significant role in the promotion of human health and disease prevention, due to antibacterial, antiviral, antifungal, antioxidative, anti-inflammatory and antitumor properties. Algal products can be used for performance improvement and the reduction of pathogenic bacteria. Algae can be served in different forms – whole seaweed meal, powder, extract, homogenate or fermented. In this chapter, the advantages (nutritional value, accessibility, etc.) and disadvantages (toxic metals, sensory perception, etc.) of the application of seaweeds as a component of food are discussed. Special attention is paid to their application in cereals, dairy and meat products. Seaweeds can also serve as a natural salt, as well as a source of hydrocolloids, especially in the confectionery industry. Macroalgae can also deliver biologically active compounds to food products of plant and animal origin indirectly, through their application in plant cultivation and animal feeding. It is predicted that seaweeds as food and seaweed-derived food flavours, colours and nutrients will continue to attract considerable commercial attention.

I. Michalak (✉) · K. Chojnacka
Faculty of Chemistry, Department of Advanced Material Technologies, Wrocław University of Science and Technology, Wrocław, Poland
e-mail: izabela.michalak@pwr.edu.pl

Keywords
Seaweeds · Nutritional value · Cereals · Dairy and meat products · Natural salt · Hydrocolloids

6.1 Introduction

The recent trend of an increasing demand for algal products results from a greater consumer focus on health ("functional foods" and "nutraceuticals", which are known to have a number of health-specific advantages) and the wider use of food additives (Wells et al. 2017). In 2014, about 28.5 million tonnes of seaweeds (macroalgae) and other algae were harvested for direct consumption or further processing for food, traditionally in Japan, Korea and China (FAO 2016). The major seaweeds that are of dietary importance in these countries belong to the genus *Undaria* (brown seaweed, Phaeophyta), commonly called wakame; *Pyropia* (red seaweed, Rhodophyta), commonly referred to as nori; and *Saccharina* (brown seaweed), commonly known as kombu (Prabhasankar et al. 2009). Another example of seaweed that is highly appreciated by consumers in Asia (Indo-Pacific region) is *Caulerpa*, with the species *C. lentillifera* and *C. racemosa* var. *turbinata* commonly known as "green caviar" (de Gaillande et al. 2017).

Worldwide, at least 145 species of macroalgae are used as food (Zemke-White and Ohno 1999). For example, in Japan, seaweeds are served at approximately 21% of meals (Yoshinaga et al. 2001). In Japanese cuisine, the red seaweed, nori, is a traditional wrapping for sushi and is also used in soups (FAO 2016). Nori and preserved kombu are also used for the wrapping of Japanese rice balls called "onigiri". Wakame is used in noodles, soups, salads, pickles, etc. Another example of traditional Japanese food is "Kobumaki" – a simmered food (often salmon or herring) wrapped in kombu, usually prepared for the New Year's holidays (Wells et al. 2017). The genus of seaweeds added as an

© Springer International Publishing AG, part of Springer Nature 2018
K. Chojnacka et al. (eds.), *Algae Biomass: Characteristics and Applications*, Developments in Applied Phycology 8,
https://doi.org/10.1007/978-3-319-74703-3_6

ingredient influences the morphology of the food produced. For example, Cofrades et al. (2008) showed that meat products formulated with the brown seaweeds (*Himanthalia elongata*, *Undaria pinnatifida*) had different characteristics (water- and fat-binding properties, hardness, chewiness of the cooked products, lower springiness and cohesiveness, colour changes) from the products made with the red seaweeds (*Pyropia umbilicalis*). Better properties were obtained in the case of brown algae, especially *Undaria pinnatifida* (Cofrades et al. 2008).

Seaweeds are not a traditional food in the Western diet, despite occasional use as a conventional ingredient, especially in coastal areas (Bouga and Combet 2015). For example, *Fucus spiralis*, *Pyropia* sp. and *Osmundea pinnatifida* are consumed as food on some of the Azorean Islands, Portugal (Paiva et al. 2014); *Saccharina digitata*, *Saccharina latissima*, *Undaria pinnatifida*, *Himanthalia elongata*, *Ulva* sp., *Gracilaria verrucosa*, *Palmaria palmata* and *Chondrus crispus* in France (MacArtain et al. 2007); and *Palmaria palmata*, *Saccharina latissima*, *Saccharina digitata* and *Alaria esculenta* in Norway (Chapman et al. 2015; Mouritsen et al. 2012). France was the first European country where seaweeds were approved for human consumption (as vegetables and condiments), which opened new opportunities for the food industry. Among them, brown seaweeds, *Ascophyllum nodosum*, *Fucus serratus*, *Fucus vesiculosus*, *Himanthalia elongata* and *Undaria pinnatifida*; red seaweeds, *Pyropia umbilicalis*, *Palmaria palmata*, *Chondrus crispus* and *Gracilaria verrucosa*; and green seaweed, *Ulva* spp. (Mabeau and Fleurence 1993) can be distinguished. With the recent trend towards the increasing popularity of Asian dishes (e.g. sushi) across Europe, seaweeds are being further introduced (Chapman et al. 2015). In other parts of the world, including all of the countries in tropical Latin America and most of Africa, seaweeds are an unappreciated resource that can be used for food production (FAO 2016). There are some records of the use of algae as an important ingredient in the diet of early humans in South America. In southern Chile, macroalgae have been used by the native people in food preparations like soups, stews and salads, for example, *Macrocystis pyrifera*, known as "huiro", which is eaten separately; *Pyropia columbina*, known as "luche", which is used in soups; and *Durvillaea antarctica*, known as "cochayuyo", which is eaten as a salad and in stews together with meat (Astorga-España et al. 2017). The use of Latin American seaweeds as an economic resource can be crucial for the wealth and sustainable livelihoods of coastal communities (Rebours et al. 2014).

Very interesting research was performed by Edwards et al. (2012), who fashioned a short questionnaire during the fourth ISAP (International Society for Applied Phycology) conference in Halifax (2011) to gather some information on the algal eating habits of the participants. It was found that 93% of the conference members (mainly phycologists, N = 273, 35 nationalities) had previously eaten algae, as well as 64% of the general public (N = 104, 25 nationalities). The frequency of consumption of algae was as follows: conference vs. general public (%) – daily 7 vs. 3, weekly 18 vs. 17, monthly 31 vs. 25 and more rarely 45 vs. 5. Among the 27 genera of macroalgae and microalgae recorded as having been eaten by the questionnaire participants, the most popular were red seaweeds (60% of phycologists and 71% of the general public). *Pyropia* spp. were the most commonly preferred. Interestingly, "lack of availability" was the main reason for not eating algae. It was concluded that greater information on algae and their wider availability could encourage people to include them as part of a balanced and healthy diet. It is also important to change peoples' mindset to consider seaweeds not as weeds or wracks but rather as a very versatile and tasty kind of food source (Mouritsen 2017).

The problem for people who do not consume algae on a historical basis is often the lack of enzymes required to thoroughly digest dietary seaweeds. The digestive flora of a particular person may take up to 4 months to produce the dedicated enzymes. Therefore, it is beneficial to eat a small amount of seaweed daily rather than larger amounts occasionally (Dhargalkar 2014). However, the popularization of the East Asian diet worldwide has gradually increased public interest and acceptance of seaweed as a food source, partly due to their suggested health benefits. Consequently, consumption of seaweed and seaweed-based products is rising, similar to the trend observed with fresh fruits and vegetables (Sharifuddin et al. 2015). In European countries, it is proposed that seaweeds be indirectly introduced into the human food chain through their addition to the diet of marine animals (e.g. fish, shrimp) produced by aquaculture (Fleurence et al. 2012). However, algae are also supplemented directly into the human diet in different forms: whole seaweed meal (fresh, dried – including a dry-ground meal) and processed seaweed (seaweed extracts, bioactive seaweed compounds, homogenates, fermented seaweeds).

Nowadays, due to continuous exposure to radiation, metal poisoning, gaseous emission, radioactive substances and other pollutions, it is recommended that seaweeds be used on a daily basis, so as to supplement the diet with nutrients and biologically active compounds, detoxify the body and reduce the negative effect of environmental pollution on human health (Dhargalkar 2014). Seaweed products are also used as a substitute for the addition of salt, artificial colours and flavours (Pedersen et al. 2013). According to data presented by Edwards et al. (2012), the main reasons given for eating algae by phycologists (N = 273) and the general public (N = 104) are health 36% vs. 13%, tradition 18% vs. 15% and taste 44% vs. 57%.

However, in the case of algae consumption, the risk related to the content of toxic elements (e.g. arsenic), as well as certain sensory properties (not acceptable to all consumers), should also be taken into account. The unavailability of algae is the greatest reason for most people for its not having been previously consumed (50% of phycologists, 49% of the general public), followed by taste (22% vs. 30%) and lack of culinary knowledge (11% vs. 35%) (Edwards et al. 2012). It should be noted that seaweeds are characterized by highly variable composition, depending on species, collection time and habitat.

In this chapter, special attention is paid to the advantages and disadvantages of seaweeds as a component of food, as the source of algae for the food industry, forms of algae that are used in the production of food, as well as some examples of algae-based meals.

6.2 Advantages of Seaweeds As a Food Source

The utilization of seaweeds as a component of a diet is mainly related to their unique composition and multifunctional properties. These marine organisms have high nutritional value, taking into account their content of proteins and amino acid profiles; lipids, including polyunsaturated fatty acids (PUFA); soluble carbohydrates; ash; minerals; vitamins; pigments; as well as polyphenols and their antioxidant properties. This issue has been described in detail in many review papers, for example, MacArtain et al. (2007), Holdt and Kraan (2011), Dhargalkar (2014), Paiva et al. (2014), Michalak and Chojnacka (2015a), Sharifuddin et al. (2015), Chu and Phang (2016), Cornish et al. (2017), de Gaillande et al. (2017) and Wells et al. (2017).

Seaweeds produce new ingredients that are called "functional" and are used in "health-promoting foods" that have a somewhat medicinal effect in the treatment or prevention of certain diseases (Mohamed et al. 2012; Paiva et al. 2014; Cornish et al. 2017). They possess antioxidant, anti-infective (antiviral, antimicrobial activity: bacterial and fungal), antitumor, anticancer, antiproliferative, anti-inflammatory, anticoagulant, phytoestrogenic and endocrine modulating properties (thyroid) (Mohamed et al. 2012). Therefore, seaweeds have the potential to be exploited in human health applications, for example, in the prevention and therapy of certain metabolic syndromes (obesity, dislipidemia, diabetes, hypertension) (Mohamed et al. 2012; Sharifuddin et al. 2015; Chu and Phang 2016), cardiovascular diseases, weight management, cancer (Brown et al. 2014), repair and protection (liver, kidneys, lungs, skin: wound healing), degenerative diseases (nerves and brain), etc. (Mohamed et al. 2012; Cornish et al. 2017). Epidemiological studies have shown an association between the dietary intake of seaweeds and a reduced prevalence of chronic diseases (Brown et al. 2014). Seaweed extracts, due to their antimicrobial properties, can also play another important role in the food industry – as natural antimicrobial agents for food preservation. Gupta et al. (2012) showed that seaweed extracts of *Himanthalia elongata* at a concentration of 6% inhibited the growth of four common causes of food spoilage (*Pseudomonas aeruginosa* and *Enterococcus faecalis*) and food pathogenic microorganisms (*Listeria monocytogenes* and *Salmonella abony*).

A seaweed diet is mainly recommended for people who are trying to control their weight (they are high in fibre, which makes a person feel full when eating a meal) and also for vegetarians, because they are a good source of amino acids (Dhargalkar 2014). Fibres of seaweed origin are used in the production of low-calorie foods, due to the texturing and bulking properties that depend on their capacity to absorb and retain water (Rupérez and Saura-Calixto 2001).

Seaweeds are usually added to food as functional ingredients. In small quantities, they enhance the nutritional value of traditional food and increase the content of dietary fibres and trace elements (Dhargalkar 2014; Astorga-España et al. 2017). As an example, Prabhasankar et al. (2009) replaced pasta ingredients with *Undaria pinnatifida* powder. It was found that 10% of seaweed-incorporated pasta had considerably improved protein and fat content, as well as increased levels of ash and fibre. Pigments present in seaweeds can influence food product colour, depending on the seaweed/extract type and the concentration added (López-López et al. 2009; Choi et al. 2012; Moroney et al. 2013). Astorga-España et al. (2017) presented results of the nutritional value of dishes (hamburgers, fettuccine, fritters and bread – staples of the Chilean diet and huiro breadsticks and luche-parsley pesto – recipes for cocktails and snacks not as commonly consumed) prepared from three dried sub-Antarctic seaweeds, *Macrocystis pyrifera*, *Pyropia columbina* and *Durvillaea antarctica* (content from 3% to 28% in the prepared dishes). Authors compared the composition of dishes made with macroalgae to commonly consumed foods like white bread, hamburgers, fettuccine made of dough enriched with eggs, fritters, breadsticks and green pesto. It was shown that the dishes prepared with algae have similar or slightly elevated levels of moisture and ash.

6.3 Disadvantages of Seaweeds As a Food Source

The promotion of algal consumption should also take into account the potential harm to consumers (Wells et al. 2017). The main risk is possible toxicity from high iodine levels in seaweeds, accumulation of arsenic and toxic metals (e.g. Cd, Cr, Hg, Ni and Pb) (Holdt and Kraan 2011; Bouga and Combet 2015; Wells et al. 2017), secondary metabolites (e.g.

prostaglandins – PGE2 in *Gracilaria vermiculophylla*, kainoids) as well as potential contamination with pathogens, radioisotopes and toxic synthetic compounds. They can also be responsible for allergic properties (Wells et al. 2017). In Japan, iodine intake from this food source is among the highest in the world. High iodine content may negatively affect individuals with underlying thyroid disorders (Zava and Zava 2011; Bouga and Combet 2015). However, before consumption, seaweeds and their components must meet certain consumer safety regulations, e.g. the level of toxic minerals in edible seaweeds (Holdt and Kraan 2011).

Consumption of seaweeds (e.g. *Palmaria palmata*, *Saccharina latissima*, *Saccharina digitata* and *Alaria esculenta*) is also strongly impacted by sensory perception (Chapman et al. 2015). Seaweeds should not exceed 20% of the dry weight of a given dish, because higher content/concentrations either "overwhelmed" traditional recipes or at least influenced the taste to the point that the dish was rejected by the consumers participating in the test (Radulovich et al. 2015). For example, Prabhasankar et al. (2009) showed that pasta with the addition of 10% *Undaria pinnatifida* was acceptable when considering sensory characteristics. However, a higher content of wakame powder (20% and 30%) in pasta was not preferred when considering sensory properties, as indicated by the lower scores given by panellists in regard to appearance and mouth feel. They complained of excessive saltiness, as well as a similar sensory effect to that of eating wakame itself. Mouritsen (2017) emphasized that when introducing seaweeds into Western cuisine, we should keep in mind that, in the case of seaweeds, "taste comes first": "No matter how nutritious and healthy a particular kind of foodstuff is, no one is going to consume it for long if it is not tasty".

6.4 Sources of Seaweeds for Food Production

Seaweeds for the food industry can be harvested from the wild sea or freshwaters with low pollution levels, from aquaculture, or it can be specially cultivated, for example, in photobioreactors (Andersen 2005; Paiva et al. 2014; Rebours et al. 2014; Radulovich et al. 2015). In the case of wild environments, seaweeds should be collected during low tide from waters that are pure and free from harmful chemicals. It is recommended that whole seaweeds not be collected but rather that the upper portions of the fronds be cut, leaving the lower portion to regenerate (Dhargalkar 2014). Today, seaweeds for food originate mainly from farming rather than natural sources (FAO 2003). Aquaculture has great potential for meeting the challenges of increasing demand for food, because it accounts for almost 50% of aquatic resources (FAO 2016; de Gaillande et al. 2017). After freshwater fish and molluscs, algae are the third-largest aquaculture crop in the world. Due to the high cost of industrial cultivation of microalgae, the largest sector of this industry is represented by macroalgal production for human food (Wikfors and Ohno 2001). Additionally, there is a possibility of applying integrated multitrophic aquaculture systems in which seaweeds can be utilized for removal of excess nutrients that come from fish farming. By this approach, the negative environmental impact of intensive fish aquaculture is reduced (Chopin et al. 2001). The main seaweed species that are specially cultivated for food production are wakame (*Undaria pinnatifida*), Japanese kelp (*Saccharina japonica*) and mozuku seaweed (*Nemacystus* spp.) (FAO 2016). The main farming systems of seaweeds involve the bottom-planting method, the off-bottom cages method, the off-bottom tray method, floating long lines, submerged rafts, land-based raceways, etc. (de Gaillande et al. 2017).

It is estimated that if 10% of all ocean areas were eventually farmed, i.e. through mariculture, including through the use of extensive methods such as assisted fisheries and seaweed ranching, total yield could equal that currently produced in agriculture, and the eutrophic potential of waters would be greatly reduced (Radulovich 2011). It was also found that the cultivation of seaweeds in aquaculture rapidly attracts biodiversity, which is manifested by a significantly larger number of fish species and individuals than in nearby control areas (Rebours et al. 2014; Radulovich et al. 2015). Moreover, the production of seaweeds in marine and freshwater environments for food purposes could be an alternative or supplementation of food production in traditional agriculture that depends on freshwater used for irrigation and rainfall variability, which is aggravated by climate change. It is predicted that water scarcity may be the most limiting factor for increasing global food production. Food production at sea, both fished and farmed, as well as both plants and animals, requires neither freshwater nor land. Therefore, the water needed for the production of food in terrestrial environments could be preserved, as equivalent amounts of food can be farmed or fished in the sea or ocean (Radulovich 2011; Stévant et al. 2017a). Seaweed farming is recognized as one of the most environmentally friendly types of aquaculture, since it uses no fertilizers, helps preserve coral reefs and is sustainable (Bixler 2017; Porse and Rudolph 2017).

The main disadvantage of harvesting algae from natural resources is their possible contamination with toxic metals and other substances. The chemical composition can also be affected by external factors such as geographic location, environment, season and sampling conditions (Mohamed et al. 2012). Additionally, high water content in seaweeds, which ranges from 70% to 90%, constitutes a challenge for conserving and transporting large amounts of biomass from harvesting to processing sites (e.g. food production) (Stévant et al. 2017a).

6.5 Forms of Seaweeds in Food

Seaweeds that are used as a component of food can be supplemented as a whole seaweed meal (fresh, dried), as well as processed seaweeds: powder, extracts, bioactive compounds, fermented, homogenized, etc. The choice of the proper form of seaweeds is crucial, because processing of seaweeds can lead to an unfavourable change in appearance, as well as to the loss of some biological properties (e.g. in extraction of biologically active compounds). For example, Radulovich et al. (2015) found that green seaweed, *Chaetomorpha* sp., was best when used directly or cooked fresh, being added whole or chopped into a variety of dishes, because drying caused undesirable changes. Seaweeds can also lose their properties during cooking. Prabhasankar et al. (2009) showed that the incorporation of commercial seaweeds (wakame) as a powder into pasta resulted in the retainment of phenolic compounds in the product upon cooking when compared to the control pasta. However, uncooked seaweed pasta exhibited higher total antioxidant content than the control pasta. It was also found that the main functional component of *Undaria pinnatifida*, fucoxanthin, was not affected by the rigorousness of the pasta-making process or by the cooking steps involved thereafter. The mode of consumption of seaweeds after harvesting is also specific to region and/or country. Usually, seaweeds (e.g. *Caulerpa*) are eaten fresh and raw (e.g. as salads), seasoned with lemon and coconut milk (de Gaillande et al. 2017).

It is important to mention that in the case of food of animal and plant origin, bioactive compounds from seaweeds in different forms may be incorporated into animal or plant products by the supplementation of animal diet or by fertilization, spraying or direct addition during food processing (Uchida and Miyoshi 2013; Michalak and Chojnacka 2015a, c; Michalak et al. 2015, 2016). Different forms of seaweed and their application in food products are presented in Fig. 6.1.

6.5.1 Whole Seaweed Meal

Whole macroalgae are commonly used, without any processing or extraction, as a vegetable from the sea in many areas around the world (e.g. in the new Nordic cuisine) (Pedersen et al. 2013). The most popular application of whole seaweeds is the production of sushi or rice balls called "onigiri", in which species of red algae, *Pyropia*, including *Pyropia yezoensis* and *Pyropia tenera*, are used as wrapping material. Fresh (raw), whole seaweeds can also be used as an ingredient in salads and soups (Radulovich et al. 2015; FAO 2016; Wells et al. 2017). Whole or chopped seaweeds can also be cooked into a variety of food preparations, baked until crispy and fried in a variety of manners. They can also be blended with fruits and vegetables (Radulovich et al. 2015). The main disadvantage of this form of algae regards their storage – specifically, their susceptibility to deterioration (Radulovich et al. 2015). Therefore, the main postharvest treatments that aim to maintain the quality of fresh seaweed biomass include seawater storage, cold storage, silage and freezing (Stévant et al. 2017a). Stévant et al. (2017b) examined the effect of short-term storage (22 h) of brown macroalgae (*Alaria esculenta* and *Saccharina latissima*) in seawater, prior to further processing, on their nutritional value. It was found that the storage treatments resulted in a rapid decrease in dry weight during the first 2 h, an increase in the ash and sodium contents and a reduction in the content of carbohydrates (mannitol and glucose (laminarin)) in *S. latissima* and fucose (fucoidan) in *A. esculenta* and polyphenols. The protein content remained relatively stable.

6.5.2 Grounded Seaweeds/Seaweed Powder

Seaweeds dried and ground to different levels of coarseness can be used as a partial substitute for wheat and maize flour in the preparation of a variety of recipes, like cookies, fried chips and spaghetti. Seaweeds in the form of powder can be sprinkled on or into different recipes, including fruit juices. They can also be scrambled or encapsulated to be consumed as a dietary supplement (Radulovich et al. 2015). For example, Mamat et al. (2014) incorporated the powder of red seaweed, *Kappaphycus alvarezii*, into wheat flour, and Mouritsen et al. (2012) used the powder of dulse (*Palmaria palmata*) to produce bread. Cofrades et al. (2011) used milled seaweeds (*Undaria pinnatifida*, *Himanthalia elongata* and *Pyropia umbilicalis*) in the preparation of post-rigour pork meat and fresh pork back fat. The main advantage of using seaweeds in the form of powder is the ease of their storage – there is no decay, and they apparently keep their properties for up to at least 9 months in the darkness, at room temperature (Radulovich et al. 2015). On the other hand, drying of seaweeds can affect their chemical composition, which can influence the product's nutritional value, as well as the extraction yield of bioactive compounds during further processing (Stévant et al. 2017a).

6.5.3 Seaweed Extraction/Seaweed Bioactive Compounds

Interest in the use of "natural green" extracts in the food industry (e.g. in various foods and beverages) is increasing. O'Sullivan et al. (2014, 2016) tested algal extracts

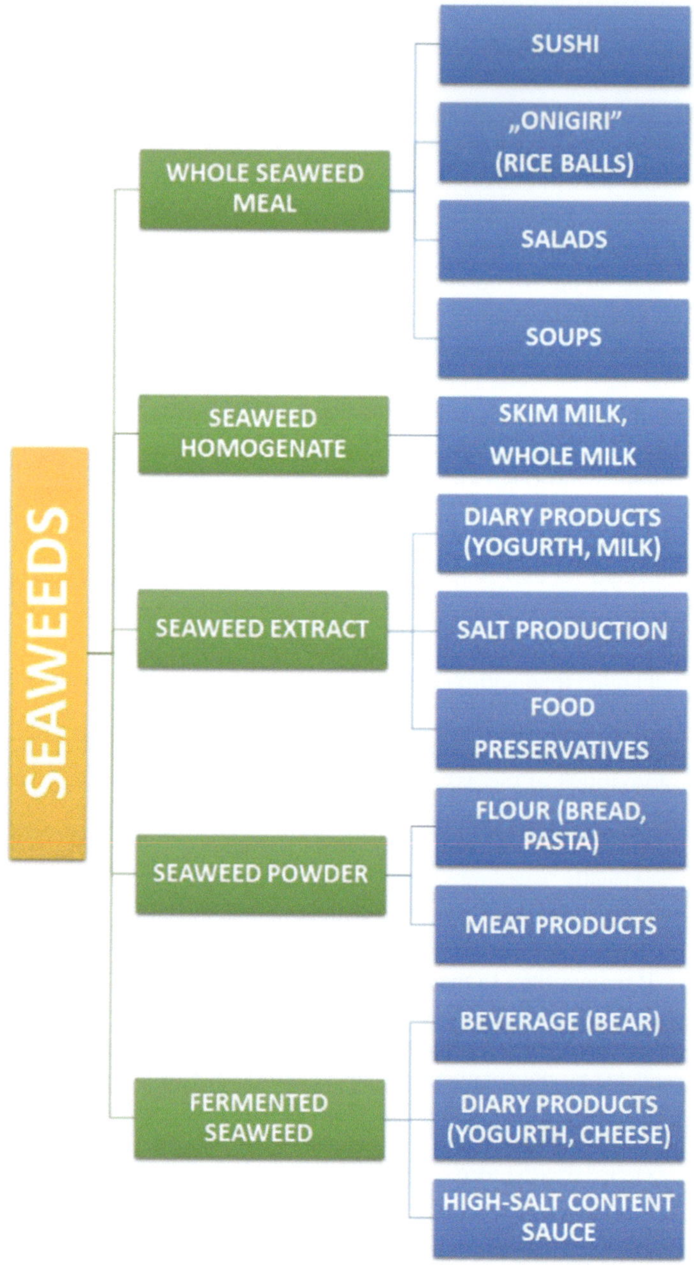

Fig. 6.1 Forms of seaweed used in food production and their exemplary applications

obtained from *Ascophyllum nodosum* and *Fucus vesiculosus* through extraction with water, 80% ethanol, 20% H_2O, and 60% ethanol, 40% H_2O as components of yoghurts and milk. Magnusson et al. (2016) produced natural macroalgal salts by washing the marine green macroalgae *Ulva ohnoi* and *Ulva tepida* with freshwater and crystallizing the resulting minerals. Washing seaweeds for 30 min at 40 °C or for 24 h at 25 °C provided an optimized yield of salt from both species while minimizing time and temperature inputs. The acid extraction technique was used to isolate polysaccharides, laminarin and fucoidan, from brown seaweed *Saccharina digitata* that were then added directly to minced pork (Moroney et al. 2013). Some seaweed extracts, due to their antibacterial properties, can be used as natural agents for food preservation (against major food spoilage and food pathogenic microorganisms) (Gupta and Abu-Ghannam 2011; Gupta et al. 2012). For example, extracts from *Himanthalia elongata*, obtained in a process with methanol under a nitrogen atmosphere in a shaker incubator, inhibited the growth of microorganisms that cause food spoilage (*Pseudomonas aeruginosa* and *Enterococcus faecalis*) and food pathogenic microorganisms (*Listeria monocytogenes* and *Salmonella abony*) (Gupta et al. 2012). This approach can help to reduce the use of chemical preservatives.

6.5.4 Seaweed Homogenization/Seaweed Homogenates

Seaweed homogenates are usually obtained from fresh seaweeds. For this purpose, homogenization using a blender and mortar and pestle is applied (Michalak and Chojnacka 2015c). In the work of Gallaher et al. (2005), concentrated oil-in-water emulsions were prepared by mixing either algae oil or Tween 20 with the remaining aqueous phase consisting of a sodium acetate imidazole buffer (pH 7). The lipid and aqueous phase was blended for 30 s using a high-speed mixer. The obtained in-water emulsions were used as components of water, skim milk, 2% milk and whole milk.

6.5.5 Seaweed Fermentation/Fermented Seaweed Products

Fermentation is an efficient method that enables the stabilization and transformation of biomass of diverse origins, including seaweeds, into a more sanitized quality product (Ennouali et al. 2006). Fermented products from algae, obtained by lactic acid or ethanol fermentation, can be ingredients in functional food (Michalak and Chojnacka 2015b). Such an example can be a fermented seaweed beverage obtained from *Gracilaria fisheri* using the bacteria *Lactobacillus plantarum* (Prachyakij et al. 2008; Hayisama-ae et al. 2014; Ratanaburee et al. 2011) or algal extracts produced from *Ascophyllum nodosum* and *Fucus vesiculosus* added to yoghurt (O'Sullivan et al. 2016). Uchida et al. (2017) produced high-salt content seaweed sauces using nori (*Pyropia yezoensis*) and halophilic lactic acid bacteria (*Tetragenococcus halophilus*) in fermentation tanks. Use of this type of bacteria creates the possibility of producing algal fermented products containing high (>10%) salt content and capable of long-term preservation (Uchida and Miyoshi 2013).

6.6 Algae-Based Products in the Human Diet

Before use as food, seaweeds harvested from wild habitats or from aquaculture should be thoroughly washed with freshwater. Debris, small fauna and epiphytes should be washed away, and unwanted parts like holdfasts and damaged tissues should be eliminated. After that, excess surface water should be removed by decantation (Radulovich et al. 2015).

Seaweeds are available in a diversified range of food products. Various product categories containing seaweeds can be distinguished, for example, bread and confectionery (bread, cakes, pizza bases, biscuits, shortbread), condiments (seaweed flakes, salad boosters, salt), drinks (gin, whisky, super shakes, smoothies), noodles and pasta (sea spaghetti, kelp noodles), salads (seaweed salad, sea salad), seaweed (whole seaweed, seaweed sheets), snacks (crackers, rice crackers, oatcakes), soup (miso soup), supplements (tablets) and sushi (sushi platters) (Bouga and Combet 2015). In a culinary context, seaweeds can be used as a vegetable, salt and spice all in one (McHugh 2003). For example, in European cuisine, marinated kelp (*Alaria esculenta*) is used in a cannelloni bean salad, laver/nori (*Pyropia umbilicalis*/*Pyropia yezoensis*) are used in chocolate molasses meringues, and dulse (*Palmaria palmata*) is used in traditional Welsh laverbread cakes and dulse-cheese scones (Wells et al. 2017). Some of the food products that were enriched with seaweeds in different form are presented below.

6.6.1 Cereal Products

In the baking industry, seaweeds are mainly used because of their hydrocolloid content, which can be applied to improve the bread-making process. Hydrocolloids improve dough handling properties, increase the quality of the fresh bread and extend the shelf life for its storage (Mamat et al. 2014). In the work of Mamat et al. (2014), the powder of red seaweed, *Kappaphycus alvarezii*, was incorporated (2–8%) into wheat flour and used for the baking of bread. The addition of seaweed powder increased the water absorption of the dough, decreased the stickiness properties and increased the firmness of the produced bread. Mouritsen et al. (2012) proposed using another red seaweed, dulse (*Palmaria palmata*), for the production of bread. It was obtained by mixing ølandshvede flour (a spelt-like wheat species that is high in gluten and protein – 13.5%), spelt flour dulse *dashi*, sourdough starter, fresh baker's yeast and salt. The added dulse supported the savoury flavour. Prabhasankar et al. (2009) produced pasta enriched with seaweeds. This is an important dish from the nutritional and gastronomic points of view. Moreover, it is a staple food in many countries and pasta products are well-accepted worldwide. Wakame (*Undaria pinnatifida*) was used to enrich pasta with functional compounds such as fucoxanthin, fucosterol, fatty acid (including n-3 and n-6 polyunsaturated fatty acids) and amino acid. Inclusion of 10% *Undaria pinnatifida* in pasta did not negatively affect the product's acceptability to consumers. Chang and Wu (2008) also produced Chinese noodles formulated with green seaweed (*Monostroma nitidum*) powder (addition of 4%, 6% and 8%). The main result was an increase in the crude fibre content of raw fresh noodles. The content of fibre and polysaccharides in the seaweeds influenced higher cooking yields due to water absorption by these components during cooking. Chapman et al. (2015) examined fish cakes, in which herbs were substituted with 5% dry sugar kelp (*Saccharina *latissima*), in consumer tests. It was found that both varieties of cake were equally attractive to consumers with regard to flavour and appearance.

The addition of seaweeds to the respective dishes (including cereal products) not only increases their nutritional value and texture but can also help alleviate many human ailments (Dhargalkar 2014). Combet et al. (2013) proposed using brown seaweed, *Ascophyllum nodosum*, as an ingredient in a nutritionally balanced pizza that was developed as a functional meal. Product reformulation of very popular unhealthy foods (e.g. fast food that is calorie dense and nutrient poor) aimed to meet nutritional guidelines for public health in a ready meal. Use of a small amount of *A. nodosum* enabled the achievement of the required target for Fe. In addition, the seaweed was also a source of vitamin A, calcium and iodine (a nutrient lacking in the British diet). Most of the adults and children tested rated it "as good as" or "better than" their usual choice.

6.6.2 Dairy Products

Dairy products provide a wide range of meals that can be enriched with seaweeds. O'Sullivan et al. (2016) tested yoghurts containing extracts (0.25% and 0.5% (w w^{-1})) prepared from the brown seaweeds *Ascophyllum nodosum* and *Fucus vesiculosus* by water or alcohol extraction. Yoghurt and related products are the most commonly manufactured and consumed food products worldwide. Sensory analysis indicated that colour, flavour and texture determined the overall acceptability of seaweed extract-enriched yoghurt. The most acceptable yoghurts, from a sensory perspective, were yoghurts with the addition of water extracts of *Ascophyllum nodosum*, regardless of the dose. It was also found that seaweed extracts added to yoghurt did not affect shelf-life characteristics – there was no influence on pH, microbiology or whey separation. Chee et al. (2005) enriched yoghurt with algal oil emulsion as a source of polyunsaturated fatty acids – n-3. Enrichment of foods with n-3 PUFA is difficult, because these fatty acids are highly susceptible to oxidative deterioration and rancidity (Chee et al. 2005; Gallaher et al. 2005). When n-3 PUFA are flavourless, they readily undergo radical oxidation, forming highly undesirable off-flavour compounds (Gallaher et al. 2005). Another drawback is that algae oil contains similar n-3 to those found in fish oils, and therefore a fishy aroma is a common defect in supplemented dairy foods. The results showed that the concentration of hydroperoxides increased for supplemented yoghurt samples over the 21-day storage period and was greater than in the non-supplemented samples. In the sensory analysis, the trained panel distinguished a stronger fishy flavour in the supplemented yoghurts, but the consumer panel rated both the control and supplemented samples similarly, as "moderately liked". The fishy aroma was also scored by a trained panel who evaluated algae oil-supplemented water and skim milk samples. This aroma did not differ statistically over the 15-day storage period. Gallaher et al. (2005) also evaluated the oxidative deterioration of the refrigerated algae oil-in-water emulsions (2, 5 and 10 wt. %) in supplemented water and skim milk samples. It was found that all of the supplemented milk samples were oxidatively stable, as compared to the supplemented water samples, after 15 days of storage. Milk was recognized to be an ideal candidate for supplementation with n-3 oil emulsion.

O'Sullivan et al. (2016) fortified milk with water and ethanolic extracts obtained from *Ascophyllum nodosum* and *Fucus vesiculosus* (doses 0.25% and 0.5%). In their work, they examined the effect of extract addition on milk quality and shelf-life attributes. The examined extracts increased the greenness and yellowness of fortified milk samples. Ethanolic extracts from *Fucus vesiculosus* in particular showed antioxidant functionality. Seaweed extract type or concentration did not affect the quality parameters, such as milk microorganisms. Sensory analysis indicated that water-prepared extracts were more acceptable than ethanolic extracts as functional ingredients in milk. However, it was also found that milk enriched with seaweed extracts did not exhibit cellular antioxidant activity, which indicated reduced biological activity of the extracts in milk.

Mouritsen et al. (2012) proposed using dulse (*Palmaria palmata*) as a promising, novel ingredient in Nordic cuisine. Seaweeds can be infused into a range of different dishes with umami taste, such as ice cream or fresh cheese. For the preparation of ice cream, dulse-infused milk (20 g of dulse per litre of milk) was mixed with cream, trimoline (inverted sugar syrup), sugar and ColdSwell cornstarch. The obtained ice creams had a very pleasing, light mauve colour. The flavour was delicate, light and floral, with the taste being compared to Japanese green tea ice cream, which is an indicator of an acceptable flavour profile. The texture of the dulse-infused ice cream was creamier and smoother (due to the polysaccharides released from *P. palmata*) than the same ice recipe without seaweed. Fresh cheese with *P. palmata* was obtained from dulse-infused milk (20 g of dulse per litre of milk), cream, buttermilk and rennet. The addition of seaweeds favourably improved the texture – producing a more viscous texture that is likely due to the carrageenan released from *P. palmata*, which also reduces the cooking time by half.

6.6.3 Meat Products

López-López et al. (2009), Moroney et al. (2013) and Cofrades et al. (2011) proposed meat containing seaweeds or seaweed extracts as a potential functional food for consumers.

In this case, bioactive compounds from seaweeds may be incorporated into the meat through the supplementation of animal diets or by direct addition during processing (e.g. Moroney et al. 2013; Michalak et al. 2015). Meat was chosen as a functional food, since it is one of the most important commonly consumed foods and offers excellent ways to promote intake of functional ingredients without any radical changes in eating habits. On the other hand, a negative image often attaches to meat products as a source of fat, saturated fatty acids, cholesterol, sodium and other substances that, in inappropriate amounts, may produce negative physiological effects (Cofrades et al. 2011). Therefore, one of the strategies is to design food that will contain reduced levels of certain unhealthy compounds, for example, fat or elements such as sodium.

López-López et al. (2009) examined low-fat, low-calorie and low-sodium frankfurters enriched with n-3 polyunsaturated fatty acids and with a well-balanced n-6 or n-3 ratio. Frankfurters were produced through the addition of seaweeds – 5.5% of *Himanthalia elongata*. Algal biomass improved the lipid profile of frankfurters through supplementation of significant amounts of DHA. The addition of seaweed also caused the production of calcium-rich, low-sodium frankfurters with better Na/K ratios and with various healthy compounds, such as fibre.

Moroney et al. (2013) used a spray-dried seaweed extract containing polysaccharides, laminarin and fucoidan, isolated from brown seaweed (*Saccharina digitata*) applied directly to minced pork (*M. longissimus dorsi*) at levels of 0.01%, 0.1% and 0.5% (w w^{-1}). It was found that the extract containing polysaccharides at a level of 0.01% can be incorporated into the pork without any adverse effect on the colour, lipid oxidation, texture or sensorial acceptance of pork patties. The extract in this dose also had no effect on the microbiological status, pH, water-holding capacity or cook loss of patties. Seaweed extract significantly decreased lipid oxidation in the cooked patties. Diaz-Rubio et al. (2011) tested the effect of antioxidant dietary fibre from brown seaweed, *Fucus vesiculosus*, added to minced horse mackerel (*Trachurus trachurus*), at doses of 1% and 2% during frozen storage. Dietary fibre from this macroalga is composed of polysaccharides such as alginates, cellulose, fucans, fucoidan and laminarins. The antioxidant activity is attributed not only to the listed polysaccharides but also to the content of phenolic compounds (mainly phloroglucinol and phlorotannins), vitamin E and certain carotenoids. It was found that samples with algal dietary fibre had lower lipid oxidation than those without it. Fish samples supplemented with 1% of dietary fibre did not have a different flavour from the control sample. Samples supplemented with 2% of fibre from algae were found to have a different flavour from the control sample, but in spite of this difference, a pleasant fresh seafood flavour was described by the panel. Choi et al. (2012) examined reduced-fat pork patties produced with the addition of *Saccharina japonica* powder (1%, 3% and 5%). This meat had significantly higher moisture, ash, carbohydrate content, yellowness and springiness than the control sample. The addition of 1% and 3% of *L. japonica* to the pork patties had the greatest overall acceptability in the sensory evaluation. Cofrades et al. (2011) examined the effect of edible seaweeds *Himanthalia elongata*, *Undaria pinnatifida* and *Pyropia umbilicalis* (2.5% and 5% dry matter) on the physicochemical and morphological characteristics of meat emulsion-based products (post-rigour pork meat and fresh pork back fat). The seaweeds influenced the properties of gel/emulsion meat systems, favouring the formation of harder and chewier structures with better water- and fat-binding properties, especially in the case of *Undaria pinnatifida*. This effect was observed using much smaller amounts of NaCl (0.5%) than are commonly used in the preparation of meat products. Chun et al. (1999) studied the addition of 1%, 2%, 3%, 4% and 5% powdered *Sargassum thunbergii* or *Gelidium amansii* to hamburger patties. *Gelidium amansii* increased the water content of the patty. With the increase in the amount of seaweed powder added, cooking yield increased. Among minerals, Na, Ca, Mg, Fe and Zn, sodium was the most abundant in seaweed patties. The best sensory scores of colour and overall acceptability were attained by the addition of 3% of *Gelidium amansii* (similar to the control group).

6.6.4 Natural Seaweed Salt

Marine algae are known to contain a healthy balance of minerals for human nutrition, including high levels of K and Mg and favourable Na to K ratios. Therefore, they constitute an attractive solution for reducing Na consumption in the daily diet. Magnusson et al. (2016) proposed taking the minerals (salt) washed from marine macroalgal biomass for the production of salt for human consumption. This will constitute the first step in a seaweed biorefinery, which concerns biomass pretreatment. This step aims to remove salt from biomass, because high salt concentrations could inhibit fermentation and corrode common alloys, such as stainless steel, used in fermentation vessels (van Hal et al. 2014). The residual biomass from the washing facility, after processing, can be used as compost for crop production, as a feed mill in animal production or as biocrude in a refinery for the production of renewable fuels (Magnusson et al. 2016). It is also suggested that seaweeds, rich in mineral elements, be used to reduce the amount of NaCl added in meat processing. Nowadays, sodium intake generally exceeds nutritional recommendations in industrialized countries (Cofrades et al. 2011). Uchida et al. (2017) proposed using nori (*Pyropia yezoensis*) for the production of high-salt content seaweed sauces in the fermentation process. Due to its richness in total nitrogen compounds, potassium and a unique free amino acid composition, the nori sauce has high potential as a novel nutritional source for humans. Seaweeds, due to their high content of minerals, can be also used as a salt substitute in the food industry (Cofrades et al. 2011).

6.7 Hydrocolloids As Food Additives

Seaweeds are also industrially processed so as to extract thickening and texturing agents and stabilizers such as agar, alginate and carrageenan (Bixler and Porse 2011; FAO 2003, 2016; Bixler 2017). The seaweed hydrocolloid industry continues to grow on the order of 2–3% per year in the Asia-Pacific region (Porse and Rudolph 2017). The production of seaweed hydrocolloids consumes nearly the largest amount of macroalgae annually, second only to Japanese nori. Carrageenan is the largest consumer of this group (Bixler 2017). These hydrocolloids have commercial applications (Bixler and Porse 2011). Agar market segmentation involves: confections/water gels, baking, retail (gel powder), meat, dairy and bacto/pharma/agarose; alginate, technical grades, food/pharma, animal feed and propylene glycol alginate (PGA); and carrageenan, meat, dairy, liquid infant formula, water gels, toothpaste and pet food (Bixler and Porse 2011; Bixler 2017). These hydrocolloids are permitted for use as food additives by major food regulatory agencies worldwide (Bixler 2017).

In the human diet, alginates are most often used for food production (e.g. restructured meat products for human foods (Bixler and Porse 2011) or in the baking industry (Mouritsen et al. 2012; Mamat et al. 2014)). The addition of alginates in food applications ranges between 0.5% and 1.5%, because their usage level is cost-driven (Brownlee et al. 2005). Alginates and their salts have emulsifying, gelling, thickening and stabilizing properties and the capacity to retain water as well (Khotimchenko et al. 2001). Consumption of alginates affects the feeling of satiety, which is an important factor that regulates food intake, and thus has great significance in the control of obesity (Chu and Phang 2016). These findings were confirmed in the work of Solah et al. (2010), who compared the satiety effect or the reduction of hunger after consumption of a whey protein-based drink versus an alginate-based drink. It was found that the physical characteristics of the drinks tested, such as viscosity and/or gel strength and protein content, reduced hunger. Alginate-based drinks with high viscosity had a more pronounced effect in reducing hunger than alginate-based drinks with low viscosity. Additionally, it was noted that viscosity reduced hunger more than the protein effect. The presence of hydrocolloids in seaweeds used as a component of food (e.g. pasta) could increase the weight of the cooked dishes. This could possibly be due to the hydration afforded by the hydrocolloids present in the seaweed powder (Prabhasankar et al. 2009). Moreover, the addition of hydrocolloids to the processed food not only improves texture and stability but also reduces food waste through its longer shelf life (Bixler 2017).

6.8 Sensory Analysis of Seaweeds

Considerable commercial attention has been paid to seaweeds as food and seaweed-derived food flavours (Bixler and Porse 2011). Seaweeds have a salty flavour (and thus can be used instead of salt, which can reduce the chances of developing high blood pressure) and are somewhat mildly spicy (Dhargalkar 2014). Further research is also exploring the use of seaweed as an alternative to salt (FAO 2016). Seaweeds are often cooked to flavour dishes or soup stocks before consumption (Zava and Zava 2011). For example, brown seaweed kombu (*Saccharina japonica*) is known to be an excellent source of umami flavour, caused by the high content of free amino acid monosodium glutamate. Umami has an industrial application as a flavour enhancer (Mouritsen et al. 2012, 2017). However, initially, the unique flavour of the seaweeds will not be readily accepted by the consumers who are not already used to it when it comes to taste (Dhargalkar 2014). In the work of Chapman et al. (2015), it was found that the seaweeds *Palmaria palmata*, *Saccharina latissima*, *Saccharina digitata* and *Alaria esculenta* were easily combined with Nordic dishes and food items such as fish (clipfish, cod, brandade, saithe, haddock, marinated halibut, ling), fish cakes, meat (cassoulet), seaweed pizza, tagliatelle pasta, camembert, chocolate ice cream, pancakes and vodka shots. The addition of seaweeds did not appear to dominate the flavour spectrum. It affected texture, appearance and colour, as well as the consistency of a variety of dishes. Many seaweed-flavoured foods (including ice creams) and drinks are being launched, with the Asia and Pacific region as the main market but with increasing interest also being shown in Europe and America.

6.9 Indirect Enrichment of Food of Animal and Plant Origin with Seaweeds

Products of animal and plant origin can be enriched with seaweeds indirectly. In the case of animal products, seaweeds in different forms can be supplemented in the animal feed or in drinking water (Holdt and Kraan 2011; Evans and Critchley 2014; Makkar et al. 2016; Michalak et al. 2017a). In plant cultivation, agro-technical treatments such as fertilization or the foliar and soil application of biostimulants of plant growth provide valuable compounds from seaweeds to plants

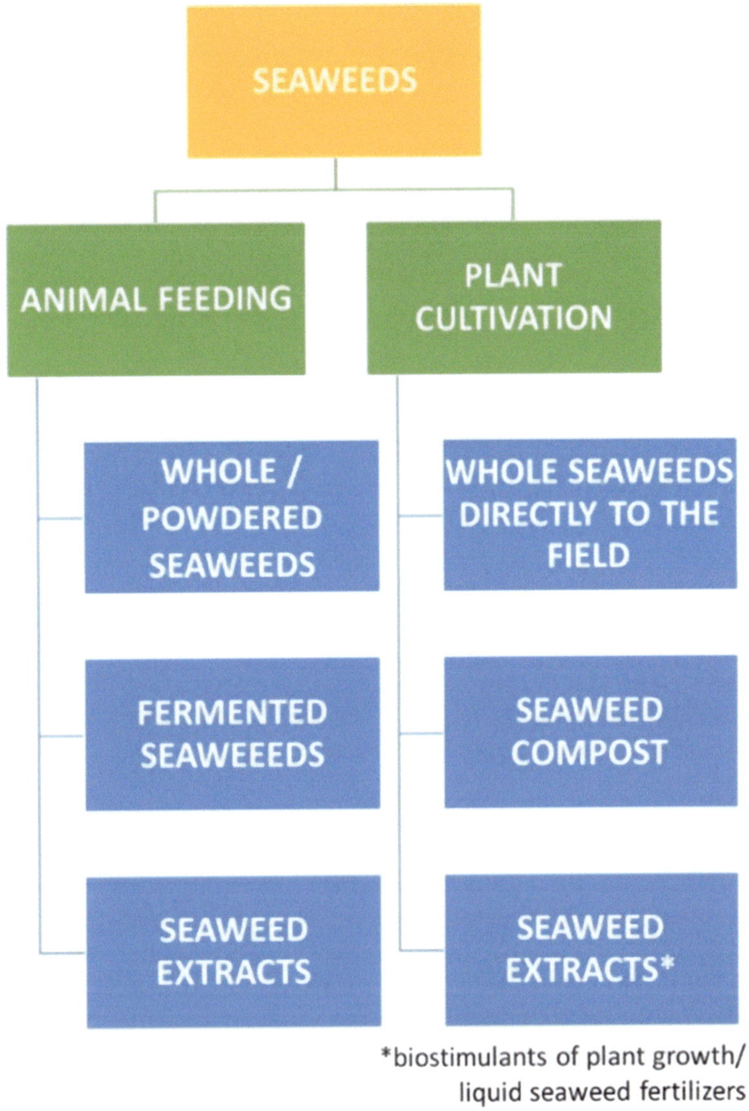

Fig. 6.2 Indirect enrichment of food of animal and plant origin with compounds from seaweeds

(Khan et al. 2009; Michalak et al. 2017a, b). Indirect enrichment of food of animal (meat, eggs, milk, etc.) and plant origin (vegetables, fruits, cereals, etc.) with compounds from seaweeds is shown in Fig. 6.2.

6.9.1 Effect of Seaweeds on the Nutritional Value of Animal Products

Among different groups of seaweeds, Phaeophyta, Rhodophyta and Chlorophyta, brown macroalgae are more studied and more exploited in terms of their application in animal feed than any other alga types. This is the result of their large size and ease of harvesting (Makkar et al. 2016). In animal feeding, the following genera and species are used: *Ascophyllum nodosum*, *Saccharina* species, *Macrocystis pyrifera*, *Sargassum* species, *Palmaria palmata* and *Ulva* species. They can be included in the diet of ruminants, pigs, poultry, rabbits, etc. (Evans and Critchley 2014; Makkar et al. 2016). The nutritional value of seaweeds in the diet of animals is mainly attributed to minerals, trace elements and vitamins, because most of the carbohydrates and proteins are not easily digestible (Holdt and Kraan 2011).

The addition of seaweeds to animal feed improves animal health and productivity and also results in an increase in mineral and pigment content and a decrease in the fat and cholesterol levels in animal products. Some examples of the influence of seaweeds supplemented into animal diets on the nutritional value of the animal products are presented below. Hong et al. (2010) examined the effect of fermented brown seaweed (1% and 2%) as a functional feed on the composition of milk. It was found that the content of fat and protein was not affected by the seaweed additive, but the level of calcium significantly increased. *Ascophyllum nodosum* supplied to the diet of cattle had no influence on the fat content of milk but did increase its iodine content (100 in control to

600 μg L^{-1} in the experimental group) (Jensen et al. 1968). Rey-Crespo et al. (2014) showed that the addition of seaweeds from the Galician coast (*Ulva rigida*, *Saccharina ochroleuca*, *Saccharina latissima*, *Saccorhiza polyschides*, *Mastocarpus stellatus* and *Sargassum muticum*) as a mineral supplement for organic dairy cattle significantly improved the animals' mineral status, particularly the concentration of iodine in milk – 290 μg L^{-1} (experimental) vs. 136 μg L^{-1} (control group). The increase was also noted for Cr – 3.4 μg L^{-1} (experimental, increase by 27%) vs. 2.5 μg L^{-1} (control group), Se – 20 μg L^{-1} vs. 17.6 (increase by 13%) and Zn – 3690 μg L^{-1} vs. 3280 (increase by 13%). It can be concluded that algae used as mineral supplements biofortify milk with microelements essential for humans.

In the work of Michalak et al. (2011, 2015), the biomass of green macroalgae was enriched with microelement ions in the biosorption process. Then, the biomass was applied as a feed additive for laying hens and pigs (instead of inorganic salts). It was found that supplementation of biological feed additives to the diet of laying hens resulted in higher microelement transfer to eggs and enhanced the colour of the yolks. In the case of the feeding experiments on pigs, after 3 months, the meat was biofortified with Cr, Mn, Fe, Cu and Zn. Dierick et al. (2009) examined the effect of the brown seaweed *Ascophyllum nodosum* on gut health and performance and the enrichment of porcine tissues with iodine. It was found that the addition of seaweeds to the feed of pigs had a depressive effect on gut flora, especially on *Escherichia coli*. Additionally, porcine tissues were enriched with iodine. These tissues can be used in a feeding strategy that aims to alleviate the actual iodine deficiencies in many communities. Also, He et al. (2002) found that meat products (e.g. from pigs) with a higher iodine content can be produced through feeding animals with a diet supplemented with algae rich in iodine, for example, *Saccharina digitata*. Doses of 5 or 8 mg of iodine in the form of macroalgae per kg of feed significantly increased the content of this element in the fresh muscle, adipose tissue and the heart but also in the liver and kidneys.

Along with ruminants and non-ruminants (e.g. pigs), macroalgae are also used in the diet of poultry. Strand et al. (1998) showed that macroalgae (brown seaweeds – *Fucus serratus* and *Fucus vesiculosus*) can enhance the colour of the egg yolks of laying hens thanks to the increase in carotenoid content, e.g. fucoxanthin metabolites. Abudabos et al. (2013) reported that the green seaweed *Ulva lactuca*, used as an alternative ingredient in broiler chicken diets (substitution 1 or 3% of corn with *Ulva*), significantly reduced total serum lipid and cholesterol concentrations when compared to the control group (diet without macroalgae). Similar results were obtained by Ginzberg et al. (2000), who tested *Porphyridium* sp. (5% and 10%) as a component of chicken diets. It was found that chickens fed with algal biomass had significantly lower serum cholesterol levels, and egg yolks from the experimental group tended to have reduced cholesterol levels and increased levels of linoleic and arachidonic acids. In addition, the colour of the yolk was darker as a result of the higher carotenoid content. Summarizing, algae in animal diets influence the content of minerals, pigments and lipids in the obtained animal products. Biofortified milk, meat and eggs with active compounds from algae can serve as functional foods.

6.9.2 Effect of Seaweeds on the Nutritional Value of Plant Products

There are different approaches to the use of seaweeds for agricultural purposes. Part of algal biomass is used directly as fertilizer in nearby land; part of it is used for composting purposes (production of organic fertilizers) or can be converted into liquid seaweed fertilizers using extraction/homogenization processes (Michalak et al. 2017b). Seaweeds are recommended for the growers for several reasons – they not only promote better germination, growth (shoot and root length, fresh and dry weight) and yield of the cultivated plants but also improve their nutritional value (Sivasankari et al. 2006; Rathore et al. 2009). The effect of the seaweed extracts on the content of nutrients and biologically active compounds was discussed earlier in this chapter. Sivasankari et al. (2006) found that aqueous seaweed extracts from brown seaweed, *Sargassum wightii* and *Caulerpa chemnitzia*, influenced chlorophyll, carotenoid, protein and amino acid content and reduced the sugar content of shoots and roots of *Vigna sinensis*. Rathore et al. (2009) showed that the application of seaweed extract, prepared from *Kappaphycus alvarezii*, in a field experiment on soybeans (*Glycine max*) improved the uptake of nutrients such as N, P, K and S. It is worth mentioning that seaweeds are able to concentrate minerals from seawater, and they contain 10–20 times more minerals than land plants (Makkar et al. 2016). Also, Michalak et al. (2017b) found that the compost and the extract from this compost, prepared from the drifting marine macroalgae collected from the Baltic Sea, increased the content of microelements such as B, Cu, Mn, Si and Zn and macroelements such as Ca, K and Mg when compared with the control group of garden cress (*Lepidium sativum*) and radish (*Raphanus sativus*). Crouch et al. (1990) also reported that the algal product Kelpak (prepared from *Ecklonia maxima*) significantly increased the amounts of Ca, K and Mg in the leaves of lettuce that were receiving an adequate supply of nutrients. Seaweed-treated plants can also have an increased level of carbohydrates, proteins, free amino acids, polyphenols, nitrogen and photosynthetic pigments, as was shown in the work of Pise and Sabale (2010), in which the concentrates of *Ulva fasciata*, *Sargassum ilicifolium* and *Gracilaria corticata* were applied in the cultivation of fenugreek. Also, Zodape et al. (2010) reported enhanced nutritional quality (increased content of proteins, carbohydrates and elements – P and Mo) in green gram treated with seaweed extract obtained from *Kappaphycus alvarezii*. The edible part of plants, enriched with biologically active compounds from seaweeds, can be consumed directly or can constitute an ingredient in many dishes.

6.10 Conclusions

Awareness of healthy eating plays a crucial role in modern society. Algal products can have an essential meaning for conscious consumers. Algae can provide not only high nutritional value (e.g. improvement of the lipid profile due to high content of polyunsaturated fatty acids) but can also be used as a source of salt, natural colours and flavours. Moreover, they can protect the food from the loss of its properties (decrease of lipid oxidation due to the presence of polysaccharides such as alginates, cellulose, fucans, fucoidan and laminarins) and can extend the shelf life (due to antimicrobial properties). Seaweeds in the production of food can also replace some of the traditionally used ingredients, for example, replacing fat in the production of meat products.

Algae-based diets are also crucial for human health. This food ingredient can be used in different forms – whole seaweed meal, fresh or dried, as well as processed seaweed in the form of seaweed extracts, seaweed bioactive compounds, homogenates and fermented seaweeds. This guarantees the possibility of multiple uses. Algae can be supplemented into the human diet directly. Alternatively, foods of animal and plant origin can be enriched with the biologically active compounds from algae indirectly, through their supplementation into animal diets or in the cultivation of plants. Algae in hen diets beneficially influence the colour of yolks, as well as reduce total serum lipids and cholesterol concentrations, while its use in pig and cattle diets increases the content of microelements in meat and the content of calcium in milk, respectively. Plants cultivated with the support of algae preparations have increased levels of minerals, carbohydrates, proteins, free amino acids, polyphenols, nitrogen and photosynthetic pigments.

Nevertheless, in order to successfully implement seaweed-enhanced food all over the world, there is a need to combine several aspects, such as innovation, cultivation and niche markets, which will lead to a greater acceptance of the seaweed products. There is also a need to establish best practices concerning harvesting, processing (which can increase or decrease the nutritional quality) and storage of seaweeds that will assure high-quality seaweed-based products. It is necessary to organize the harvesting of raw materials in order to exploit the potential of the marine areas to produce food.

Finally, it is important to provide greater information on algae and their wider availability in order to encourage people to include them as part of a balanced and healthy diet.

Acknowledgements This project was financed in the framework of a grant entitled "The effect of bioactive algae enriched by biosorption on the certain minerals such as Cr(III), Mg(II) and Mn(II) on the status of glucose in the course of metabolic syndrome horses. Evaluation *in vitro* and *in vivo*" (2015/18/E/NZ9/00607), attributed by the National Science Centre in Poland.

References

Abudabos AM, Okab AB, Aljumaah RS, Samara EM, Abdoun KA, Al-Haidary AA (2013) Nutritional value of green seaweed (Ulva lactuca) for broiler chickens. Ital J Anim Sci 12:e28. https://doi.org/10.4081/ijas.2013.e28

Andersen RA (2005) Algal culturing techniques, 1st edn. Elsevier Academic Press, New York

Astorga-España MS, Mansilla A, Ojeda J, Marambio J, Rosenfeld S, Mendez F, Rodriguez JP, Ocaranza P (2017) Nutritional properties of dishes prepared with sub-Antarctic macroalgae – an opportunity for healthy eating. J Appl Phycol 29:2399–2406

Bixler HJ (2017) The carrageenan controversy. J Appl Phycol 29:2201–2207

Bixler HJ, Porse H (2011) A decade of change in the seaweed hydrocolloids industry. J Appl Phycol 23:321–335

Bouga M, Combet E (2015) Emergence of seaweed and seaweed-containing foods in the UK: focus on labeling, iodine content, toxicity and nutrition. Foods 4:240–253

Brown ES, Allsopp PJ, Magee PJ, Gill CI, Nitecki S, Strain CR, McSorley EM (2014) Seaweed and human health. Nutr Rev 72:205–216

Brownlee IA, Allen A, Pearson JP, Dettmar PW, Havler ME, Atherton MR, Onsøyen E (2005) Alginate as a source of dietary fiber. Crit Rev Food Sci Nutr 45:497–510

Chang HC, Wu L-C (2008) Texture and quality properties of Chinese fresh egg noodles formulated with green seaweed (*Monostroma nitidum*) powder. J Food Sci 73(8):S398–S404

Chapman AS, Stévant P, Larssen WE (2015) Food or fad? Challenges and opportunities for including seaweeds in a Nordic diet. Bot Mar 58(6):423–433

Chee CP, Gallaher JJ, Djordjevic D, Faraji H, McClements DJ, Decker EA, Hollender R, Peterson DG, Roberts RF, Coupland JN (2005) Chemical and sensory analysis of strawberry flavoured yogurt supplemented with an algae oil emulsion. J Dairy Res 72:311–316

Choi YS, Choi JH, Han DJ, Kim HY, Kim HW, Lee MA, Chung HJ, Kim CJ (2012) Effects of *Laminaria japonica* on the physico-chemical and sensory characteristics of reduced-fat pork patties. Meat Sci 91:1–7

Chopin T, Buschmann A, Halling C, Troell M, Kautsky N, Neori A, Kraemer GP, Zertuche-González JA, Yarish C, Neefus C (2001) Integrating seaweeds into marine aquaculture systems: a key toward sustainability. J Appl Phycol 37:975–986

Chu WL, Phang SM (2016) Marine algae as a potential source for anti-obesity agents. Mar Drugs 14:222. https://doi.org/10.3390/md14120222

Chun SS, Park JR, Park JC, Suh JS, Ahn CB (1999) Quality characteristics of hamburger patties added with seaweed powder. J Korean Soc Food Sci Nutr 28(1):140–144

Cofrades S, López-López I, Solas MT, Bravo L, Jiménez-Colmenero F (2008) Influence of different types and proportions of added edible seaweeds on characteristics of low-salt gel/emulsion meat systems. Meat Sci 79:767–776

Cofrades S, Lopez-Lopez I, Jimenez Colmenero F (2011) Applications of seaweed in meat-based functional foods. In: Kim S-K Handbook of marine macroalgae: Biotechnol Appl Phycol, Wiley, pp 491–499

Combet E, Jarlot A, Aidoo KE, Lean MEJ (2013) Development of a nutritionally balanced pizza as a functional meal designed to meet published dietary guidelines. Pub Health Nutr 17(11):2577–2586

Cornish ML, Critchley AT, Mouritsen OG (2017) Consumption of seaweeds and the human brain. J Appl Phycol 29:2377–2398

Crouch IJ, Beckett RP, van Staden J (1990) Effect of seaweed concentrate on the growth and mineral nutrition of nutrient-stressed lettuce. J Appl Phycol 2:269–272

de Gaillande C, Payri C, Remoissenet G, Zubia M (2017) *Caulerpa* consumption, nutritional value and farming in the Indo-Pacific region. J Appl Phycol 29:2249–2266

Dhargalkar VK (2014) Uses of seaweeds in the Indian diet for sustenance and well-being. Sci Cult 80:192–202

Diaz-Rubio ME, Serrano J, Borderias J, Saura-Calixto F (2011) Technological effect and nutritional value of dietary antioxidant *Fucus* fiber added to fish mince. J Aquat Food Prod Technol 20:295–307

Dierick N, Ovyn A, De Smet S (2009) Effect of feeding intact brown seaweed *Ascophyllum nodosum* on some digestive parameters and on iodine content in edible tissues in pigs. J Sci Food Agricult 89(4):584–594

Edwards MD, Holdt SL, Hynes S (2012) Algal eating habits of phycologists attending the ISAP Halifax conference and members of the general public. J Appl Phycol 24:627–633

Ennouali M, Ouhssine M, Ouhssine K, Elyachioui M (2006) Biotransformation of algal waste by biological fermentation. Afr J Biotechnol 5:1233–1237

Evans FD, Critchley AT (2014) Seaweeds for animal production use. J Appl Phycol 26:891–899

FAO (2003) In: DJ MH (ed) A guide to the seaweed industry, FAO Fisheries Technical Paper – 441. FAO, Rome. http://www.fao.org/3/a-y4765e.pdf. Accessed 28 June 2017

FAO (2016) The state of world fisheries and aquaculture 2016, Contributing to food security and nutrition for all. FAO, Rome. www.fao.org/3/a-i5555e.pdf. Accessed 22 May 2017

Fleurence J, Morançais M, Dumay J, Decottignies P, Turpin V, Munier M, Garcia-Bueno N, Jaouen P (2012) What are the prospects for using seaweed in human nutrition and for marine animals raised through aquaculture? Trends Food Sci Technol 27:57–61

Friis Pedersen S, Meland M, Rebours C (2013) Macroalgae for an increasing organic market. Bioforsk Fokus 8(2):337–338

Gallaher JJ, Hollender R, Peterson DG, Roberts RF, Coupland JN (2005) Effect of composition and antioxidants on the oxidative stability of fluid milk supplemented with an algae oil emulsion. Int Dairy J 15:333–341

Ginzberg A, Cohen M, Sod-Moriah UA, Shany S, Rosenshtrauch A, Arad S (2000) Chickens fed with biomass of the red microalga *Porphyridium* sp. have reduced blood cholesterol level and modified fatty acid composition in egg yolk. J Appl Phycol 12:325–330

Gupta S, Abu-Ghannam N (2011) Recent developments in the application of seaweeds or seaweed extracts as a means for enhancing the safety and quality attributes of foods. Innov Food Sci Emerg Technol 12(4):600–609

Gupta S, Cox S, Rajauria G, Jaiswal A, Abu-Ghannam N (2012) Growth inhibition of common food spoilage and pathogenic microorganisms in the presence of brown seaweed extracts. Food Bioprocess Technol 5(5):1907–1916

Hayisama-ae W, Kantachote D, Bhongsuwan D, Nokkaew U, Chaiyasut C (2014) A potential synbiotic beverage from fermented red seaweed (*Gracilaria fisheri*) using *Lactobacillus plantarum* DW12. Int Food Res J 21:1789–1796

He ML, Hollwich W, Rambeck WA (2002) Supplementation of algae to the diet of pigs: a new possibility to improve the iodine content in the meat. J Anim Physiol Anim Nutr 86:97–104

Holdt SL, Kraan S (2011) Bioactive compounds in seaweed: functional food applications and legislation. J Appl Phycol 23:543–597

Hong ZS, Lee ZH, Xu CX, Yin JL, Jin YC, Lee HJ, Lee SB, Choi YJ, Lee HG (2010) Effect of fermented brown seaweed waste (FBSW) on milk production, composition and physiological responses in Holstein dairy cows. J Animal Sci Technol 52:287–296

Jensen A, Nebb H, Sæter EA (1968) The value of Norwegian seaweed meal as a mineral supplement for dairy cows. Norwegian Institute of Seaweed Research, TAPIR. Report 32, pp 1–35

Khan W, Rayirath UP, Subramanian S, Jithesh MN, Rayorath P, Hodges DM, Critchley AT, Craigie JS, Norrie J, Prithiviraj B (2009) Seaweed extracts as biostimulants of plant growth and development. J Plant Growth Regul 28:386–399

Khotimchenko YS, Kovalev VV, Savchenko OV, Ziganshina OA (2001) Physical–chemical properties, physiological activity, and usage of alginates, the polysaccharides of brown algae. Russ J Mar Biol 27:S53–S64

López-López I, Cofrades S, Ruiz-Capillas C, Jiménez-Colmenero F (2009) Design and nutritional properties of potential functional frankfurters based on lipid formulation, added seaweed and low salt content. Meat Sci 83:255–262

Mabeau S, Fleurence J (1993) Seaweed in food products: biochemical and nutritional aspects. Trends Food Sci Technol 4:103–107

MacArtain P, Gill CIR, Brooks M, Campbell R, Rowland IR (2007) Nutritional value of edible seaweeds. Nutr Rev 65:535–543

Magnusson M, Carl C, Mata L, de Nys R, Paul NA (2016) Seaweed salt from *Ulva*: a novel first step in a cascading biorefinery model. Algal Res 16:308–316

Makkar HPS, Tran G, Heuzé V, Giger-Reverdin S, Lessire M, Lebas F, Ankers P (2016) Seaweeds in livestock diets: a review. Anim Feed Sci Technol 212:1–17

Mamat H, Matanjun P, Ibrahim S, Md Amin SF, Hamid MA, Rameli AS (2014) The effect of seaweed composite flour on the textural properties of dough and bread. J Appl Phycol 26:1057–1062

McHugh DJ (2003) A guide to the seaweed industry. FAO Fisheries Technical Paper, T441

Michalak I, Chojnacka K (2015a) Algae as production systems of bioactive compounds. Eng Life Sci 15:160–176

Michalak I, Chojnacka K (2015b) Functional fermented food and feed from seaweed. In: Montet D, Ray RC (eds) Fermented foods: biochemistry & biotechnology, vol 12. CRC Press (Taylor & Francis Group), Boca Raton, pp 231–247

Michalak I, Chojnacka K (2015c) Production of seaweed extracts by biological and chemical methods. In: Kim S-K, Chojnacka K (eds) Marine algae extracts: processes, products, applications, vol 7. WILEY–VCH Verlag GmbH & Co. KGaA, Weinheim, pp 121–143

Michalak I, Chojnacka K, Dobrzański Z, Górecki H, Zielińska A, Korczyński M, Opaliński S (2011) Effect of enriched with microelements macroalgae on egg quality parameters and mineral content of eggs, eggshell, blood, feathers and droppings. J Anim Physiol Anim Nutr 95:374–387

Michalak I, Chojnacka K, Korniewicz D (2015) New feed supplement from macroalgae as the dietary source of microelements for pigs. Open Chem 13:1341–1352

Michalak I, Chojnacka K, Dmytryk A, Wilk R, Gramza M, Rój E (2016) Evaluation of supercritical extracts of algae as biostimulants of plant growth in field trials. Front Plant Sci 7:1591. https://doi.org/10.3389/fpls.2016.01591

Michalak I, Chojnacka K, Saeid A (2017a) Plant growth biostimulants, dietary feed supplements and cosmetics formulated with supercritical CO_2 algal extracts. Molecules 22:66. https://doi.org/10.3390/molecules22010066

Michalak I, Wilk R, Chojnacka K (2017b) Bioconversion of Baltic Seaweeds into organic compost. Waste Biomass Valor 8(6):1885–1895

Mohamed S, Hashim SN, Rahman HA (2012) Seaweeds: a sustainable functional food for complementary and alternative therapy. Trends Food Sci Technol 23:83–96

Moroney NC, O'Grady MN, O'Doherty JV, Kerry JP (2013) Effect of a brown seaweed (*Laminaria digitata*) extract containing laminarin and fucoidan on the quality and shelf-life of fresh and cooked minced pork patties. Meat Sci 94:304–311

Mouritsen OG (2017) Those tasty weeds. J Appl Phycol 29:2159–2164

Mouritsen OG, Williams L, Bjerregaard R, Duelund L (2012) Seaweeds for umami flavour in the New Nordic Cuisine. Flavour 1:4. https://doi.org/10.1186/2044-7248-1-4

O'Sullivan AM, O'Callaghan YC, O'Grady MN, Waldron DS, Smyth TJ, O'Brien NM, Kerry JP (2014) An examination of the potential of seaweed extracts as functional ingredients in milk. Int J Dairy Technol 67:182–193

O'Sullivan AM, O'Grady MN, O'Callaghan YC, Smyth TJ, O'Brien NM, Kerry JP (2016) Seaweed extracts as potential functional ingredients in yogurt. Inn Food Sci Em Technol 37:293–299

Paiva L, Lima E, Ferreira Patarra R, Neto AI, Baptista J (2014) Edible Azorean macroalgae as source of rich nutrients with impact on human health. Food Chem 164:128–135

Pise NM, Sabale AB (2010) Effect of seaweed concentrates on the growth and biochemical constituents of *Trigonella Foenum-Graecum L.* J Phytology 2(4):50–56

Porse H, Rudolph B (2017) The seaweed hydrocolloid industry: 2016 updates, requirements, and outlook. J Appl Phycol 29:2187–2200

Prabhasankar P, Ganesan P, Bhaskar N, Hirose A, Stephen N, Gowda LR, Miyashita K (2009) Edible Japanese seaweed, wakame (*Undaria pinnatifida*) as an ingredient in pasta: chemical, functional and structural evaluation. Food Chem 115:501–508

Prachyakij P, Charernjiratrakul W, Kantachote D (2008) Improvement in the quality of a fermented seaweed beverage using an antiyeast starter of *Lactobacillus plantarum* DW3 and partial sterilization. World J Microbiol Biotechnol 24:1713–1720

Radulovich R (2011) Massive freshwater gains from producing food at sea. Water Policy J 13:547–554

Radulovich R, Umanzor S, Cabrera R, Mata R (2015) Tropical seaweeds for human food, their cultivation and its effect on biodiversity enrichment. Aquaculture 436:40–46

Ratanaburee A, Kantachote D, Charernjiratrakul W, Penjamras P, Chaiyasut C (2011) Enhancement of γ-aminobutyric acid in a fermented red seaweed beverage by starter culture Lactobacillus plantarum DW12. Electron J Biotechnol 4:717–3458. https://doi.org/10.2225/vol14-issue3-fulltext-2

Rathore SS, Chaudhary DR, Boricha GN, Ghosh A, Bhatt BP, Zodape ST, Patolia JS (2009) Effect of seaweed extract on the growth, yield and nutrient uptake of soybean (*Glycine max*) under rainfed conditions. S Afr J Bot 75:351–355

Rebours C, Marinho-Soriano E, Zertuche-González JA, Hayashi L, Vásquez JA, Kradolfer P, Soriano G, Ugarte R, Abreu MH, Bay-Larsen I, Hovelsrud G, Rødven R, Robledo D (2014) Seaweeds: an opportunity for wealth and sustainable livelihood for coastal communities. J Appl Phycol 26:1939–1951

Rey-Crespo F, López-Alonso M, Miranda M (2014) The use of seaweed from the Galician coast as a mineral supplement in organic dairy cattle. Animal 8(4):580–586

Rupérez P, Saura-Calixto F (2001) Dietary fibre and physicochemical properties of edible Spanish seaweeds. Eur Food Res Technol 212(3):349–354

Sharifuddin Y, Chin Y-X, Lim P-E, Phang S-M (2015) Potential bioactive compounds from seaweed for diabetes management. Mar Drugs 13:5447–5491

Sivasankari S, Venkatesalu V, Anantharaj M, Chandrasekaran M (2006) Effect of seaweed extracts on the growth and biochemical constituents of *Vigna sinensis*. Bioresour Technol 97:1745–1751

Solah VA, Kerr DA, Adikara CD, Meng X, Binns CW, Zhu K, Devine A, Prince RL (2010) Differences in satiety effects of alginate-and whey protein-based foods. Appetite 54:485–491

Stévant P, Rebours C, Chapman A (2017a) Seaweed aquaculture in Norway: recent industrial developments and future perspectives. Aquacult Int. https://doi.org/10.1007/s10499-017-0120-7

Stévant P, Marfaing H, Rustad T, Sandbakken I, Fleurence J, Chapman A (2017b) Nutritional value of the kelps *Alaria esculenta* and *Saccharina latissima* and effects of short-term storage on biomass quality. J Appl Phycol 29:2417–2426

Strand A, Herstad O, Liaaen-Jensen S (1998) Fucoxanthin metabolites in egg yolks of laying hens. Comp Biochem Physiol A 119:963–974

Uchida M, Miyoshi T (2013) Algal fermentation – the seed for a new fermentation industry of foods and related products. JARQ 47(1):53–63

Uchida M, Kurushima H, Ishihara K, Murata Y, Touhata K, Ishida N, Niwa K, Araki T (2017) Characterization of fermented seaweed sauce prepared from nori (*Pyropia yezoensis*). J Biosci Bioeng 123:327–332

van Hal JW, Huijgen WJJ, López-Contreras AM (2014) Opportunities and challenges for seaweed in the biobased economy. Trends Biotechnol 32:231–233

Wells ML, Potin P, Craigie JS, Raven JA, Merchant SS, Helliwell KE, Smith AG, Camire ME, Brawley SH (2017) Algae as nutritional and functional food sources: revisiting our understanding. J Appl Phycol 29:949–982

Wikfors GH, Ohno M (2001) Impact of algal research in aquaculture. J Phycol 37:968–974

Yoshinaga J, Morita M, Yukawa M, Shiraishi K, Kawamura H (2001) Certified reference material for analytical quality assurance of minor and trace elements in food and related matrixes based on a typical Japanese diet: interlaboratory study. J AOAC Int 84:1202–1208

Zava TT, Zava DT (2011) Assessment of Japanese iodine intake based on seaweed consumption in Japan: a literature based analysis. Thyroid Res 4:14. http://www.thyroidresearchjournal.com/content/4/1/14

Zemke-White WL, Ohno M (1999) World seaweed utilisation: an end-of-century summary. J Appl Phycol 11:369–376

Zodape ST, Mukhopadhyay S, Eswaran K, Reddy MP, Chikara J (2010) Enhanced yield and nutritional quality in green gram (*Phaseolus radiata L*) treated with seaweed (*Kappaphycus alvarezii*) extract. J Sci Ind Res 69:468–471

7 Algae and Their Extracts in Medical Treatment

Karolina Korzeniowska, Bogusława Górka, Jacek Lipok, and Piotr P. Wieczorek

Abstract

Scientific efforts undertaken in recent decades have undoubtedly proven that algae are an inestimable and, what is more, important renewable source of hundreds of chemical compounds characterized by a wide spectrum of biological activity. Polysaccharides, phenolic compounds and their derivatives, pigments, proteins, lipids, and fatty acids are the substances that can be isolated from seaweeds, possessing antimicrobial, antiviral, antifungal, antilipidemic, antitumor, antidiabetic, anticoagulant, antioxidant, and antiallergic properties, which can be successfully utilized for human needs. The majority of attention has been focused on marine polysaccharides and their application in medicine and pharmacology. Biodegradability into environmentally harmless products, excellent biocompatibility, the lack of toxicity, and physiological indifference make them bioactive compounds of huge therapeutic potential in the pharmaceutical and biomedicinal fields. Many of them are key components for the production of medical devices and pharmaceuticals and play an important role in biomedical applications, such as wound healing/dressing, drug delivery, and controlled release. A relatively high content of phlorotannins and fatty acids, especially omega-3 acids, makes algae a natural and functional food or diet supplement, rich in vitamins that support the nervous system and lower the glucose level in blood. Although several of these properties have been proven by significant amounts of research or in many clinical trials, there are still many possibilities for using biologically active compounds from seaweeds in ways that improve human health and wellness.

Keywords

Algae in medicine · Seaweed polysaccharides · Bioactive compounds · Antioxidant activity · Phlorotannins · Fatty acids

K. Korzeniowska (✉) · B. Górka · J. Lipok · P. P. Wieczorek
Faculty of Chemistry, Department of Analytical and Ecological Chemistry, Opole University, Opole, Poland
e-mail: kkucab@uni.opole.pl

7.1 Introduction

Both low and high molecular compounds exhibiting a range of interesting biological properties can be isolated from many living organisms. One rich and unappreciated source of biologically active compounds is marine plants, including algae. For many years, algae, sometimes referred to as seaweeds, have been considered a huge reservoir of bioactive compounds, exhibiting a number of biological, nutritional, and functional properties (Agatonovic-Kustrin et al. 2016; Bruno de Sousa et al. 2017). This chemical diversity makes algae and their extracts an interesting research object, but first and foremost, their potential for use in the medical, pharmaceutical, and cosmetic industries deserves exploration (Thomas and Kim 2013; Bajpai et al. 2014).

Among the compounds most often isolated from algae are polysaccharides, pigments, lipids and fatty acids, proteins and amino acids, polyphenols, minerals, vitamins, lectins, terpenes, phytohormones, betaines, polyamines, and sterols, whose content and biological activity vary significantly within and between taxonomic groups (Stengel et al. 2011; Michalak and Chojnacka 2015). Many of these compounds are of special interest due to their medicinal properties. Literature shows that algae and their compounds exhibit antibiotic (Shanab 2007), antioxidant (Shanab 2007; Gupta and Abu-Ghannam 2011; Wijesekara et al. 2011; Ngo and Kim 2013; Thomas and Kim 2013; Wang et al. 2013, 2014; Fang et al. 2015; Michalak and Chojnacka 2015), anti-inflammatory (Ponce et al. 2003; Thomas and Kim 2013; Wang et al. 2014; Fang et al. 2015; Michalak and Chojnacka 2015), antidiabetic (Thomas and Kim 2013; Fang et al. 2015), antitumor (Gupta and Abu-Ghannam 2011; Thomas and Kim 2013; Wang et al. 2014; Fang et al. 2015; Michalak and Chojnacka 2015), antihypertensive (Thomas and Kim 2013), antiallergic (Ngo and Kim 2013; Thomas and Kim 2013), antifungal (Thomas and Kim 2013; Cheung et al. 2014; Fang et al. 2015; Michalak and Chojnacka 2015), anticoagulant (Ponce et al. 2003; Gupta and Abu-Ghannam 2011; Wijesekara et al. 2011; Ngo and Kim 2013; Wang et al. 2014), antiviral (Ponce et al. 2003; Gupta and Abu-Ghannam 2011; Wijesekara et al. 2011; Cheung et al. 2014;

© Springer International Publishing AG, part of Springer Nature 2018
K. Chojnacka et al. (eds.), *Algae Biomass: Characteristics and Applications*, Developments in Applied Phycology 8,
https://doi.org/10.1007/978-3-319-74703-3_7

Wang et al. 2014; Michalak and Chojnacka 2015), antimicrobial (Gupta and Abu-Ghannam 2011; Ibrahim and Lim 2015; Mashjoor et al. 2016), antilipidemic (Wang et al. 2014), and cytotoxic properties (Fang et al. 2015; Mashjoor et al. 2016), along with many others. In addition, some reports indicate that some bioactive compounds derived from algae may be useful for anti-HIV (Fang et al. 2015), antihepatotoxic (Wang et al. 2014), antinociceptive (Wang et al. 2014), and antileishmanial activities (Bruno de Sousa et al. 2017).

This review presents the groups of compounds isolated from different species of alga that exhibit particular therapeutic potential. In addition, their biological activity, together with possible applications of products already introduced into the market, is described.

7.2 Possibilities for Using Algae Extracts with Therapeutic Benefits

The healing properties of algae and their extracts have been known for centuries. Prior to 1950, algae were used exclusively in traditional and folk medicine. They have long been considered a source of valuable ingredients and have been used in the treatment of a variety of ailments and illnesses. For thousands of years, algae have been used in traditional Chinese medicine. To date, the most popular algae used as a cure are brown algae *Sargassum*, in particular, the four algal species *S. fusiforme*, *S. pallidum*, *S. horneri*, and *S. thunbergii*. It is believed that these species contribute to the treatment of diseases such as atherosclerosis, hyperlipidemia, hypertension, cancer, and diseases of the thyroid. One example of a folk medicine was a beverage based on red algae of the species *Chondrus crispus* and *Mastocarpus stellatus*. This drink was consumed for ailments such as colds, sore throats, and chest infections, including tuberculosis. Algae cooked in milk or water have also been used in the treatment of burns and kidney problems. Another example of the use of algae in traditional medicine is the preparation of red algae *Porphyra umbilicalis* juice. The consumption of this juice every day for 3 weeks, at a dose of three spoonfuls, proved to be effective against cancer. In particular, positive effects have been observed in the fight against breast cancer. The same algal species has been used to treat constipation and indigestion. In the 1980s and 1990s, interest in algae did not increase due to the raw material itself but rather to bioactive compounds isolated from it. Studies have shown that algae contain a number of compounds of pharmacological potential. The assumption is made that algae have become an interesting object of study as a result of the search for compounds with therapeutic activity. Moreover, in 1977–1987, algae provided about 35% of newly discovered biologically active compounds. To date, they are being used as medicinal ingredients in a wide variety of formulations and supplements (Dias et al. 2012; Pereira and Costa-Lotufo 2012; Chen et al. 2016; Tannoury et al. 2017).

Today, more and more people are aware of the benefits of healthy and safe eating. Most consumers are looking for unprocessed foods without synthetic additives and products with "bio" in their name. Therefore, interest in natural products is constantly increasing in the food industry. That is why raw materials, whose extracts can act as healthy and safe food supplements, are being sought. And most importantly, they can be a safe alternative to "artificial" food additives (Gupta and Abu-Ghannam 2011). Algae were, and still are, used as components in healthy food. An important group of bioactive compounds derived from algae are sulfated polysaccharides. One type of such a compound is agarans, and its known representative is a compound called porphyran. Porphyran is, first and foremost, a good-quality food fiber and is the main ingredient in seaweed called "nori," one of the major nutrients in Japanese cuisine (Bhatia et al. 2008). Another type of sulfated polysaccharide is carrageenans. Carrageenans were first isolated in 1837 and have long been used as gelling agents in the food industry (Liu et al. 2015b). Furthermore, they have many medical applications, having for years been used in Ireland as a traditional therapeutic tea and as components of antitussives (Kraan 2012). In addition to polysaccharides, algae contain a number of other phytochemicals that provide healthy properties for food. Such compounds include, inter alia, phenolic compounds and fatty acids. Many studies have reported that these compounds are responsible for antioxidant, antimicrobial, anticancer, and antiviral activity. However, the main component of marine algae, constituting 25–75% of dry matter, is dietary fiber. It is primarily responsible for the proper functioning of the digestive system and, together with minerals, helps in reducing salt content. An example of algae that are commonly used in Asian cuisine is *Undaria pinnatifida*. Under the name of miyok, it is being served in the form of soups, salads, and side dishes (Gupta and Abu-Ghannam 2011; Kılınç et al. 2013).

Nowadays, cosmeceuticals of natural origin are very popular. Cosmeceuticals are described as cosmetic products containing a number of bioactive compounds and providing therapeutic benefits. The healing potential of algae and their constituents is commonly used in the cosmetic industry (Wijesinghe and Jeon 2011). Skin is exposed to a number of harmful factors due to constant contact with the external environment. In particular, solar radiation can be responsible for poor skin conditions. Exposure to ultraviolet light is associated with a variety of harmful effects, ranging from photoreception to skin cancer. Skin lesions include, inter alia, pigmentation, telangiectasia, erythema, wrinkles, actinic keratosis, squamous cell carcinoma, basal cell carcinoma, and many others (Bowszyc-Dmochowska 2010). Many studies describe algae phytochemicals as important cosmeceutical ingredients that can be effective in preventing these diseases. This group of compounds primarily includes phenolic compounds, inter alia, phlorotannins, sulfate polysaccharides, n-3 fatty acids, and carotenoids. In addition, algae extracts are rich in vitamins A, B, C, and E, amino acids, and minerals, which are also essential in the production of many cosmetics. The physiological activity of these compounds used in the production of cosmeceuticals includes antioxidant, anti-inflammatory, antiallergic, antiaging, anti-wrinkle, cytoprotective, and whitening properties. In addition, they protect against oxidative stress and ultraviolet radiation and participate in processes such as reduction of melanin synthesis and tyrosinase and matrix metalloproteinases inhibitory activity. The compounds responsible for these properties were identified in the brown algae extracts and, in particular, from species such as *Ecklonia cava*, *Sargassum siliquastrum*, *Sargassum marginatum*, *Padina tetrastromatica*, *Trochomorpha conoides*, *Fucus vesiculosus*, *Fucus serratus*, *Ishige okamurae*, and *Ecklonia stolonifera* (Wijesinghe and Jeon 2011; Thomas and Kim 2013).

7.3 Bioactive Compounds from Algae with Various Functional Properties as Medicinal Ingredients

7.3.1 Polysaccharides

One of the most important groups of algae compounds, commonly described in the literature as important therapeutic constituents, are polysaccharides, in particular, sulfated ones. Polysaccharides are carbohydrate biopolymers that are composed of repeating units, which are simple sugars joined in glycosidic linkages with varying degrees of polymerization. Due to the presence of carboxylate, amino, and hydroxyl groups in the composition, these polymers are usually hydrophilic (Stengel et al. 2011; Silva et al. 2014; Stadnik and de Freitas 2014).

In marine and terrestrial organisms, polysaccharides are mainly structural and storage elements and energy conveyors. They also play an important role in the structural integrity and mechanical strength of the tissue. In addition, they participate in the identification of cells, the regulation of signaling, cell proliferation, cell differentiation, and immune reactions. Many polysaccharides are able to regulate water mobility, and the water itself significantly affects the physical and functional characteristics of these sugars. Because of their exceptional biological and physicochemical properties, these compounds exhibit a wide range of applications (Reddy et al. 2011; Wang et al. 2013; Ahmed et al. 2014; Stadnik and de Freitas 2014).

Typically, there are two groups of algae polysaccharides: storage and cell wall polysaccharides. The high content of polysaccharides and the range of their properties are precisely the reason why these compounds attract considerable attention and are widely used. The total concentration of algae polysaccharides depends on many factors, such as environmental conditions, time of the year of harvest, and the species of alga. It has been found that it can even reach 76% (Stengel et al. 2011; Silva et al. 2012a; Senthilkumar et al. 2013; Herrero et al. 2015). The main algae polysaccharides include alginates, laminarans, and sulfated polysaccharides (carrageenans, agarans, fucoidans, ulvans) (Mayakrishnan et al. 2013). Table 7.1 shows the groups of polysaccharides isolated from different algal species with examples of their biological activities.

One of the most important properties of these biopolymers is their hydrophilicity and the formation of hydrogels. As a result, these algae colloids are used in many industries, in particular, in the food and cosmetic industries (Stengel et al. 2011; Vo et al. 2015). The variety of polysaccharides and their ability to form complexes with other important biomolecules result in these compounds having a number of advantageous biological properties, which, in turn, makes them desirable compounds in the pharmaceutical industry and medicine.

In recent years, the potential of certain marine polysaccharides extracted from seaweed and shellfish to strengthen the immune system and lower the activity of an allergic reaction has been described. These potential components are primarily alginates, fucoidans, porphyran, and chitin and its derivatives (Vo et al. 2015). These biopolymers are increasingly being studied for use in tissue engineering. Because of their many advantages, including biocompatibility, good availability, and the fact that they can be adapted to the particular purpose or function, they show promise as potential materials for regeneration of almost all tissues. The most studied and utilized polysaccharides in this area include glycosaminoglycans, starch, cellulose, pectin, chitosan, alginates, agar, pullan, gellan, and xanthan gum. Tissue-engineered materials occur in various forms, such as porous and fibrous scaffolds and injectable hydrogels. In addition, in most cases, they are connected with other polymers, natural or synthetic, or with inorganic nanoparticles. They have been tested, among other things, in regard to the engineering of blood vessels, heart valves, cartilage and intervertebral disks, bones, skeletal muscle, skin, liver, and nervous tissue. They are also used to encapsulate and deliver cells (Bačáková et al. 2014). Due to the great number of the previously mentioned features and benefits, polysaccharides play an important role in biomedical applications such as wound healing/dressing, drug delivery, and controlled release. In oral drug delivery applications, polysaccharides act as excipients to increase the solubility and bioavailability of the active substance and to ensure stability of the final drug products (Liu et al. 2015a).

7.3.1.1 Alginates

Alginates are a group of anionic copolymers isolated mainly from brown algae and, in particular, from species

Table 7.1 List of some types of algae polysaccharide with examples of their possible biological activity

Source	Type of polysaccharide	Species	Possible biological effect	References
Green algae (Chlorophyta)	Ulvans	*Ulva pertusa*	Antioxidant, antihyperlipidemic	Silva et al. (2012b)
		Ulva lactuca	Antioxidant, anticoagulant, antihyperlipidemic, antiviral	Silva et al. (2012b)
		Ulva rigida	Immunostimulating	Silva et al. (2012b)
		Ulva fasciata	Anticoagulant	Silva et al. (2012b)
		Ulva reticulata	Anticoagulant	Silva et al. (2012b)
Brown algae (Phaeophyta)	Alginates	*Macrocystis pyrifera*	Antiallergic	Vo et al. (2015)
		Sargassum vulgare	Anticancer	Pereira and Costa-Lotufo (2012)
	Laminarans	*Laminaria* sp.	Anticoagulant, hypercholesterolemic, hypolipidemic	Kraan (2012), and Mayakrishnan et al. (2013)
	Fucoidans	*Saccharina japonica*	Anticoagulant, anti-inflammatory, antiangiogenic, antiviral, antithrombotic, antioxidant, antitumor	Xu et al. (2017) Senthilkumar et al. (2013), and Zorofchian Moghadamtousi et al. (2014)
		Undaria pinnatifida	Immunomodulatory, antitumor, anticancer	Senthilkumar et al. (2013) Senthilkumar et al. (2013), and Zorofchian Moghadamtousi et al. (2014)
		Laminaria japonica	Antiviral, antithrombotic, anticoagulant, antioxidant, antilipidemic	Senthilkumar et al. (2013), and Zorofchian Moghadamtousi et al. (2014)
		Laminaria saccharina	Anti-inflammatory, anticancer, antitumor	Senthilkumar et al. (2013), and Zorofchian Moghadamtousi et al. (2014)
		Ascophyllum nodosum	Antithrombotic, anticancer, antitumor	
		Fucus vesiculosus	Antioxidant, anticancer, antitumor	Senthilkumar et al. (2013)
		Ecklonia kurome	Antioxidant	Chale-Dzul et al. (2015)
		Turbinaria tricostata	Hepatoprotective, antioxidant	Chale-Dzul et al. (2017)
		Dictyota ciliolata	Antioxidant	Chale-Dzul et al. (2017)
		Padina sanctae-crucis	Antioxidant	Chale-Dzul et al. (2017)
		Sargassum fluitans	Antioxidant	
Red algae (Rhodophyta)	Carrageenans	*Tichocarpus crinitus*	Antiviral	Cheung et al. (2014)
		Kappaphycus striatum	Anticancer	Pereira and Costa-Lotufo (2012)
		Chondrus crispus	Antiviral	Prajapati et al. (2014)
		Phyllophora brodiaei	Anticoagulant	Prajapati et al. (2014)
	Porphyran	*Porphyra tenera*	Antiallergic	Vo et al. (2015)
		Porphyra yezoensis	Antiallergic	Vo et al. (2015)
		Porphyra haitanensis	Antioxidant, anticoagulant	Liu et al. (2015a)

such as *Laminaria hyperborea*, *Laminaria digitata*, *Laminaria japonica*, *Ascophyllum nodosum*, and *Macrocystis pyrifera*, representing 17–45% of the algal dry weight (George and Abraham 2006; Vera et al. 2011; Goh et al. 2012; Kraan 2012; Lee and Mooney 2012). Alginates have the ability to create gels that are responsible for the mechanical and structural features enabling the flexibility and participation in the ion exchange of brown seaweeds (Silva et al. 2012a). These biopolymers have a number of important advantages, including the fact that hydrogels form under mild conditions of pH and temperature and are biodegradable, biocompatible, cheap, non-toxic, and susceptible to sterilization and storage (Mayakrishnan et al. 2013). Many articles have reported that alginates mainly have an antioxidant effect, and their biological activity depends on the molecular weight and content of the anionic groups. The unique properties of alginates and their biological activity can be adapted to many applications (Ahmed et al. 2014).

Alginates are valuable and important polysaccharides, as they can be applied to the engineering and the regeneration of almost all tissues. In the form of injection hydrogel, they are being used to encapsulate and deliver bubbles ovarian follicle maturation, islets of Langerhans and stem cells. Alginate hydrogel has been tested in nerve tissue engineering, bone tissue engineering and cartilage, intervertebral disks, and the regeneration of skeletal muscle. In conjunction with poly(ethylene glycol) molecules and antibodies, it participates in the capture of intraepithelial human blood progenitor cells, while, when associated with hydroxyapatite, it takes part in the regeneration of the bone interface. As a porous skeleton, it creates deposits of capillaries in tissues, and as a nanofiber skeleton, it is used in engineering skin tissue and building blood vessel replacements. It also plays an important role in cardiac tissue engineering (Bačáková et al.

2014). Alginates are present in the pharmaceutical industry as useful encapsulate matrices and drug carriers. In addition, when included in oral tablets, they can improve the bioadhesive properties of said tablets (Goh et al. 2012). The ability of alginic acid to bind divalent metal causes the heavy metals to be gelled by alginic acid, and consequently, they become insoluble in the intestine and cannot be absorbed into the body tissue. Therefore, both it and its derivatives are used in the treatment of gastritis and as antiulcer substances. Examples of such drugs containing alginates are "Gaviscon," "Algitec," and "Gastralgin." Besides, alginates have an antitoxic effect on hepatitis through the normalizing of lipid and glycogen in the liver, an example of such a drug being "Detoxal" (Kraan 2012). Alginate hydrogels show a structural resemblance to the extracellular matrix of the living tissue, so that they play an important role in the process of wound healing/dressing, delivery of bioactive substances, and transplantation of cells. Due to their adsorbent and hemostatic properties, they have many uses in the protection of wounds. Alginate-based wound dressings have many advantages, like the support of granulation tissue formation, swift epithelialization, and healing. They can also maintain a physiologically moist microenvironment and reduce bacterial wound infections. Some of the commonly known, commercially available dressings are Algicell™, AlgiSite M™, Comfeel Plus™, Kaltostat™, Sorbsan™, and Tegagen™ (Lee and Mooney 2012).

7.3.1.2 Laminarans

Laminarans are polysaccharides present in brown marine algae, mainly of the *Laminaria* species. These polysaccharides can also be isolated in smaller quantities from *Ascophyllum*, *Fucus*, and *Undaria*. The content of laminarans in brown algae depends on the season and the environment and can provide up to 35% of the dry algal weight. They are divided into two types: soluble and insoluble (Vera et al. 2011; Kraan 2012; Mayakrishnan et al. 2013). In comparison with other algae polysaccharides, they create neither gels nor viscous solutions. After structural modifications, such as sulfonation, oxidation, and reduction, laminarans exhibit anticoagulant activity (Kraan 2012).

The fact that laminarans obtained from brown algae create neither gels nor viscous solutions would seem to reduce their pharmaceutical and biomedical potential. However, studies indicate that, like other polysaccharides, they have a number of beneficial properties for human health. The anticoagulant activity of these polysaccharides has been described as one that can be increased by structural modifications increasing sulfation. The anticoagulant activity of brown algae *Laminaria* was first described in 1941, which stated that antithrombotic action was contained in the biologically active substance. Another paper presented laminarans as safe ingredients for surgical dusting powder and potential antitumor agents. Together with other polysaccharides, they are components of preparations for strengthening the immune system. Laminarans play a protective role against bacterial pathogens and severe radiation and increase the level of B cells and helper T cells (Kraan 2012). It has been reported that laminarans exhibit hypercholesterolemic and hypolipidemic activity. They also play a key role in reducing cholesterol absorption in the gut, reducing the level of total and free cholesterol, triglycerides, and phospholipids in the liver. Besides, they lower systolic blood pressure by producing antihypertensive responses (Mayakrishnan et al. 2013).

7.3.1.3 Sulfated Polysaccharides

A number of the biological activities and special properties associated with algae extracts are largely attributed to sulfated polysaccharides. They can also be found in animals and plants, but the most important and most common source is marine algae. Sulfated polysaccharides are anionic macromolecules, in which the hydroxyl group is replaced by a sulfate group, and they are considered to be one of the most important components of the algal cell wall. In addition, they play an important role in the regulation of ions and exhibit different biological functions in living marine organisms. The chemical structure of these biopolymers varies, depending on the species of alga (Mayakrishnan et al. 2013; Ahmed et al. 2014).

7.3.1.3.1 Carrageenans

Carrageenans are anionic sulfated polysaccharides extracted from the line of red seaweed, in particular, *Kappaphycus alvarezii*, *Chondrus*, *Eucheuma*, *Gigartina*, and *Hypnea* species. These natural linear polymers are the main structural component of red seaweed cell walls, which are from 30% to 75% of the dry weight (Vera et al. 2011; Li et al. 2014; Stadnik and de Freitas 2014).

Carrageenans may have a promising future in the pharmaceutical industry as excipients in drug delivery systems. Studies have described their immunomodulatory, anticoagulant, antitumor, antihyperlipidemic, and antiviral action. These polysaccharides are used as excipient tablets for controlled drug release systems. Due to their strong development of emulsion and improved uniformity in colloidal suspensions, acting as a gelling agent in acid-neutralizing gels, they play an important role in pharmaceutical preparations. The physicochemical properties of carrageenans may enable these compounds to be used as the matrix for the preparation of oral extended-release tablets with limited drug loading capacity. For example, one study investigated the application of two commercially available carrageenans, Gelcarin® GP-379 and Viscarin®, for the preparation of tablets with tripelennamine HCl. They are responsible for the stability of the storage and extend the durability of the suspension of antibiotics (Reddy et al. 2011; Li et al. 2014). Carrageenans as sulfated polysaccharides are capable of scavenging free radicals, showing the same antioxidant effects. It was therefore concluded that there is a relationship between sulfate content and antioxidant activity. The biological action of these biopolymers has been studied for protection against viruses, fungi, and bacteria. It has been tested in the treatment of respiratory weakness against flu viruses but also other viruses, such as hepatitis A, dengue, and herpes. In addition, they may act as inhibitors for sexually transmitted diseases. One good example of antiviral use is a carrageenan-based microbicidal vaginal gel trade-named Carraguard™, which is widely used in African countries to combat HIV and other sexually transmitted diseases (Hayashi and Reis 2012). Other articles present carrageenans as immunomodulators and compounds of anticancer activity. In recent years, their anticoagulant potential in lowering cholesterol and triglyceride levels has been tested. They also play an important role in tissue engineering as excipients in the transmission of drugs, as well as some immobilizing enzymes. In addition, carrageenan hydrogel was used to encapsulate cells, including chondrocyte cells of the nose and fat cells, and is a promising material for the regeneration of the cartilage (Silva et al. 2012b; Liu et al. 2015a).

7.3.1.3.2 Agarans (Porphyran)

Another type of sulfated polysaccharide is agarans. One of the better known agarans is porphyran. Porphyran is a complex galactan isolated from red algae cell walls and, in particular, of the genus *Porphyra*, including *Porphyra capensis*, *P. haitanensis*, *P. tenera*, *P. suborbiculata*, and *P. yezoensis*. This high-molecular-weight, sulfated polysaccharide is a water-soluble polymer with high viscosity and gelling, stabilizing, and thickening properties. Moreover, this biopolymer is a potential biologically active agent for health applications, because it exhibits, inter alia, antiallergic, antioxidant, and anticoagulant action (Bhatia et al. 2008; Jiao et al. 2011; Stadnik and de Freitas 2014).

Porphyran is primarily a high-quality dietary fiber. Studies performed in rats indicate its function in preventing carcinogenesis in the intestines and antitumor activity against Meth-A fibromas. It may also exhibit antihypertensive activity and reduce blood cholesterol levels. This describes their potential role in the pharmaceutical industry as compounds with antioxidant, immunostimulating, anticoagulant, antihypertensive, anti-fatigue, anticoagulant, antiviral, antibacterial, antihyperlipidemic, antiallergic, antiaging, hepatoprotective, hypocholesterolemic, hypoglycemic, and antiulcer properties. It has been shown that porphyran may be able to inhibit contact hypersensitivity reaction by reducing the IgE level in serum. In addition, porphyran fractions obtained from *P. yezoensis* stimulated macrophages in vivo and in vitro. Oral application of *P. vietnamensis* porphyran contributed to an increase in the weight of organs such as the spleen or thymus and an increase in total lymphocytes and leukocytes. These are just some of the polysaccharide activities examined. A selection of other porphyran activities have been described in a report by S. Bhatia et al. (Bhatia et al. 2008; Kraan 2012; Vo et al. 2015).

7.3.1.3.3 Ulvans

Ulvans are heteropolysaccharides occurring in green algae, mainly of the genus *Ulva*, including *Ulva pertusa* and *Ulva lactuca*, known as "sea lettuce," and their content in green seaweed cell walls is from 8 to 29% of the algal dry weight (Vera et al. 2011; Kraan 2012; Stadnik and de Freitas 2014). This group of sulfated polysaccharides is soluble in water, and their solubility increases with rising temperature. Ulvans form weak gels in deionized water with low internal viscosity. Other factors that affect gelling properties of ulvans are also divalent cations and pH. In addition, gels are thermoreversible without heat treatment. Ulvans are used in the synthesis of chemicals and, in particular, are a source of such rare uronic acids as iduronic acid, which is used in the synthesis of analogs of antithrombotic heparin. Thus, these algae polysaccharides can be applied in the synthesis of compounds with therapeutic action (Kraan 2012; Alves et al. 2013).

From the perspective of activity within the biomedical and pharmaceutical industries, polysaccharides derived from green algae ulvans deserve special attention. A report by Silva et al. presented a series of examined properties of these polymers. They emphasized their antioxidant, antitumor, antiproliferative, immunostimulating, anticoagulant, antihyperlipidemic, and antiviral activity. In relation to the antioxidant activity, several papers have described their ability to chelate metals and scavenge superoxide and hydroxyl radicals. As with other polysaccharides, ulvans also find use in regenerative medicine and tissue engineering. They are used in the form of nanofibers, membranes, particles, and hydrogels (Silva et al. 2012b). It is believed that ulvans, as a natural source of dietary fiber, have a beneficial effect on digestion and improve absorption of nutrients from the small intestine. The study describes their effect on reducing effective total cholesterol and LDL cholesterol through decomposition to bile acids. Hydrogel based on ulvans complexed with calcium ions increases the viscosity of the intestinal lumen content, thus facilitating the absorption of bile acids from the ileum (Mayakrishnan et al. 2013).

7.3.1.3.4 Fucoidans

Fucoidans are the most abundant polysaccharides derived from various species of brown macroalgae, such as *Ascophyllum nodosum* and *Fucus vesiculosus*, and their contents in the brown seaweeds are from 5% to 20% of the algal dry weight. These are biopolymers present as a structural component of the fiber cell wall and intercellular space that strengthen, protect against dehydration, and prevent drying. This type of sulfated polysaccharide was isolated from brown algae for the first time by Kylin in 1913 and was subsequently given the name "fucoidin." According to IUPAC rules, it is currently called fucoidan, but it is also known under the names fucan and fucosan. The biological activity of fucoidans varies with their molecular weight (Jiao et al. 2011; Vera et al. 2011; Senthilkumar et al. 2013; Ahmed et al. 2014; Stadnik and de Freitas 2014). First of all, an antitumor effect of these compounds has been described. An interesting overview of the antitumor activity of fucoidans was presented in the work by Moghadamtousi and others. It describes the effects of fucoidans derived from various species of brown algae on cancer cells (Zorofchian Moghadamtousi et al. 2014).

The biopharmaceutical industry has found use for fucoidans due to their antiviral, antimicrobial, antithrombotic, and immunological effects. In recent years, their potential as natural components of cosmeceuticals has been described. Studies suggest that these algae biopolymers exhibit antiaging and antioxidant activity, creating a protective effect on the skin. In the cosmetics industry, they are used as antiaging products, due to increased activity of the metalloproteinase-1 enzyme in human skin. They have also been investigated in the field of biomedicine. In this area, their potential for nanomedicine, tissue engineering, and drug delivery systems has been indicated. It has been reported that fucoidans stimulate the production of hepatocyte growth factor (HGF). In the delivery of biologically active agents, fucoidans are often complexed with chitosan. This new microsphere delivery system based on a combination of fucoidans and chitosan is called "fucosphere." This cross-linking has been proposed for the controlled release of drug concentration-dependent polymers and proteins and the development of nanoparticles for the delivery of anticoagulants. There is also a proposal to use a hydrogel based on a fucoidan-chitosan combination as an accelerator for burn injuries. In this complex, fucoidan was intended to provide an anticoagulant effect, and chitosan, with its excellent properties, was intended to form a hydrogel dressing to facilitate wound healing. Other polymers joined with fucoidan are, inter alia, alginates and polyesters such as polycaprolactone (PCL). Development of a fucoidan complex of chitosan and alginate was meant to stimulate rapid wound healing. A fucoidan mixture of polycaprolactone was used for regeneration of bone tissue in the form of a nanofibrous scaffold. In this context, fucoidan is to be responsible for cell proliferation, alkaline phosphatase activity, expression of I-type collagen, and mineral seat (Silva et al. 2012b; Ahmed et al. 2014). In the review, Senthilkumar and others presented fucoidan as a potential agent for tumor therapy. They describe its role in processes such as apoptosis, invasion, angiogenesis, and metastasis. According to studies conducted, fucoidan inhibits proliferation of cancer cells via cell cycle arrest, induction of apoptosis, inhibition of metastasis and angiogenesis, and growth regulation of signaling molecules (Senthilkumar et al. 2013). Fucoidans are now considered promising algal compounds for the treatment of disorders of the cardiovascular system and protection of the myocardium. There is a hypothesis that fucoidans cause detoxification of isoproterenol through antioxidant activity and changes in the lipid profile. Isoproterenol is responsible for causing myocardial infarction. This compound induces an increase in the concentration of enzymes, such as keratin kinase, dehydrogenase, and serum transaminase. Furthermore, it increases LDL cholesterol and decreases HDL cholesterol. Studies show that fucoidans can provide protection against isoproterenol, reversing its effects and preventing damage caused by the compound (Mayakrishnan et al. 2013).

7.3.2 Phenolic Compounds

Antioxidant activity is one of the most commonly described properties of compounds isolated from natural sources. It is particularly important in the fight against the free radicals that are responsible for many diseases, especially skin diseases, including cancer. Therefore, raw materials from which natural antioxidants can be isolated are highly sought after. In many products, they may serve as therapeutic agents, and in other products, they may be added to protect other bioactive compounds from the harmful effects of the external environment (Thomas and Kim 2013; Wang et al. 2013; Ahmed et al. 2014).

As was previously mentioned, antioxidant activity is associated with protection against free radicals. Therefore, it is important to clarify what free radicals are. A free radical is defined as any ion, atom, or molecule that contains an unpaired electron and is capable of independent existence. Many such entities occur in unstable, and thus also highly reactive, forms, and they can act as reductants or oxidants. One of the major types of free radical is the reactive oxygen species (ROS). Oxygen-containing free radicals include hydroxyl radical species, superoxide anion radicals, hydrogen peroxide, single oxygen, hypochlorite, peroxynitrite radicals, and nitric oxide radicals. Free radicals may be generated inside the human body during various metabolic transformations and processes or may be derived from the external environment (ultraviolet and X-ray radiation, cigarette smoke, environmental pollution, chemicals, ozone). The presence of unpaired electrons causes free radicals to be more reactive than other molecules. Worst of all, they can adversely affect biologically important molecules (lipids, proteins, carbohydrates, nucleic acids) by damaging them. The damage of these important molecules and the imbalance between the production of free radicals and antioxidant activity are the causes of oxidative stress. Free radicals and the formation of oxidative stress are the cause of many illnesses and diseases, such as atherosclerosis, inflammatory diseases, cancers, skin diseases including aging, ischemic diseases, AIDS, emphysema, haemochromatosis, cardiovascular diseases, neurological disorders, hepatotoxicity, asthma, gastric ulcers, and many others. These examples show how dangerous free radicals are and how important it is to find protection against them (Huamantupa et al. 2011; Wang et al. 2013; Hayyan et al. 2016).

Algae, as well as plants, are exposed to a number of factors, both biotic and abiotic. To protect against stressful conditions (e.g., high oxygen concentration, light, temperature, pathogen attack, high pressure, salinity), algae produce many important metabolites that are antioxidants. One of the most commonly known and important groups of compounds with antioxidant properties is the phenolic compounds (Gupta and Agrawal 2007; Dai and Mumper 2010; Stengel et al. 2011; Taş et al. 2015; Pérez et al. 2016; Agatonovic-Kustrin and Morton 2018). The antioxidant activity of phenolic compounds is primarily related to the stabilization and binding of free radicals, reducing properties, the inhibition of certain enzymes, and the ability to chelate metal ions (Li et al. 2011). Compared to red and green algae, brown algae are characterized by a higher content of phenolic compounds (Heffernan et al. 2015). The largest proportion of phenolic compounds contained in brown algae is phlorotannins, while in green and red algae, the most commonly occurring polyphenols are bromophenols, phenolic acids, and flavonoids (Pérez et al. 2016; Sanz-Pintos et al. 2017).

Phenolic compounds (polyphenols) are a huge and diverse group of phytochemicals with a wide range of biological properties. It is estimated that a group of polyphenols consists of more than 8000 naturally occurring compounds. The primary source of these valuable organic compounds is plants (all parts of a plant, including roots, leaves, seeds, flowers, and fruits) and products made from them (e.g., cereals, wine, olive oil, tea, and coffee). A common feature of all phenolic compounds is the presence of one or more hydroxyl groups bonded directly to the aromatic rings. Because of the carbon skeleton construction and the presence of functional groups, phenolic compounds are divided into two main groups: phenolic acids (derivatives of hydroxybenzoic and hydroxycinnamic acids) and flavonoids (derivatives of 2-phenylchromane, consisting of six major subclasses: flavonols, flavanols, flavanones, flavones, isoflavones, and anthocyanins). Furthermore, the phenolic compounds may include more complex biomolecules, such as lignans, stilbenes, and tannins. Due to the diversity of polyphenolic compounds, their occurrence in many natural raw materials, and their valuable biological properties (antioxidant, antidiabetic, anti-inflammatory, anticarcinogenic, antiviral, antimicrobial, antiproliferative, hemolytic, and antimalarial activity), they are of interest to many scientists (Capozzi et al. 2001; Dai and Mumper 2010; Onofrejová et al. 2010; Bahadoran et al. 2013; Roselló-Soto et al. 2015; Agatonovic-Kustrin et al. 2016; Pérez et al. 2016; Agatonovic-Kustrin and Morton 2018; Vimala and Poonghuzhali 2017).

Brown algae are a type of alga that is commonly described in the literature as a source of antioxidant compounds, in particular, phenolic compounds. The properties of these algae have been known and studied for many years. However, due to the availability of biomass of different species of brown alga and the advancement of technology, including the development of analytical methods, they are still an interesting research object for the analysis of phenolic compounds. One group of phenolic compounds commonly found in brown algae is phlorotannins. In addition, these compounds have been tested for their biological properties. Many studies describe the great potential of phlorotannins in the treatment of many diseases associated with oxidative stress, in particular, cancer (Zenthoefer et al. 2017).

Phlorotannins are biopolymers composed of phloroglucinol (1,3,5-trihydroxybenzene) linked to each other in various ways with a wide range of molecular sizes (126–650 kDa). These polymers are formed in the acetate-malonate (polyketide) pathway. In algae, they occur in free form or in association with other cell wall components. The total concentration of algae phlorotannins is within the range from 5% to 30% of dry weight. Phlorotannins are strongly hydrophilic, and their physiological activity depends on the degree of polymerization. In addition, these polyphenols are described as strong heavy metal chelators and inhibitors of many enzymes. Depending on the combination of monomer units, phlorotannins are divided into fucols (phlorotannins with phenyl linkages), phlorethols (phlorotannins with ether linkages), fuhalols (phlorotannins with ether linkages and additional hydroxyl groups), eckols (phlorotannins with dibenzodioxin linkage and phenoxyl substitution), fucophlorethols (phlorotannins with both ether and phenyl linkages), and carmalols (phlorotannins with a dibenzodioxin linkage) (Li et al. 2011, 2017; Eom et al. 2012; Heffernan et al. 2015; Lopes et al. 2016; Pérez et al. 2016; Tenorio-Rodriguez et al. 2017; Zenthoefer et al. 2017). Literary references indicate that the only source of these polyphenolic secondary metabolites is algae, in particular, brown algae. This is a further reason why it can be stated that algae are a source of valuable and unique ingredients (Eom et al. 2012). Phlorotannins have a number of biological properties. Their therapeutic potential is broadly described in the literature. It has been shown that these compounds may be active ingredients of many drugs, supplements, nutraceuticals, and cosmetics. Examples of individual phlorotannins, the species of algae from which they are isolated, and their healing potential are presented below and in Table 7.2.

Zenthoefer and his co-workers investigated Baltic Sea brown algae *Fucus vesiculosus* and their cytotoxic potential against human pancreatic cancer cells. Those studies confirmed the cytotoxic activity of the examined algae extracts against tumor cells. Structural analysis and characterization of phenolic compounds were performed to identify compounds responsible for such significant antitumor activity. Determination of the total content of phenolic compounds (the Folin-Ciocalteu method) noted the presence of polyphenols in the tested extracts. In addition, structural analysis using spectroscopic and chromatographic methods confirmed the presence of two active polyphenolic compounds, belonging to phlorotannins (Zenthoefer et al. 2017). Another example of the therapeutic effect of phlorotannins is their potential in the treatment of hypertension. It has been shown that brown algae (*Ecklonia cava*, *Ecklonia stolonifera*, *Pelvetia siliquosa*, *Undaria pinnatifida*) rich in phlorotannins demonstrate the ability to inhibit the angiotensin-I-converting enzyme (ACE). ACE is an enzyme that plays an important role in regulating blood pressure. Development of drugs with the addition of compounds capable of inhibiting ACE can be effective in treating cardiovascular diseases. The ability of enzyme inhibition was also demonstrated for phlorotannins isolated from the algae *Ecklonia cava*. The impact of these bioactive compounds on matrix metalloproteinase (MMP) inhibition has been reported. MMPs play an important role in diseases such as chronic inflammation, wrinkles, osteoporosis, arthritis, periodontal disease, and cancer (Li et al. 2011).

Everyone wants to look beautiful, slow down the aging process, and look for the "elixir of youth." There are many reports in the literature about the health effects of algae extracts and their bioactive compounds on the skin. Phenolic compounds belong to this group of compounds that may be potential components of the "elixir of youth." Thomas and co-authors portray polyphenols as active compounds in the protection of skin and the treatment of skin diseases. They have shown that phlorotannins can be potential natural inhibitors of enzymes involved in the processes that take place in skin. Wrinkles are one of the effects of oxidative stress and are a symptom of skin aging. One of the causes of wrinkles is collagen degradation caused by increased regulation of matrix metalloproteinase (MMP). In vitro studies of methanolic extracts from algae *Corallina pilulifera* showed that the phenolic compounds derived from these extracts showed

Table 7.2 Examples of phlorotannins, their sources, and potential biological effects (Li et al. 2011; Eom et al. 2012; Thomas and Kim 2013; Lopes et al. 2016)

Phlorotannin	Species of algae	Potential biological effects
Eckol	*Ecklonia cava* *Eisenia arborea, Ecklonia kurome, Eisenia bicyclis, Ecklonia stolonifera*	Skin whitening, antihypertensive, antimicrobial, antiallergic, antioxidant, enzyme inhibitory effect, anticancer, anti-inflammatory, photoprotective (skin treatment), antidiabetic
Dieckol	*Ecklonia cava, Ecklonia kurome, Eisenia bicyclis, Ecklonia stolonifera*	Skin whitening, antihypertensive, antimicrobial, antioxidant, enzyme inhibitory effect, anticancer, anti-inflammatory, antiallergic, bactericidal activity, photoprotective (skin treatment), antidiabetic
Phloroglucinol	*Ecklonia cava, Ishige okamurae, Ecklonia kurome, Ecklonia stolonifera,* *Eisenia bicyclis*	Skin whitening, antihypertensive, antimicrobial, anticoagulant, antidiabetic
6,6′-bieckol	*Eisenia arborea, Ecklonia cava, Ishige okamurae*	Antiallergic, enzyme inhibitory effect (inhibition of cholinesterases in the treatment of Alzheimer's disease), anti-HIV
6,8′-bieckol	*Eisenia arborea*	Antiallergic
8,8′-bieckol	*Ecklonia cava, Ecklonia kurome, Eisenia bicyclis, Eisenia arborea*	Antiallergic, antimicrobial, antioxidant, enzyme inhibitory effect, anti-inflammatory, anticancer, bactericidal activity, anti-HIV, antidiabetic
Phlorofucofuroeckol	*Eisenia arborea, Ecklonia cava, Ecklonia kurome, Eisenia bicyclis, Ecklonia stolonifera*	Antiallergic, antimicrobial, enzyme inhibitory effect, anti-inflammatory, anticancer, antidiabetic
2-phloroeckol	*Ecklonia stolonifera*	Antidiabetic
7-phloroeckol	*Eisenia bicyclis, Ecklonia cava*	Antimicrobial, skin whitening, enzyme inhibitory effect (tyrosinase inhibition in the treatment of pigmentation disturbances)
Hydroxybenzodioxin	*Eisenia bicyclis*	Antidiabetic
Dioxinodehydroeckol	*Eisenia bicyclis, Ecklonia cava, Ecklonia stolonifera*	Antimicrobial, anticancer, antidiabetic
Fucofuroeckol	*Eisenia bicyclis*	Antidiabetic
Diphlorethohydroxycarmalol	*Ishige okamurae*	Antidiabetic

photoprotective activity against UV radiation and prevented expression of MMPs in human dermal fibroblast cells. It has also been shown that phlorotannins can be inhibitors of tyrosinase, which is an enzyme that catalyzes the rate-limiting pigmentation stage, thus phlorotannins may play a significant part in the skin pigmentation process. The tyrosinase inhibitory activity of phenolic compounds is related to their ability to chelate the copper ions in this enzyme. In addition, in vitro studies on mouse skin models have confirmed the efficacy of brown alga phenolic compounds in the protection against harmful UV radiation and their potential anticancer activity. These studies have shown that both algae feeding and topical algae treatment have resulted in suppression of cyclooxygenase-2 (COX-2) expression and cell proliferation. One of the known skin diseases that many people suffer from is atopic dermatitis. It is a pruritic inflammatory skin disease that can occur at any age. It is a very onerous disease, because it manifests itself with itchy skin and excessive scratching. Furthermore, it is speculated that atopic dermatitis can cause diseases such as asthma, food intolerance, and allergic rhinitis. The mediator of allergic and inflammatory processes is histamine, and the enzyme involved in allergic reactions is hyaluronidase. In vitro studies have shown that extracts rich in phlorotannins isolated from brown algae *Eisenia arborea*, *Eisenia bicyclis*, *Ecklonia kurome*, and *Ecklonia cava* can prevent allergic and inflammatory reactions, because they are capable of inhibiting hyaluronidase. It is interesting to note that some crude phlorotannins have a stronger inhibitory effect on hyaluronidase than some known inhibitors, such as catechins, sodium cromoglycate, and epigallocatechin gallate (Thomas and Kim 2013).

There were also studies performed on the contribution of phlorotannins in the treatment of sleep disorders. Cho and others have shown that phlorotannins may be promising drugs for insomnia and sedation and may be a source of novel hypnotic drugs. The positive effects of these phenolic compounds have been observed in mice that had sleep disorders (Cho et al. 2014).

One of the major diseases affecting civilization is diabetes. The World Health Organization (WHO) estimates that by 2030, diabetes will be the seventh leading cause of death worldwide (Mathers and Loncar 2006). Metabolic disorders associated with this disease and the resulting complications can lead to other dangerous diseases. The increase in the incidence of diabetes shows that current treatments are insufficient. For this reason, it may be necessary to introduce additional methods involving the use of supplements or nutraceuticals, which may be effective in the fight against diabetes and its complications. Some in vitro and in silico studies have shown that these potential additives may be phenolic compounds and other bioactive compounds isolated from the brown algae *Sargassum wightii*. It was found that ethanolic extracts from this algal species had strong α-amylase and α-glucosidase inhibitors. In addition, these extracts have been shown to possess anti-inflammatory activity that may decrease the complications of diabetes. The antidiabetic potential of phlorotannins has also been demonstrated in algae extracts of the species *Fucus vesiculosus*, *Ecklonia cava*, *Sargassum aquifolium*, *Ecklonia stolonifera*, *Fucus distichus*, *Sargassum ringgoldianum*, *Sargassum polycystum*, *Spatoglossum asperum*, *Padina pavonica*, *Turbinaria ornata*, *Ishige okamurae*, and *Ascophyllum nodosum* (Kim and Kim 2012; Bahadoran et al. 2013; Lopes et al. 2016; Fazeela Mahaboob Begum and Hemalatha 2017).

These examples show how valuable biomolecules are phenolic compounds, including algae phlorotannins. Hence, algae extracts have enormous potential for use in medicine and as ingredients in pharmaceuticals, functional foods, and cosmetics. It can be said that they are strong antioxidants and inhibitors of many enzymes responsible for a number of diseases.

7.3.3 Fatty Acids

Algae are also a source of fatty acids, which are important components in the proper functioning of the human body. Fatty acids are defined as aliphatic carboxylic acids with a number of carbon atoms from 4 to 28. These compounds are generally divided depending on their construction (straight or branched), degree of saturation (saturated or unsaturated), and number of double bonds (monounsaturated (MUFA) or polyunsaturated (PUFA)). Depending on the position of the first double bond from the methyl end, polyunsaturated fatty acids can be classified as n-3 and n-6. Algal fatty acids are derived from phospholipids, glycolipids, and nonpolar glycerolipids, and the lipid content of algae ranges from 0.12% to 6.73% of dry weight (Pérez et al. 2016; Vyssotski et al. 2017). Considering the healing potential of fatty acids, polyunsaturated fatty acids (PUFAs), in particular, omega-3 PUFAs, are most important. Polyunsaturated fatty acids are essential fatty acids (EFA), because they cannot be synthesized in the human body (Olasehinde et al. 2017). They are the main components of the brain and retina. Therefore, they play an important role in the proper functioning of the nervous and visual systems. In addition, it is described that omega-3 fatty acids have anti-inflammatory, antiarrhythmic, antithrombotic, and anticoagulant effects. This activity can be used to treat cardiovascular and inflammatory diseases (Uauy et al. 2001; Lee 2013; Papenfus et al. 2013). In addition, studies that involve topical preparations with n-3 fatty acids have confirmed their effectiveness in the prevention of periodontal disease (Pérez et al. 2016). The literature on the importance of fatty acids in medicine mainly concerns polyunsaturated fatty acids and their important role in the development of every human being. Because of their many biological properties, it is important that these acids be supplied. In particular, pregnant women should take omega-3 acids to ensure proper development of the baby. They are supplied to the body in the form of supplements, pharmaceuticals, functional food ingredients, and cosmetic products (Innis 2007; Papenfus et al. 2013; Michalak and Chojnacka 2015; van der Wurff et al. 2017). Examples of the biological activity of fatty acids and their health properties are shown below and in Table 7.3. In addition, examples of algae that are sources of these valuable biologically active compounds are presented.

Algae are capable of metabolizing many fatty acids through various oxidative pathways (Smit 2004). High levels of saturated and unsaturated fatty acids, both mono- and polyunsaturated, were reported in ethyl acetate extracts of red algae such as *Gracilaria vermiculophylla* and *Porphyra dioica*. Regarding polyunsaturated fatty acids, green algae contain mainly C18 PUFAs (e.g., α-linolenic, stearidonic, linoleic acid), and red algae are rich in C20 PUFAs (e.g., arachidonic, eicosapentaenoic acid), whereas brown algae contain both C18 and C20 PUFAs. However, the largest reservoir of fatty acids, including important omega-3 acids, is microalgae (*Arthrospira* sp., *Isochrysis* sp., *Odontella* sp., *Pavlova* sp., *Porphyridium* sp., *Crypthecodinium* sp., *Nannochloropsis* sp., and *Phaeodactylum* sp.) (Pérez et al. 2016; Olasehinde et al. 2017). The content and profile of fatty acids, both saturated and unsaturated, largely depend on the algae's habitat conditions and the seasons of harvesting (Papenfus et al. 2013).

The n-3 and n-6 polyunsaturated acids play an important role in medical treatment (Chen et al. 2016). Polyunsaturated fatty acids, especially those of the n-3 group, are crucial to the treatment of heart disease. Their source is primarily fish oil, which also contains seaweed oil. Omega-3-based medicines are available on the market. These drugs, called Epanova and Lovaza, are approved by the Food and Drug Administration (FDA) and are intended for people with coronary heart disease and dyslipidemia (Lee 2013; Ito 2015; Chen et al. 2016). An example of algae rich in fatty acids that have shown potential in the treatment of cardiovascular disease are algae of the genus *Sargassum* (*S. fusiforme*, *S. pallidum*, *S. horneri*, and *S. thunbergii*). The extracts of these algae contain both monounsaturated and polyunsaturated PUFAs, including important n-3 PUFAs (Chen et al. 2016). It was found that polyunsaturated fatty acids are not the only ones to have therapeutic potential in regard to coronary and cardiovascular diseases. Preliminary studies carried out by Yang and his co-workers presented monounsaturated fatty acids as potential compounds for the prevention of atherosclerosis. They also found that supplementation with long-chain MUFAs altered lipoprotein proteomes in the LDLR-deficient (Papenfus et al. 2013; Yang et al. 2017). In addition, similar results have been observed by Chen and others describing monounsaturated and polyunsaturated fatty acids as potential biomarkers for hyperlipidemia (Chen et al. 2017).

PUFAs, as major components of the brain, strongly affect the functioning of the nervous system. Therefore,

Table 7.3 Examples of algal fatty compounds with their biological activity (Terés et al. 2008; Michalak and Chojnacka 2015; Kalaiselvan et al. 2016; Pérez et al. 2016; Thanigaivel et al. 2016; Vieira et al. 2017)

Algae source	Species	Fatty compounds	Biological activity
Green algae (Chlorophyta)	*Ulva fasciata*	PUFAs	Antibacterial
	Ulva reticulata	Palmitic acid	Antibacterial
	Chlorococcum humicola	Fatty acids	Antibacterial, antifungal
	Enteromorpha linza	Unsaturated fatty acids (stearidonic acid, gamma-linolenic acid)	Antimicrobial
	Anadyomene saldanhae, *Caulerpa cupressoides*	Fatty acids	Antiprotozoal
Brown algae (Phaeophyta)	*Himanthalia elongata*	Fatty acids (palmitic acid, palmitoleic acid, oleic acid)	Antibacterial, antifungal, antioxidant
	Padina sp., *Canistrocarpus cervicornis*, *Dictyota* sp.	Fatty acids	Antiprotozoal
	Lobophora variegata	Fatty acids	Antibacterial, pupicidal
Red algae (Rhodophyta)	*Gracilaria corticata*	PUFAs	Antibacterial
	Jania corniculata, *Laurencia papillosa*	Fatty acids (tetradecanoic acid, hexadecanoic acid, octadecanoic acid, 9-octadecenoic acid, tetracosanoic acid)	Antipathogenic
	Gelidiella acerosa	Saturated fatty acids (palmitic acid)	Antioxidant, antimicrobial
	Ochtodes secundiramea, *Bostrychia tenella*	Fatty acids	Antiprotozoal
	Asparagopsis taxiformis	Fatty acids	Antimicrobial
Microalgae (microphyte)	*Dunaliella salina*	PUFAs	Antibacterial, antifungal
	Haematococcus pluvialis	Fatty acids	Antibacterial, antifungal

many studies have focused on finding sources of essential fatty acids that can be effective in the treatment of neurodegenerative diseases. One such disease is Alzheimer's disease. This disease is a serious neurological disorder and the most common form of dementia. Alzheimer's disease causes brain damage, cognitive and memory impairment, and loss of synapses and neurons and ultimately leads to cell death. There are indications that the imbalance of omega-3 and omega-6 fatty acids may affect the development of this disease. The most important fatty acid of the brain is docosahexaenoic acid (DHA, omega-3 fatty acid). Along with other, slightly less important acids, such as eicosapentaenoic acid (EPA, omega-3 fatty acid) and arachidonic acid (AA, omega-6 fatty acid), DHA is involved in neurological processes such as synaptic transmission, cognitive development, memory function, and neuronal plasticity. Reports in the literature indicate that microalgae are a rich source of omega-3 fatty acids. High levels of DHA and EPA are found in microalgal species such as *Arthrospira* sp., *Isochrysis* sp., *Odontella* sp., *Pavlova* sp., *Porphyridium* sp., *Crypthecodinium* sp., *Nannochloropsis* sp., and *Phaeodactylum* sp. (Papenfus et al. 2013; Briffa et al. 2017; Olasehinde et al. 2017). Clinical studies indicate that one of the causes of Alzheimer's disease is the effect of aluminum as a neurotoxic metal. This harmful metal can be delivered to the body through the consumption of food or through exposure to a polluted environment. Animal studies have shown the neuroprotective effects of extracts from brown algae *Sargassum ilicifolium* through inhibition of aluminum induction, improving memory, protecting against free radicals, and increasing the level of neurotransmitters responsible for memory functions. These studies have confirmed the potential therapeutic effect of algae extracts in the treatment of neurodegenerative diseases such as Alzheimer's disease (Sumithra et al. 2016). Other brown algae that have been found to have important omega-3 acids such as DHA are *Cystoseira usneoides*, *Stypocaulon scoparium*, and *Sargassum vulgare* (Papenfus et al. 2013).

Clinical studies have shown that high levels of DHA and arachidonic acid affect visual acuity. These experiments have been carried out on newborns whose food was supplemented with polyunsaturated fatty acid oil. Improvement of visual function was primarily related to the actions of the acids involved in modulating gene expression of the developing retina and brain. The major acid responsible for these processes is DHA, which, inter alia, affects photoreceptor membranes, rod and cone development, activation of rhodopsin and increase the opsin expression. This is a nother example of a health action of fatty acids that is necessary for proper human development, as well as the importance of supplementation with essential fatty acids throughout life, particularly in its initial period (Uauy et al. 2001).

Taking into account the essential fatty acids, docosahexaenoic acid (DHA) is of particular interest. This long-chain highly unsaturated n-3 fatty acid possesses a number of biological and functional properties that can be used in medical treatment. DHA is metabolically related to other important healthy omega-3 fatty acids (Calder 2016). Therefore, it is important to look for raw materials that can be a rich source of these important acids. Many studies indicate that this valuable source that we are looking for is algae, both macro- and microalgae.

7.4 Conclusions

Currently, the aspiration to use natural products is gaining importance in all human activities, especially those related to health, well-being, and beauty. Natural sources can be successfully utilized in some branches of bio-based industry. Over the years, algae have already found application in many fields of life, among which the most important seem to be medicine, pharmacology, and the cosmetic industry. Algae extracts are an inestimable and, more importantly, renewable source of hundreds of chemical compounds that offer a wide spectrum of desired physiological activities.

References

Agatonovic-Kustrin S, Morton DW (2018) Quantification of polyphenolic antioxidants and free radical scavengers in marine algae. J Appl Phycol 30:113–120

Agatonovic-Kustrin S, Morton DW, Ristivojević P (2016) Assessment of antioxidant activity in Victorian marine algal extracts using high performance thin-layer chromatography and multivariate analysis. J Chromatogr A 1468:228–235

Ahmed ABA, Adel M, Karimi P, Peidayesh M (2014) Chapter ten – pharmaceutical, cosmeceutical, and traditional applications of marine carbohydrates. In: Kim S-K (ed) Advances in food and nutrition research. Academic, Cambridge, pp 197–220

Alves A, Sousa RA, Reis RL (2013) A practical perspective on ulvan extracted from green algae. J Appl Phycol 25:407–424

Bačáková L, Novotná K, Parzek M (2014) Polysaccharides as cell carriers for tissue engineering: the use of cellulose in vascular wall reconstruction. Physiol Res 63:S29–S47

Bahadoran Z, Mirmiran P, Azizi F (2013) Dietary polyphenols as potential nutraceuticals in management of diabetes: a review. J Diabetes Metab Disord 12:1–9

Bajpai VK, Rather IA, Lim J, Park YH (2014) Diversity of bioactive polysaccharide originated from marine sources: a review. Indian J Mar Sci 43:1857–1869

Bhatia S, Sharma A, Sharma K, Kavale M, Chaugule BB, Dhalwal K, Namdeo AG, Mahadik KR (2008) Novel algal polysaccharides from marine source: porphyran. Pharmacogn Rev 2:271–276

Bowszyc-Dmochowska M (2010) Działanie promieniowania ultrafioletowego na skórę. Ostre i przewlekłe uszkodzenie posłoneczne. Homines Hominibus 6:29–42

Briffa M, Ghio S, Neuner J, Gauci AJ, Cacciottolo R, Marchal C, Caruana M, Cullin C, Vassallo N, Cauchi RJ (2017) Extracts from two ubiquitous Mediterranean plants ameliorate cellular and animal models of neurodegenerative proteinopathies. Neurosci Lett 638:12–20

Bruno de Sousa C, Gangadhar KN, Morais TR, Conserva GAA, Vizetto-Duarte C, Pereira H, Laurenti MD, Campino L, Levy D, Uemi M, Barreira L, Custódio L, Passero LFD, Lago JHG, Varela J (2017) Antileishmanial activity of meroditerpenoids from the macroalgae *Cystoseira baccata*. Exp Parasitol 174:1–9

Calder PC (2016) The DHA content of a cell membrane can have a significant influence on cellular behaviour and responsiveness to signals. Ann Nutr Metab 69:8–21

Capozzi G, Lo Nostro P, Menichetti S, Nativi C, Sarri P (2001) Easy synthesis of polyphenolic 4-thiaflavans with a 'double-faced' antioxidant activity. Chem Commun:551–552

Chale-Dzul J, Moo-Puc R, Robledo D, Freile-Pelegrín Y (2015) Hepatoprotective effect of the fucoidan from the brown seaweed *Turbinaria tricostata*. J Appl Phycol 27:2123–2135

Chale-Dzul J, Freile-Pelegrín Y, Robledo D, Moo-Puc R (2017) Protective effect of fucoidans from tropical seaweeds against oxidative stress in HepG2 cells. J Appl Phycol 29:2229–2238

Chen Z, Xu Y, Liu T, Zhang L, Liu H, Guan H (2016) Comparative studies on the characteristic fatty acid profiles of four different Chinese medicinal *Sargassum* seaweeds by GC-MS and chemometrics. Mar Drugs 14:1–8

Chen Y-L, Xiao C-H, Hu Z-X, Liu X-S, Liu Z, Zhang W-N, Zhao X-J (2017) Dynamic lipid profile of hyperlipidemia mice. J Chromatogr B 1055–1056:165–171

Cheung RCF, Wong JH, Pan WL, Chan YS, Yin CM, Dan XL, Wang HX, Fang EF, Lam SK, Ngai PHK, Xia LX, Liu F, Ye XY, Zhang GQ, Liu QH, Sha O, Lin P, Ki C, Bekhit AA, Bekhit AED, Wan DCC, Ye XJ, Xia J, Ng TB (2014) Antifungal and antiviral products of marine organisms. Appl Microbiol Biotechnol 98:3475–3494

Cho S, Yoon M, Pae AN, Jin Y-H, Cho N-C, Takata Y, Urade Y, Kim S, Kim J-S, Yang H, Kim J, Kim J, Han J-K, Shimizu M, Huang Z-L (2014) Marine polyphenol phlorotannins promote non-rapid eye movement sleep in mice via the benzodiazepine site of the GABAA receptor. Psychopharmacology 231:2825–2837

Dai J, Mumper RJ (2010) Plant phenolics: extraction, analysis and their antioxidant and anticancer properties. Molecules 15:7313–7352

Dias DA, Urban S, Roessner U (2012) A historical overview of natural products in drug discovery. Metabolites 2:303–336

Eom SH, Kim YM, Kim SK (2012) Antimicrobial effect of phlorotannins from marine brown algae. Food Chem Toxicol 50:3251–3255

Fang HY, Chokkalingam U, Chiou SF, Hwang TL, Chen SL, Wang WL, Sheu JH (2015) Bioactive chemical constituents from the brown alga *Homoeostrichus formosana*. Int J Mol Sci 16:736–746

Fazeela Mahaboob Begum SM, Hemalatha S (2017) Characterization, in silico and in vitro determination of antidiabetic and anti-inflammatory potential of ethanolic extract of *Sargassum wightii*. Asian J Pharm Clin Res 10:297–301

George M, Abraham TE (2006) Polyionic hydrocolloids for the intestinal delivery of protein drugs: alginate and chitosan – a review. J Control Release 114:1–14

Goh CH, Heng PWS, Chan LW (2012) Alginates as a useful natural polymer for microencapsulation and therapeutic applications. Carbohydr Polym 88:1–12

Gupta S, Abu-Ghannam N (2011) Recent developments in the application of seaweeds or seaweed extracts as a means for enhancing the safety and quality attributes of foods. Innov Food Sci Emerg Technol 12:600–609

Gupta S, Agrawal SC (2007) Survival and reproduction in some algae under stress conditions. Folia Microbiol 52:603–617

Hayashi L, Reis RP (2012) Cultivation of the red algae *Kappaphycus alvarezii* in Brazil and its pharmacological potential. Braz J Pharmacog 22:748–752

Hayyan M, Hashim MA, Alnashef IM (2016) Superoxide ion: generation and chemical implications. Chem Rev 116:3029–3085

Heffernan N, Brunton NP, FitzGerald RJ, Smyth TJ (2015) Profiling of the molecular weight and structural isomer abundance of macroalgae-derived phlorotannins. Mar Drugs 13:509–528

Herrero M, Sánchez-Camargo ADP, Cifuentes A, Ibáñez E (2015) Plants, seaweeds, microalgae and food by-products as natural sources of functional ingredients obtained using pressurized liquid extraction and supercritical fluid extraction. Trends Analyt Chem 71:26–38

Huamantupa I, Cuba M, Urrunaga R, Paz E, Ananya N, Callalli M, Pallqui N, Coasaca H (2011) Richness, use and origin of expended medicinal plants in the markets of the Cusco City. Rev Peru Biol 18:283–291

Ibrahim D, Lim S-H (2015) In vitro antimicrobial activities of methanolic extract from marine alga *Enteromorpha intestinalis*. Asian Pac J Trop Biomed 5:785–788

Innis SM (2007) Fatty acids and early human development. Early Hum Dev 83:761–766

Ito MK (2015) A comparative overview of prescription omega-3 fatty acid products. P T 40(826–836):857

Jiao G, Yu G, Zhang J, Ewart HS (2011) Chemical structures and bioactivities of sulfated polysaccharides from marine algae. Mar Drugs 9:196–233

Kalaiselvan I, Senthamarai M, Kasi PD (2016) 2,3,7,8-TCDD-mediated toxicity in peripheral blood mononuclear cells is alleviated by the antioxidants present in *Gelidiella acerosa*: an in vitro study. Environ Sci Pollut Res 23:5111–5121

Kılınç B, Cirik S, Turan G, Tekogul H, Koru E (2013) Seaweeds for food and industrial applications. In: Muzzalupo I (ed) Food industry. InTech, Rijeka, pp 735–748

Kim M-J, Kim HK (2012) Insulinotrophic and hypolipidemic effects of *Ecklonia cava* in streptozotocin–induced diabetic mice. Asian Pac J Trop Med 5:374–379

Kraan S (2012) Algal polysaccharides, novel applications and outlook. In: Chang C-F (ed) Carbohydrates – comprehensive studies on glycobiology and glycotechnology. InTech, Rijeka, pp 489–532

Lee J-H (2013) Polyunsaturated fatty acids in children. Pediatr Gastroenterol Hepatol Nutr 16:153–161

Lee KY, Mooney DJ (2012) Alginate: properties and biomedical applications. Prog Polym Sci 37:106–126

Li Y-X, Wijesekara I, Li Y, Kim S-K (2011) Phlorotannins as bioactive agents from brown algae. Process Biochem 46:2219–2224

Li L, Ni R, Shao Y, Mao S (2014) Carrageenan and its applications in drug delivery. Carbohydr Polym 103:1–11

Li Y, Fu X, Duan D, Liu X, Xu J, Gao X (2017) Extraction and identification of phlorotannins from the brown alga, *Sargassum fusiforme* (Harvey) Setchell. Mar Drugs 15:1–15

Liu J, Willför S, Xu C (2015a) A review of bioactive plant polysaccharides: biological activities, functionalization, and biomedical applications. Bioact Carbohydr Diet Fibre 5:31–61

Liu J, Zhan X, Wan J, Wang Y, Wang C (2015b) Review for carrageenan-based pharmaceutical biomaterials: favourable physical features versus adverse biological effects. Carbohydr Polym 121:27–36

Lopes G, Andrade PB, Valentão P (2016) Phlorotannins: towards new pharmacological interventions for diabetes mellitus type 2. Molecules 22:1–21

Mashjoor S, Yousefzadi M, Esmaeili MA, Rafiee R (2016) Cytotoxicity and antimicrobial activity of marine macro algae (*Dictyotaceae* and *Ulvaceae*) from the Persian Gulf. Cytotechnology 68:1717–1726

Mathers CD, Loncar D (2006) Projections of global mortality and burden of disease from 2002 to 2030. PLoS Med 3:2011–2030

Mayakrishnan V, Kannappan P, Abdullah N, Ahmed ABA (2013) Cardioprotective activity of polysaccharides derived from marine algae: an overview. Trends Food Sci Technol 30:98–104

Michalak I, Chojnacka K (2015) Algae as production systems of bioactive compounds. Eng Life Sci 15:160–176

Ngo DH, Kim SK (2013) Sulfated polysaccharides as bioactive agents from marine algae. Int J Biol Macromol 62:70–75

Olasehinde TA, Olaniran AO, Okoh AI, Koulen P (2017) Therapeutic potentials of microalgae in the treatment of Alzheimer's disease. Molecules 22:1–18

Onofrejová L, Vašíčková J, Klejdus B, Stratil P, Mišurcová L, Kráčmar S, Kopecký J, Vacek J (2010) Bioactive phenols in algae: the application of pressurized-liquid and solid-phase extraction techniques. J Pharm Biomed Anal 51:464–470

Papenfus HB, Kulkarni MG, Stirk WA, Finnie JF, Van Staden J (2013) Effect of a commercial seaweed extract (Kelpak®) and polyamines on nutrient-deprived (N, P and K) okra seedlings. Sci Hortic 151:142–146

Pereira RC, Costa-Lotufo LV (2012) Bioprospecting for bioactives from seaweeds: potential, obstacles and alternatives. Braz J Pharmacog 22:894–905

Pérez MJ, Falqué E, Domínguez H (2016) Antimicrobial action of compounds from marine seaweed. Mar Drugs 14:1–38

Ponce NMA, Pujol CA, Damonte EB, Flores ML, Stortz CA (2003) Fucoidans from the brown seaweed *Adenocystis utricularis*: extraction methods, antiviral activity and structural studies. Carbohydr Res 338:153–165

Prajapati VD, Maheriya PM, Jani GK, Solanki HK (2014) Carrageenan: a natural seaweed polysaccharide and its applications. Carbohydr Polym 105:97–112

Reddy K, Mohan GK, Satla S, Gaikwad S (2011) Natural polysaccharides: versatile excipients for controlled drug delivery systems. Asian J Pharm Sci 6:275–286

Roselló-Soto E, Galanakis CM, Brnčić M, Orlien V, Trujillo FJ, Mawson R, Knoerzer K, Tiwari BK, Barba FJ (2015) Clean recovery of antioxidant compounds from plant foods, by-products and algae assisted by ultrasounds processing. Modeling approaches to optimize processing conditions. Trends Food Sci Technol 42:134–149

Sanz-Pintos N, Pérez-Jiménez J, Buschmann AH, Vergara-Salinas JR, Pérez-Correa JR, Saura-Calixto F (2017) Macromolecular antioxidants and dietary fiber in edible seaweeds. J Food Sci 82:289–295

Senthilkumar K, Manivasagan P, Venkatesan J, Kim SK (2013) Brown seaweed fucoidan: biological activity and apoptosis, growth signaling mechanism in cancer. Int J Biol Macromol 60:366–374

Shanab SMM (2007) Antioxidant and antibiotic activities of some seaweeds (Egyptian Isolates). Int J Agric Biol 9:220–225

Silva TH, Alves A, Ferreira BM, Oliveira JM, Reys LL, Ferreira RJF, Sousa RA, Silva SS, Mano JF, Reis RL (2012a) Materials of marine origin: a review on polymers and ceramics of biomedical interest. Int Mater Rev 57:276–307

Silva TH, Alves A, Popa EG, Reys LL, Gomes ME, Sousa RA, Silva SS, Mano JF, Reis RL (2012b) Marine algae sulfated polysaccharides for tissue engineering and drug delivery approaches. Biomatter 2:278–289

Silva AKA, Letourneur D, Chauvierre C (2014) Polysaccharide nanosystems for future progress in cardiovascular pathologies. Theranostics 4:579–591

Smit AJ (2004) Medicinal and pharmaceutical uses of seaweed natural products: a review. J Appl Phycol 16:245–262

Stadnik MJ, de Freitas MB (2014) Algal polysaccharides as source of plant resistance inducers. Trop Plant Pathol 39:111–118

Stengel DB, Connan S, Popper ZA (2011) Algal chemodiversity and bioactivity: sources of natural variability and implications for commercial application. Biotechnol Adv 29:483–501

Sumithra M, Arunachalam G, Chitra V (2016) *Sargassumilici folium* (Turner) C. Agardh ameliorate oxidative stress induced neuronal degeneration in the Animal Model of Alzheimer's Disease. Biosci Biotechnol Res Asia 13:1069–1074

Tannoury MY, Saab AM, Elia JM, Harb NN, Makhlouf HY, Diab-Assaf M (2017) In vitro cytotoxic activity of *Laurencia papillosa*, marine red algae from the Lebanese coast. J Appl Pharm Sci 7:175–179

Taş B, Ertürk Ö, Yılmaz Ö, Ayvaz MÇ, Ertürk EY (2015) Chemical components and biological activities of two freshwater green algae from Ordu, Turkey. Turk J Biochem 40:508–517

Tenorio-Rodriguez PA, Murillo-Álvarez JI, Campa-Cordova ÁI, Angulo C (2017) Antioxidant screening and phenolic content of ethanol extracts of selected Baja California Peninsula macroalgae. J Food Sci Technol 54:422–429

Terés S, Barceló-Coblijn G, Benet M, Álvarez R, Bressani R, Halver JE, Escribá PV (2008) Oleic acid content is responsible for the reduction in blood pressure induced by olive oil. Proc Natl Acad Sci U S A 105:13811–13816

Thanigaivel S, Chandrasekaran N, Mukherjee A, John T (2016) Seaweeds as an alternative therapeutic source for aquatic disease management. Aquaculture 464:529–536

Thomas NV, Kim SK (2013) Beneficial effects of marine algal compounds in cosmeceuticals. Mar Drugs 11:146–164

Uauy R, Hoffman DR, Peirano P, Birch DG, Birch EE (2001) Essential fatty acids in visual and brain development. Lipids 36:885–895

van der Wurff ISM, Meyer BJ, de Groot RHM (2017) A review of recruitment, adherence and drop-out rates in omega-3 polyunsaturated fatty acid supplementation trials in children and adolescents. Forum Nutr 9:1–32

Vera J, Castro J, Gonzalez A, Moenne A (2011) Seaweed polysaccharides and derived oligosaccharides stimulate defense responses and protection against pathogens in plants. Mar Drugs 9:2514–2525

Vieira C, Gaubert J, de Clerck O, Payri C, Culioli G, Thomas OP (2017) Biological activities associated to the chemodiversity of the brown algae belonging to genus *Lobophora* (*Dictyotales*, *Phaeophyceae*). Phytochem Rev 16:1–17

Vimala T, Poonghuzhali TV (2017) In vitro antimicrobial activity of solvent extracts of marine brown alga, *Hydroclathrus clathratus* (C. Agardh) M. Howe from Gulf of Mannar. J Appl Pharm Sci 7:157–162

Vo TS, Ngo DH, Kang KH, Jung WK, Kim SK (2015) The beneficial properties of marine polysaccharides in alleviation of allergic responses. Mol Nutr Food Res 59:129–138

Vyssotski M, Lagutin K, MacKenzie A, Mitchell K, Scott D (2017) Phospholipids of New Zealand edible brown algae. Lipids 52:629–639

Wang H, Liu YM, Qi ZM, Wang SY, Liu SX, Li X, Wang HJ, Xia XC (2013) An overview on natural polysaccharides with antioxidant properties. Curr Med Chem 20:2899–2913

Wang L, Wang X, Wu H, Liu R (2014) Overview on biological activities and molecular characteristics of sulfated polysaccharides from marine green algae in recent years. Mar Drugs 12:4984–5020

Wijesekara I, Pangestuti R, Kim SK (2011) Biological activities and potential health benefits of sulfated polysaccharides derived from marine algae. Carbohydr Polym 84:14–21

Wijesinghe WAJP, Jeon YJ (2011) Biological activities and potential cosmeceutical applications of bioactive components from brown seaweeds: a review. Phytochem Rev 10:431–443

Xu Y, Zhang Q, Luo D, Wang J, Duan D (2017) Low molecular weight fucoidan ameliorates the inflammation and glomerular filtration function of diabetic nephropathy. J Appl Phycol 29:531–542

Yang Z-H, Gordon SM, Sviridov D, Wang S, Danner RL, Pryor M, Vaisman B, Shichijo Y, Doisaki N, Remaley AT (2017) Dietary supplementation with long-chain monounsaturated fatty acid isomers decreases atherosclerosis and alters lipoprotein proteomes in LDLr−/− mice. Atherosclerosis 262:31–38

Zenthoefer M, Geisen U, Hofmann-Peiker K, Fuhrmann M, Kerber J, Kirchhöfer R, Hennig S, Peipp M, Geyer R, Piker L, Kalthoff H (2017) Isolation of polyphenols with anticancer activity from the Baltic Sea brown seaweed *Fucus vesiculosus* using bioassay-guided fractionation. J Appl Phycol 29:2021–2037

Zorofchian Moghadamtousi S, Karimian H, Khanabdali R, Razavi M, Firoozinia M, Zandi K, Abdul Kadir H (2014) Anticancer and antitumor potential of fucoidan and fucoxanthin, two main metabolites isolated from brown algae. Sci World J 2014:1–10

8 Application of Algae Biomass and Algae Extracts in Cosmetic Formulations

Bogusława Łęska, Beata Messyasz, and Grzegorz Schroeder

Abstract

Biomass of marine algae and their extracts are now one of the most widely used natural ingredients in cosmetics, because of the presence of a wide variety of biologically active compounds in their thalli. Mainly, it comes down to marine species, but the most current interdisciplinary research shows that freshwater macroscopic green algae species (e.g., *Chara fragilis*, *Cladophora glomerata*, *Ulva flexuosa*) may also be a rich source of macro- and micronutrients and other bioactive substances such as fatty acids, polysaccharides, pigments, polyphenols, etc. Freshwater macroalgae are a rare object of study and are practically absent within the cosmetics market. In this chapter, algae harvesting and the processing of algal biomass are discussed. This chapter shows, in particular, that the presence of bioactive substances in their thalli determines the broad biological activity and potential use of freshwater algae in the production of cosmetics.

Keywords

Algal biomass · Algae extract · Bioactive compounds · Cosmetic · Extraction methods

8.1 Introduction

Cosmetic chemistry is a relatively new area that has recently seen rapid development. This development is related to increasing public interest in taking care of one's health and appearance. The search is on for cosmetic ingredients and products that can preserve a youthful appearance and good skin condition for the longest possible time. Bioactive substances of natural origin have become particularly desirable in contemporary cosmetology. Biologically active compounds are those that have particular impact on the physiological and metabolic functions of the organism (Imhoff et al. 2011; Barbosa-Pereira et al. 2013; Hardouin et al. 2014). Many types of biological activity have been identified, for example, antioxidative, anti-inflammatory, immunostimulating, and many others. The category of bioactive substances comprises a diversity of chemical compounds, such as fatty acids, phenol compounds, carotenoids, saccharides, proteins, terpenes, alkaloids, and many others. These compounds are natural metabolites of plant processes, so they represent both organic and inorganic products of chemical transformations taking place inside of plant cells. That is why plant extracts are very popular and have been used in medicine, veterinary, pharmacy, cosmetology, and food production (Verkleij 1992; Allen et al. 2001; O'Doherty et al. 2010; Fabrowska et al. 2015a, b; Michalak and Chojnacka 2015; Pádua et al. 2016).

Many algae extracts, products, and micronized algae have been used over the years in dietary supplements and cosmetics, especially marine species. However, the path to acceptance and use of new species is long, and there are many barriers that need to be overcome. In the paper by Gellenbeck (2011), the author discusses in detail what is needed to achieve acceptance for large-scale marketing and a broad range of consumer products. The path to successful incorporation of a new algal product in cosmetics can be long and complex, but the benefits of high-quality, innovative, high-value products can support growing businesses and ensure a mutually effective partnership between suppliers and manufacturers.

Algae are known to contain a wide range of biologically active compounds, particularly those beneficial to human skin. Marine algae species are known to contain a series of bioactive substances and are widely used as ingredients in cosmetics, drugs, diet supplements, and food. The chemical composition of freshwater algae is recognized as being much poorer, as indicated by the small number of scientific reports on this subject. According to the Web of Science database of May 2017, only 33 scientific reports feature the keywords "freshwater algae and bioactive compounds," while the num-

B. Łęska (✉) · G. Schroeder
Faculty of Chemistry, Adam Mickiewicz University in Poznan, Poznan, Poland
e-mail: bogunial@amu.edu.pl

B. Messyasz
Faculty of Biology, Department of Hydrobiology, Adam Mickiewicz University in Poznan, Poznan, Poland

© Springer International Publishing AG, part of Springer Nature 2018
K. Chojnacka et al. (eds.), *Algae Biomass: Characteristics and Applications*, Developments in Applied Phycology 8,
https://doi.org/10.1007/978-3-319-74703-3_8

ber of analogous reports on "marine algae and bioactive compounds" is over 600. Results of the studies on bioactive compounds in freshwater macroalgae are very promising, and their biomass could be a potential raw product for fodder, the food industry, or the cosmetic industry. Perhaps in the near future, freshwater algae will become an alternative to their marine counterparts.

8.2 Characterization of Algae Used in Cosmetic Products

The cosmetic industry uses different algae species coming from different habitats, both macro- and microalgae from marine and freshwater environments. The diversity of algae types used is reflected in their different properties. The species most widely used in the cosmetic industry include the following:

1. Algae from the *Ulva* genus: cosmopolite green macroalgae encountered in most seas and oceans. Many species from this genus also occur in freshwater, reaching considerable biomass (Messyasz and Rybak 2009). The algae from this genus can have two types of thallus: those of corrugated structure characteristic of young organisms and those wrinkled with a blistered structure typical of mature thalli (Bliding 1963, 1968; Starmach 1972; Van den Hoek et al. 1995; Blomster et al. 2000). In Poland, the following freshwater species of this genus are known: *Ulva prolifera* O.F. Mueller, *Ulva intestinalis L.*, *Ulva compressa L.*, *Ulva flexuosa* Wulfen, and *Ulva paradoxa* Agardh. Along the Baltic Sea coast, one can find *Ulva lactuca* L.; *Ulva linza* L.; *Ulva muscoides* Clemente; *Ulva procera* (Ahlner) Hayden, Blomster, Maggs, Silva, Stanhope, and Waaland; or *Ulva radiata* (Agardh) Hayden, Blomster, Maggs, Silva, Stanhope, and Waaland (Messyasz and Rybak 2009; Messyasz et al. 2015a, b). The cosmetics industry most often uses *Ulva lactuca*, which contains vitamins A, B, C, and E, magnesium, iron, and aosaine, a protein containing the same amino acids as the human elastin (glycine, proline, lysine). For this reason, preparations based on this algae species are used in mature skin care products to reduce wrinkles (Czerpak and Jabłońska-Trypuć 2008; Gade et al. 2013; Manoylov 2014).
2. Algae from the *Cladophora* genus, also classified as green macroalgae (Van den Hoek et al. 1995). They occur in both marine and freshwater environments, but the majority of its species grow in warm seas. Their thallus is threadlike and strongly branched. The cell walls are sometimes encrusted with calcium compounds. The algae belonging to this genus contain a specific carotenoid dye – siphonoxanthin (Guiry and Guiry 2015). The most popular species representing this genus are *Cladophora glomerata* (L.) Kütz., *Aegagropila linnaei* Kütz., *Cladophora brasiliana* Martens, *Cladophora columbiana* Collins, *Cladophora dalmatica* Kütz., *Cladophora fracta* (O.F.Müller ex Vahl) Kütz., and *Cladophora ordinata* (Børgesen) Hoek (www.algaebase.org). The cosmetic industry uses mostly *Cladophora glomerata*, because of its antioxidant and antibacterial properties. The high antioxidant activity of the extract from *Cladophora glomerata* follows from a high content of phenol and polyphenol compounds. It has also been shown to have antibacterial activity toward the strains: *Salmonella enterica*, *Staphylococcus aureus*, *Bacillus subtilis*, and *Proteus mirabilis* (Soltani et al. 2011; Fabrowska et al. 2015a; Zhou et al. 2017).
3. *Arthrospira* is a microscopic alga, belonging to the kingdom of Eubacteria and phylum of Cyanobacteria (Van den Hoek et al. 1995; Komárek et al. 2014; Guiry and Guiry 2015). Because of its exceptional ability to adapt, it occurs in a number of different environments: in soil, sand, bogs, marine water, freshwater, tropical lakes, and even in hot springs (Sun et al. 2016). The most popular in the cosmetics and pharmaceutical industries are *Spirulina maxima* (Setchell and Gardner) Geitler and *Arthrospira platensis* Gomont (syn. *Spirulina platensis* (Gomont) Geitler) (Molski 2010; El-Baz et al. 2013). As additions to cosmetic products, they slow down the aging processes of skin and have an antioxidant effect (Debacq-Chainiaux et al. 2006; Czerpak and Jabłońska-Trypuć 2008; Pielesz 2010).
4. *Laminaria* belongs to brown algae and is found mostly in marine waters in cold and temperate climate zones. It looks like a great leaf, reaching sizes of up to 20 m (Serisawa et al. 2004; Pielesz 2010; Minchin and Nunn 2014). The species from this genus show great contents of iodine, potassium, alginic acid, laminarin, and other valuable ingredients. Oil extracts from *Laminaria ochroleuca* Bachelot de la Pylaie, *Laminaria digitata* (Hudson) Lamour, and *Laminaria cloustoni* Edmondston protect the skin from photoallergic reactions and activate cell metabolism. Water extracts from *Laminaria hyperborea* (Gunn.) Foslie and *Laminaria saccharina* (L.) Lamour show anti-seborrheic properties and restore the proper functioning of oily skin. The iodine contained in these brown algae stimulates the activity of lipase and accelerates decomposition of fat tissues; thus the algae from *Laminaria* genus are used in anti-cellulite cosmetic products. The species *Laminaria digitata* is particularly rich in proline, lysine, and glycine, which are the amino acids present in human elastin, so they are used in products improving skin elasticity and hair conditioners (Debacq-Chainiaux et al. 2006; Czerpak and Jabłońska-Trypuć 2008; Pielesz 2010).

5. *Fucus* also represents brown algae and, similarly to *Laminaria*, grows in marine water, mostly in cold seas, including along the Baltic Sea coast (Pielesz 2010; Guiry and Guiry 2015). *Fucus vesiculosus* L. is most often used in the cosmetics industry. Its characteristic feature is a ribbonlike thallus covered with air bladders. Algae from this species contain iodine, magnesium, manganese, zinc, vitamin C, fucoxanthin, laminarin, and fucoidan. In cosmetic products, *Fucus vesiculosus* is used as an additive to filters protecting against UV radiation and in anti-cellulite products. Other species of significance in cosmetology are *Fucus spiralis* L. and *Fucus serratus* L. (Debacq-Chainiaux et al. 2006; Czerpak and Jabłońska-Trypuć 2008; Pielesz 2010; D'Orazio et al. 2012).
6. *Pylaiella*, belonging to the group of brown algae, is characterized by a threadlike thallus and grows mainly in cold seas (Van den Hoek et al. 1995; Silberfeld et al. 2014; Guiry and Guiry 2015). The species *Pylaiella littoralis* (L.) Kjellman, *Pylaiella ochotensis* Ruprecht, *Pylaiella flexilis* Ruprecht, or *Pylaiella seriata* Kuckuck contain microelements, vitamins, polysaccharides, alginic acid, mannitol, laminarin, fucoidan, and many other bioactive substances. Thanks to the content of its thalli, *Pylaiella* shows skin-moisturizing properties, is capable of skin nutrition, shows antioxidative skin-protecting and skin-revitalizing properties, and activates the skin's metabolism (Blomster et al. 2000; Czerpak and Jabłońska-Trypuć 2008; Manoylov 2014).
7. *Ceramium* algae are red algae of multicellular, filamentous, and leaf-shaped thalli (Van den Hoek et al. 1995; Blomster et al. 2000; Guiry and Guiry 2015). They live mostly in marine waters and prefer warm climatic zones, largely remaining scarce in cold climates. They contain large amounts of amino acids, proteins, vitamins, mineral salts, and agar. A popular cosmetic raw product is *Ceramium rubrum* Agardh, which contains creatine-like substances, used for skin care for the hands and legs (Czerpak and Jabłońska-Trypuć 2008; Pielesz 2010). The extracts from this species show antibacterial properties (Ikawa et al. 1973). Other important species from this genus are *Ceramium circinatum* (Kütz.) Agardh, *Ceramium diaphanum* (Lightf.) Roth, and *Ceramium tenuicorne* (Kütz.) Waern.

8.3 Algae Collection and the Processing of Algal Biomass

Algal biomass for the cosmetic and pharmaceutical industries and for production of other preparations is obtained from two sources. Producers can either purchase the already-processed raw algae products from firms dealing with their cultivation and distribution, e.g., from algae farms, or they can collect algae from sites at which they naturally occur or from algae cultivation sites (Fabrowska et al. 2015a, b; Shah et al. 2016; Sebök et al. 2017). In order to ensure the highest quality, they are collected from the purest waters in Japan, Brittany, and Hawaii. Algae are collected manually from boats equipped with a special arm, some mainly, or even solely, in France (Surget et al. 2017).

Proper procedures for dealing with algae biomass require a great deal of care from the very beginning. Collection of algae biomass must be performed in such a way as to ensure the harvesting of a representative sample. Biomass samples must be collected from as many sites in a given area (lake, river) as possible, at appropriate depth and preferably on the same day. The biomass must be transported to the laboratory immediately after collection in tightly closed water containers, at a low temperature and with no exposure to light. In the laboratory, the biomass is washed many times with distilled water to remove all contaminants, like sand, other algae species, and other plants. The washed biomass is subjected to microscopic and DNA analysis to verify the species' composition. The first stage of algae biomass processing is its drying. In order to preserve and protect the biologically active substances, the traditional method of drying in the sun has been replaced with sublimation drying, in which water is transformed straight from the solid to the gas phase, omitting the liquid phase. This can be realized by lyophilization, which takes place in special reactors below 0 °C and under reduced pressure. The biomass is frozen at −50 °C or with the use of liquid nitrogen to −196 °C. Then, it is placed in a vacuum, where the water is sublimated. Then, the biomass is dried at 40–50 °C until it becomes a constant mass. Algae must be subjected to drying 2–4 h after their collection; as with longer times, the risk of decay and decomposition of the biologically active substances increases. After drying, the biomass is subjected to micronization (homogenization) by, for example, a laboratory grinder, to obtain the maximum possible comminution. The small particles obtained in the process can more easily penetrate into deeper layers of the skin and ensure more effective activity of the bioactive ingredients. The algae biomass must be stored in the dark and at low temperature (−20 °C), because many bioactive substances present in it (e.g., carotenoids, polyphenols) easily undergo decomposition under the effect of light and elevated temperature (Fabrowska et al. 2015a, b).

It seems that algae farming is the best way to provide valuable biomass. Optimization of growth conditions and control permit a harvesting free from contaminants and ensure a standard biomass of high quality. On a small scale, the growth of algae can also be conducted in laboratories. The first algae farms appeared in Japan, where microalgae from the genus *Chlorella* were grown, followed by farms growing *Spirulina* in Mexico. At present, algae farming also takes place in the USA, India, and Israel (Borowitzka 2013a, b). Algae farms are an ecological solution, as they eliminate excess CO_2 from the atmosphere and are cost-effective, as the production of biomass from microalgae is, on average, 5000 tons of dry mass per year (Fabrowska et al. 2015a, b; Guiry and Guiry 2015).

For production of cosmetics, algae are used in their micronized form or as extracts.

8.4 Plant Extracts

Recently, consumers have shown increasing demand for products based on substances derived from natural products. Contemporary cosmetology has also shown increasing interest in the development of products based on biologically active substances of natural origin, mostly coming from plants. The cosmetics market offers an increasing number of formulations containing plant extracts (Malinowska et al. 2014; Michalak et al. 2017; Szopa et al. 2017).

Plant extracts can be obtained from different parts of the plant, leaves, stems, roots, flowers, fruit, bark, or seeds, depending on the localization of the desired biologically active substance that needs to be extracted. Care must be taken to collect the plants at the time when they contain the highest amount of the desired substance. The extracts are obtained by extraction of maceration (flooding of the raw product in a solvent) of properly dried and comminuted plant parts. In order to ensure high-quality products, the plants should meet strict pharmacopoeian or technological norms. The raw plant products are subjected to standardization, which is a set of procedures aimed at ensuring a constant and repeatable level of biologically active components in each raw product unit. It permits obtaining standardized extracts containing components of documented biological activity and high purity, including microbiological purity (Cefali et al. 2016; Michalak et al. 2017; Szopa et al. 2017).

For application in cosmetic products, a few types of extract from dried plant parts are used:

- Water-glycol extracts, obtained by flooding plant parts in a mixture of water and propylene or butylene glycol, which are moisturizing and conserving substances. This type of extraction is used to isolate saponins, tannins, mucus, anthocyanins, flavonoids, saccharides, amino acids, and water-soluble vitamins.
- Glycol extracts, which are less popular than water-glycol ones because of the high viscosity of glycol, although they ensure higher microbiological stability and contain a greater number of biologically active components as glycols, are good solvents.
- Water-alcohol extracts, which are rarely used because of the drying effect of ethanol on the skin and because their effectiveness is limited to a small number of low-polarity substances.
- Glycerin extracts, the glycerin of which ensures additional moisturizing effect, but the extracts contain smaller numbers of biologically active components and show high viscosity.
- Oil extracts, obtained by flooding plant parts in a vegetable oil, synthetic triglyceride, or fatty ester (e.g., octyl palmitate); this method permits isolation of essential oils, carotenes, fatty acids, phospholipids, phytosterols, and lipid-soluble vitamins.
- Dry extracts, containing no more than 5% of water, obtained by evaporation of the solvent from the earlier prepared extract, usually of a water-alcohol type.
- Multicomponent extracts, mixed ones obtained by subsequent treatment of raw plant products with polar and nonpolar solvents (Martini 2014).

Plant extracts are one of the most numerous and widely used raw cosmetic products of plant origin. Their widespread use and popularity stem from the richness of the biologically active components they contain and thus the multidirectional effect they have on the skin. Extraction draws out biologically active substances from the plant parts so that they can be more effectively used by the skin. The use of an appropriate solvent for extraction, for example, a vegetable oil, can enhance the cosmetic effect, or it can play an additional conserving role. Cosmetics containing plant extracts are milder and more easily absorbable by the human skin than products based on synthetic substances; moreover they far more rarely cause irritation or allergic reaction (Malinowska et al. 2014). Of course, the marketing aspect is also important, as there is increasing demand for products that are ecological and of natural origin (Malinowska et al. 2014; De Lima Yamaguchi et al. 2015; Michalak et al. 2017; Szopa et al. 2017).

8.5 Algae Extracts

In cosmetic products, extracts from marine algae are most often used. They are usually subjected to water or lipid extraction. Water extracts contain water-soluble substances, such as proteins, polysaccharides, and water-soluble vitamins. These extracts are used in moisturizing creams. Lipid extracts contain phospholipids, glycolipids, free fatty acids, lipid-soluble vitamins, and steroids and are applied in regenerating, nourishing, and protective creams (Fabrowska et al. 2015a, b; Guiry and Guiry 2015). Depending on the type of extract and its isolated bioactive substances, algae extracts can have moisturizing, soothing, occlusive, antiaging, antioxidant, anti-acne, and anti-cellulite effects. That is why algae extracts are applied as ingredients in many cosmetic products: day and night creams; under-eye, moisturizing, antiaging, and anti-acne creams; face masks; shampoos; and hair conditioners. The extracts from *Laminaria ochroleuca*, *Laminaria hyperborea*, and *Chlorella vulgaris* Beyer are those most often encountered. Other species whose extracts are popular in cosmetic formulations are *Chondrus crispus* Stackhouse, *Corallina officinalis* L., *Ulva compressa* L., *Fucus vesiculosus*, *Ulva lactuca*, *Laminaria digitata*, *Laminaria saccharina*, and *Spirulina platensis*. The extracts from brown algae are more often used in cosmetic products than those from green and red algae (Malinowska 2011).

According to the criterion of innovation, the extraction methods can be divided into the classical and the advanced. The classical methods, for example, in the Soxhlet apparatus, involving shaking and maceration, are gradually being phased out in favor of modern methods, mainly because of their low efficiency, long processing time, and use of large volumes of solvents. Contemporary extraction methods employ additional physical or chemical agents, such as hydrolytic enzymes, microwave radiation, and supercritical solvents, which accelerate the process and increase its yield. These methods are known as supercritical fluid extraction (SFE), microwave-assisted extraction (MAE), and enzyme-assisted extraction (EAE). The methods of extraction are developed to meet the principles of "green chemistry," that is, reduction of both the labor and energy needed and the use of organic solvents, gas emissions, and production of waste and wastewater (Michalak and Chojnacka 2014; Fabrowska et al. 2015a; Esquivel-Hernández et al. 2017). A correct solvent, or a mixture of solvents, is chosen to correspond to the physicochemical properties of a given biomass sample and the target bioactive substances. The choice of solvent depends mainly on the solubility of the target substance to be extracted and on the ease of its removal from the extract. Besides the analyte's solubility in the extracting solution, the yield of the process depends on the character of the primary matrix, its comminution, and the type of bonding made with the analyte. In order to obtain algae extract of the greatest possible content of biologically active components, the method of extraction must correspond to the properties of these active components. For instance, SFE is effective for isolation of fatty acids (Messyasz et al. 2015a; Michalak et al. 2017) and carotenoids (Rastogi et al. 2010; Fabrowska et al. 2016) from freshwater algae. For isolation of sulfated polysaccharides, the most effective is hot water extraction in a Soxhlet apparatus in combination with enzymatic hydrolysis (Pankiewicz et al. 2016). The design of new methods for isolation of bioactive substances from algae biomass is complex and time-consuming. Optimization of extraction conditions is often achieved through the trial and error method, or it could be supported with the help of computer programs (e.g., Statgraphics Centurion XVI®; StatPoint Technologies, Inc., Warrenton, VA, USA) (Fabrowska et al. 2016). The programs help design the experiment and calculate the optimum conditions, which significantly shortens the process. The yield of extraction is influenced by the solvent's properties, such as the diffusion coefficient, solvent viscosity, rate of mass transfer of the extracted substance, and process conditions: time, temperature, and pressure (Herrero et al. 2010; Onofrejová et al. 2010; Sánchez-Camargo et al. 2014; Schroeder et al. 2015).

8.6 Bioactive Compounds in Algae

8.6.1 Fatty Acids

Fatty acids make up a group of aliphatic, lipophilic monocarboxylic acids, divided into those that are saturated, in which carbon atoms make only single bonds, and those that are unsaturated, in which at least one double bond is present. The group of polyunsaturated fatty acids comprises a group of essential fatty acids (EFA) that are necessary for the correct functioning of the human organism. The essential fatty acids are divided into omega-9, omega-6, and omega-3, depending on the number of carbon atoms that make up the first double bond, counted from the terminal methyl group (Fang et al. 2004; Molski 2010; Bedoux et al. 2014; Messyasz et al. 2015a).

Taking into account the biological activity of fatty acids, those most important for human organisms are the essential fatty acids. They occur in the intercellular cement of the *epidermis corneum*, in the skin lipids and subcutaneous tissue. They are substrates for the synthesis of lipid components of cell membranes, they regulate production of sebum and epidermis keratinization, and they protect against over-drying of the skin and exfoliation of the epidermis. In cosmetic products, they mainly act as emollients, that is, compounds that soften and smooth the skin and form an occlusive film on the skin's surface, protecting against excessive loss of water and reducing the transepidermal water loss (TWL) coefficient. The lipid occlusive film on the surface of the skin also protects against pathogens, which is related to the anti-inflammatory, anti-allergic, and soothing effect of EFA (Fang et al. 2004; Bedoux et al. 2014).

Fatty acids are primary metabolites and are stored in algae cells. Fatty acids are the most numerous group of lipids to occur in algae. The most widespread fatty acids in nature are unsaturated acids contained in a molecule of 16 or 18 carbon atoms. The freshwater alga *C. glomerata* was found to be a rich source of fatty acids, and its extract can contain up to 62.5% of them (Pankiewicz et al. 2016). The most abundant (17.4%) was palmitic acid (C16:0), used in cosmetic products as an emulsifier (Messyasz et al. 2015a). The extracts from *C. glomerata* also contained significant amounts of unsaturated fatty acids from the group of omega-3, omega-6, omega-7, and omega-9, showing a wider range of biological activity. The dominant fatty acid (9.3%) was oleic acid (C18:1 [n-9]), which reduces transepidermal water loss, increases skin moisturization, regenerates the epidermis lipid barrier, and alleviates inflammations (Fabrowska et al. 2015a; Messyasz et al. 2015a). Another unsaturated fatty acid found in *C. glomerata* is linoleic acid (C18:2 [n-6]), which occurs in the intercellular cement of the skin and shows anti-inflammatory properties (Fabrowska et al. 2015a; Messyasz et al. 2015a), while α-linoleic acid (C18:3 [n-3]), also present in *C. glomerata*, regulates the functioning of the circulatory system, alleviates hearth arrhythmia, and increases elasticity of blood vessels (Bedoux et al. 2014). An interesting fatty acid present in algae is palmitic-oleic acid (C16:1 [n-7]), which shows strong antioxidant properties and is well absorbed and tolerated by the skin (El-Baky et al. 2009). The other unsaturated fatty acids found in *C. glomerata* are stearic acid (C18:4 [n-3]), arachidonic acid (C20:4 [n-6]), and eicosapentaenoic acid (C20:5 [n-3]). They show skin-regenerating properties and soothing, anti-inflammatory, and anti-allergic properties; they enhance the immunological barrier of the epidermis and delay skin aging (Fang et al. 2004; Molski 2010; Bedoux et al. 2014; Fabrowska et al. 2015a). Thanks to the high content of fatty acids, the freshwater macroalgae *C. glomerata* can be used in cosmetology as components of moisturizing, regenerating, protective, or antiaging creams.

8.6.2 Carotenoids

Carotenoids are unsaturated hydrocarbons belonging to the tetraterpene group (40-carbon atom terpenoids). They are built of isoprene units containing five carbon atoms and are classified as lipids. Because of their structure, they are nonpolar and are well-soluble in lipids. Their common feature is the presence of coupled double bond systems, which determine the carotenoids' ability to absorb light in a visible range and their characteristic red-orange color (Widomska et al. 2009; Rivera and Canela-Garayoa 2012). Carotenoids show strong antioxidant properties, which means that they are able to fight free radicals and delay skin aging processes. That is why carotenoids are commonly used in the production of antiaging cosmetics and diet supplements. They are able to stimulate fibroblasts so as to synthesize collagen and elastin, which contribute to the removal of wrinkles and enhance the elasticity of skin. Moreover, they are active in the processes of epidermis keratinization, prevent excessive drying of the skin, and protect the epithelium tissue against damage. Carotenoids are also involved in stabilization of the work of sebaceous glands through regulation of lipid management in the skin, enhance the protective properties of the epidermis, and reduce the TWL coefficient. As they absorb light, they are also used in sunscreen lotions and creams. Thanks to their photoprotective activity, they enhance skin resistance to sunlight and protect the skin against cancer. Finally, they are used as natural dyes in lipsticks, face powders, nail polish, and eyeshadows (Vilchez et al. 2011; Igielska-Kalwat et al. 2012; Bedoux et al. 2014).

Green algae contain lower amounts of carotenoids than brown or red algae. In green algae, the dominant dyes are chlorophylls a and b, while among carotenoids, the dominant dyes are *β*-carotene, lutein, and zeaxanthin. Brown algae show high contents of fucoxanthin and β-carotene, while red algae contain lutein, zeaxanthin, and β-carotene (Takaichi 2011).

The content of carotenoids in freshwater green algae is poorly known, and the studies in this field are still in the initial stage. Relationships between chlorophyll a and carotenoids were examined in extracts of freshwater algal populations in 1967 (Moss 1967), but in 2016, the first results on the analysis of carotenoids in three species of freshwater green macroalgae were published (Fabrowska et al. 2016). The species *Chara fragilis* Desvaux (syn. *Chara globularis* Thuiller) is a rich source of carotenoids and contains fucoxanthin, micromonol, astacin, lutein, citranaxanthin, and mactraxanthin. The green algae species *C. glomerata* contains fucoxanthin as the dominant carotenoid along with astacin. In the species *U. flexuosa*, the presence of deinoxanthin and siphonoxanthin was detected. All of these carotenoids have been detected for the first time in freshwater algae species. Particularly interesting was the identification of fucoxanthin in *C. fragilis* and *C. glomerata*, in which this carotenoid was dominant (Fabrowska et al. 2016). The presence of fucoxanthin in freshwater green algae is rather surprising, as this carotenoid is more typical of marine brown algae, in particular, of the species *Fucus vesiculosus*, *Fucus serratus*, *Laminaria digitata*, *Ascophyllum nodosum* (L.) Le Jolis (Le Tutour et al. 1998), and *Sargassum muticum* (Yendo) Fensholt (Milledge et al. 2016). Only a few authors have reported the presence of fucoxanthin in freshwater macroalgae (Schagerl and Pichler 2000). Fucoxanthin is a valuable bioactive component, as it is a very strong antioxidant capable of free radical neutralization (Raposo et al. 2015), as well as showing both anti-inflammatory (Heo et al. 2010) and anticancer activities (Pádua et al. 2016). Fucoxanthin inhibits the processes of melanogenesis, so it can be used as an active ingredient in cosmetic products that deal with skin discolorations (Shimoda et al. 2010). Another important carotenoid detected in *C. fragilis* is lutein, which shows antioxidative (Raposo et al. 2015), anti-inflammatory and photoprotective (Bian et al. 2012), and anticancer activity (Maoka et al. 2012). It is also used in prophylactics and treatments for eye conditions such as macular degeneration, cataracts, and pigmentary retinal degeneration (Ma and Lin 2010). The other carotenoids have been less thoroughly studied, but in general, carotenoids show similar bioactivity and are good antioxidants. Because of the content of carotenoids, extracts from freshwater algae can be used in the future as components of drugs, diet supplements, and antiaging and sunscreen cosmetics. Particularly interesting is the freshwater species *C. fragilis*, which is the richest source of carotenoids from among the green algae studied, containing fucoxanthin and lutein of well-known desired activities.

8.6.3 Sulfated Polysaccharides

Sulfated polysaccharides (SPs) are made up of a large number (10–30,000) of monosaccharides linked though glycoside bonds and that additionally contain sulfate groups. These compounds belong to the group of acidic polysaccharides, as they contain acid residues such as sulfate groups and uronic acids (Ngo and Kim 2013; Martini 2014; Mišurcová et al. 2014). Different types of SP are characteristic of brown algae (Phaeophyta), red algae (Rhodophyta), and green algae (Chlorophyta), i.e., fucoidans are characteristic of brown algae, galactans of red algae, and ulvans of green algae. Ulvans are built mainly of glucose, rhamnose, xylose, glucuronic acid, and iduronic acid, besides which they can also contain mannose, arabinose, and galactose (Robic and Lahaye 2007). The composition of monosaccharides and their connections in ulvans vary depending on many factors, such as the species of alga, their environment, the season of the year (Lahaye et al. 1999; Robic et al. 2009), and the method of extraction (Siddhanta et al. 2001). Their characteristic features are repeated mers of 3-sulfate of rhamnose connected via a (1,4)-α-glycoside bond with glucuronic or iduronic acid (Lahaye and Axelos 1993). Similarly as with other polysaccharides, ulvans are semicrystalline substances of white color and hygroscopic and soluble in hot water and alkali solutions; after cooling, they swell in water, making gels (Alves et al. 2010, 2013). The biological activity of SPs, including ulvans, is diverse. They have been proven to show antioxidant (Qi and Sun 2015), antiproliferative, and apoptotic properties (Ahmed and Ahmed 2014), as well as immunostimulating (Leiro et al. 2007), anticoagulating (El-Baky et al. 2009), antihyperlipidemic (Yu et al. 2003), antibacterial (Gadenne et al. 2013), and antiviral activities (Aguilar-Briseño et al. 2015). Like other SPs, ulvans can act as humectants in cosmetic products, that is, the substances that moisturize the skin through absorption of water, prevent excessive skin drying, and protect the skin against toxic chemicals and harmful microorganisms (Martini 2014). Because of their antioxidative activity (Qi and Sun 2015), their help in regeneration of skin tissues (Chandika et al. 2015), and the induction of collagen synthesis related to a high content of rhamnose (Andrès et al. 2006), ulvans can be used in antiaging cosmetics. Taking into account their antioxidative activity (Qi and Sun 2015), antibacterial activity (Gadenne et al. 2013) and their ability toward gelation (Alves et al. 2010, 2013), ulvans can also act as stabilizers, preservatives, and densifiers of cosmetic products.

SPs are the primary metabolites of algae and are mainly used as building and storage substances. They are found in large amounts in algae cell walls, so they are often called acidic algae polysaccharides (Mišurcová et al. 2014). The most numerous group of biologically active compounds found in algae are carbohydrates, as they make up about 60% of all biologically active substances produced by algae (Martini 2014). Only about 6% of reports on polysaccharides in algae concern freshwater algae, while 94% concern marine algae. So far, ulvans have only been isolated from marine algae *Ulva* species: *U. flexuosa*; *U. lactuca* (Castelar et al. 2014); *U. armoricana* Dion, de Reviers, and Coat; *U. rotundata* Bliding; *U. scandinavica* Bliding; *U. olivascens* Dangeard; *U. gigantea* (Kütz.) Bliding (Lahaye et al. 1999); *U. rigida* Agardh (Lahaye and Ray 1996); *U. clathrata* (Roth) Agardh (Hernández-Garibay et al. 2010); and *U. pertusa* Kjellman (Pengzhan et al. 2003).

Taking into account the properties and activity of ulvans (Pankiewicz et al. 2016), their presence in freshwater algae means that the freshwater species *C. glomerata* and *U. flexuosa* could act as a new natural raw product with a wide range of potential applications. Ulvans isolated from these species can be used in medicine, pharmaceuticals, and cosmetology and serve as active ingredients with multidirectional activities (Mišurcová et al. 2014).

8.6.4 Polyphenols

Polyphenols are a complex group of compounds whose characteristic feature is the presence of an aromatic ring containing two or more hydroxyl groups. According to the number of aromatic rings in the molecule, groups of simple polyphenols containing a single ring (e.g., gallic acid and pyrocatechol) and complex polyphenols containing two or more aromatic rings (e.g., usnic acid and hypericin) are distinguished. Polyphenols include polyphenolic acids, flavonoids, isoflavonoids, anthocyanins, catechins, coumarins, and stilbenes (Martini 2014). They are usually colorless, but there are also some colorful compounds, e.g., flavonoids are usually yellow, while anthocyanins are blue violet. The presence of many hydroxyl groups in the molecules of these compounds determines their antioxidative properties (Bedoux et al. 2014).

Phenolic compounds show antioxidative and antiradical activity based on two mechanisms. The first involves inactivation of the reactive oxygen species, while the second prevents the generation of free radicals. Of particular importance for the cosmetics and pharmaceutical industries are flavonoids. They inhibit the activity of enzymes taking part in free radical generation, chelate transition metal ions, and protect vitamin C against oxidation. Apart from that, flavonoids show a protecting effect on blood vessels, preventing their brittleness and breaking, and so are used in preparations for care of couperose skin and treatment of varicose veins and atherosclerosis. Polyphenols also show anti-inflammatory, anti-allergic, and anti-swelling activity. Isoflavonoids, which are derivatives of flavonoids, alleviate skin irritation and inflammation, protect against UV radiation, and reduce sunburns. They can be used in antiaging preparations, as they show antioxidative activity and stimulate synthesis of collagen, elastin, and hyaluronic acid in the skin, which improves skin elasticity, density, and moisturization. Phenolic acids are known to have an anti-inflammatory effect and antibacterial, antifungal, and antiradical activity (Bedoux et al. 2014). Their range of biological activity makes polyphenols desirable ingredients in cosmetics for mature skin, sensitive skin, and couperose skin. Algae are rich in polyphenols, which are produced as secondary metabolites in these plants. They are synthesized in response to stress and protect against biotic factors, e.g., attack of pathogens, and abiotic factors, e.g., UV radiation and the presence of heavy metals (Molski 2010; Stengel et al. 2011). Brown algae and certain green algae from the class of Zygnematophyceae contain large amounts of complex polyphenols, e.g., fucoles, phlorethols, and fucophlorethols. These compounds belong to the group of phlorotannins and show bacteriocidic and fungicidal properties; they are antioxidants and are able to chelate metal ions (Czerpak and Jabłońska-Trypuć 2008). Particularly rich in phlorotannins and other polyphenols are the marine species of brown algae *Ascophyllum nodosum*, *Fucus vesiculosus* (Pavia and Toth 2000), and *Eisenia bicyclis* (Kjellman) Setchell (Machu et al. 2015). Red algae (e.g., *Palmaria palmata* (L.) Weber and Mohr, *Porphyra tenera* Kjellman), green macroalgae (*Ulva clathrata*, *Ulva rigida*), and green microalgae (*Chlorella pyrenoidosa* Chick, *Spirulina platensis*) also contain polyphenols, but in far lower amounts than brown algae (Peña-Rodríguez et al. 2011; Yildiz et al. 2012; Machu et al. 2015; Longo and Hay 2017). Data in the literature on polyphenols in freshwater algae are very scarce. For instance, in freshwater species *C. glomerata*, the presence of coumaric acid, benzoic acid (Kartal et al. 2009), and metoxyphenyl and phenyl ester (Amornlerdpison et al. 2011) has been established, while *Chara hispida* L. is known to contain coumaric acid (Kartal et al. 2009). Alcohol extracts of *C. fragilis*, *C. glomerata*, and *U. flexuosa* have been studied and checked for the presence of phenolic compounds, applying the Folin-Ciocalteu method. The highest total content of phenolic compounds from among the green algae studied was found in *C. fragilis*, and this species has the highest potential or applications as a natural source of polyphenols (Fabrowska et al. 2016).

8.6.5 Macroelements and Microelements

Ions of macro- and microelements are vital for correcting the functioning of the human organism. Macroelements are defined as those elements needed by the human organism in large amounts, the daily demand for macroelements reaching over 100 mg. This group of elements includes carbon, oxygen, calcium, phosphorus, sodium, potassium, chlorine, sulfur, and silicon. Microelements are those elements needed by humans in trace amounts, that is, in amounts less than 100 mg per day, and this group includes boron, zinc, fluorine, iodine, cobalt, magnesium, copper, molybdenum, selenium, vanadium, and iron (Molski 2010; Bedoux et al. 2014). Each of these elements has a specific role in the organism, and many of them are important for skin condition, so both macro- and microelements are desirable ingredients for cosmetics and diet supplements.

Algae contain many macro- and microelements, and their ions occur in these plants in a well-absorbable form. The important aspect is the biosorption process. Novel dietary feed supplements can be produced through biosorption, in which micronutrients are bonded with biological material. The process is controlled by the equilibrium between functional groups and micronutrient ions, yielding the products with controlled release properties (Michalak et al. 2015). Algae are the source of such macroelements as sodium, potassium, calcium, and chlorine and contain considerable amounts of microelements, such as zinc, copper, iodine, iron, and manganese (Martini 2014; Fabrowska et al. 2015a). The presence of these elements in the algae thalli is related to the fact that algae cell walls can be easily penetrated by compounds of low molecular mass, such as water, gases, or ions (Wang and Chen 2009). That is why the marine algae show a higher content of iodine and potassium in the biomass than the freshwater algae. Results of elemental analysis of the freshwater species *C. glomerata* have proven that this algae species is a rich source of calcium, potassium, and magnesium and, in smaller amounts, also contains sodium, iron, copper, and zinc. The particularly high content of calcium in *C. glomerata* can be related to the presence of diatoms living on the thallus of this alga. The contents of macro- and microelements in *C. glomerata* indicate the potential of application of this alga in the cosmetic industry (Messyasz et al. 2015a).

8.7 Application of Algae in the Cosmetics Industry

One of the primary aims of contemporary cosmetology is to maintain the youthful look of the skin for as long as possible. To achieve this, cosmetology proposes antiaging cosmetics that not only reduce the already present wrinkles and discolorations, but should prevent the aging of the skin that may be caused by different factors in the first place. A very important cause of aging processes is the activity of free radicals, e.g., the reactive oxygen species (ROS). The reaction of free radical generation is initiated by light, with the conducive effect of heavy metals that act as catalysts in free radical transformations. As a result of a chain reaction, the chemically reactive radicals are formed. Their characteristic feature is the presence of unpaired electrons whose reactivity may damage the cell components: proteins, lipids, DNA, and others. This damage contributes to the aging of the skin and to the development of cancer changes. To prevent these changes, cosmetology offers antioxidants or radical scavengers that are able to deactivate the free radicals (Tonnesen and Karlsen 2002; Molski 2010; Pielesz 2010; Stengel et al. 2011). Particular care is taken to offer antioxidants of natural origin, consistent with the recent tendency to search for highly effective naturally active ingredients, mainly originating in plants.

Both marine and freshwater algae are known to contain antioxidants, so their antioxidant properties have been studied by many authors (Samarkoon and Jeon 2012). The research group has studied and confirmed the antioxidant potential of red algae *Mastocarpus stellatus* (Stackhouse) Guiry by the FRAP (ferric reducing ability of plasma) method (Gómez-Ordóñez et al. 2012). Another research group has reported on the antioxidant properties of different macro- and microalgae species, studied, e.g., by the DPPH (2,2-diphenyl-1-picrylhydrazyl) method of radical reduction (Ngo et al. 2011). Algae have been shown to contain spontaneous antioxidants that directly deactivate free radicals, as well as the complex antioxidants that chelate the heavy metal cations, acting as catalysts of oxidation and converting them into non-dissociating complexes. The spontaneous antioxidants that occur in algae are vitamins C and E, polypeptides, phenolic compounds, β-carotene and other carotenes, astaxanthin, and other xanthophyllic dyes (Molski 2010; Bedoux et al. 2014). The chelating antioxidants present in algae are alginic acid and alginates, ulvans, polyphenols, fucoles, phlorethols, fucophlorethols, and certain peptides, e.g., the carnosine (β-alanyl-L-histidine) present in red algae *Acanthophora delilei* Greville (Fleurence 2004; Fabrowska and Łęska 2012). The antioxidants present in algae not only protect the skin against free radicals but also the components of cosmetic products sensitive to oxidation, e.g., vitamin E protects lipids against peroxidation (Romera et al. 2007; Molski 2010).

Algae have become one of the most attractive and widely used sources of natural bioactive substances used in contemporary cosmetic production. They are usually used in micronized form, as a powder or extract (Fabrowska et al. 2015a). As they contain a wide range of biologically active compounds, amino acids, proteins, carbohydrates, fatty acids, vitamins, polyphenols, and macro- and microelements, depending on the species and environment they grow in, algae are used in different cosmetic preparations. They are the source of many valuable active ingredients and demonstrate a number of the activities mentioned below (Stolz and Obermayer 2005; Pielesz 2010; Bedoux et al. 2014; Martini 2014):

- Provide nourishing substances for the skin and protect the skin against destructive environmental factors
- Purify the skin from toxins, alleviate inflammations, and show bacteriostatic activity
- Enhance osmosis in intercellular space and cellular metabolism
- Enhance the synthesis of collagen and elastin, improve skin elasticity and firmness, restore the correct pH
- Purify the tissues through removal of harmful products from metabolism and accelerate decomposition of any excess of lipids
- Regenerate the skin, delay the aging processes, protect against free radicals
- Protect the skin against UV radiation and photoaging
- Regulate the work of sebaceous glands and absorb sebum
- Alleviate skin irritations, help heal wounds
- Contribute to the removal of cellulite, stretch marks, psoriasis, and acne
- Protect against loss of moisture and moisturize the skin

Moreover, alginates contained in algae increase the stability of cosmetic products, regulate their consistency, act as densifying agents, facilitate spreading, and protect O/W emulsions against drying. Algal dyes, chlorophyll (green), fucoxanthin (brown), astaxanthin (pink), zeaxanthin (yellow), phycoerythrin (red), and phycocyanin (blue), are used for production of color cosmetics (lipsticks, eyeshadows, nail polish) (Molski 2010; Pielesz 2010; Bedoux et al. 2014; Martini 2014). Extracts from many algae show bacteriostatic and skin-healing properties (Rao et al. 2010; Cox et al. 2014; Akremi et al. 2017).

Algae can be used in all types of cosmetic products. Because of a wide range of activity of their active components, they are used in skin creams for dry, damaged, mature skin, mixed skin, and oily skin with a tendency toward acne. Algae extracts are used in cosmetic balms, tonics, milks, face masks, sunscreen preparations, skin-firming preparations, anti-cellulite products, shampoos, hair conditioners, soap, shower gels, peeling creams, color cosmetics, and other products (Molski 2010; Pielesz 2010; Martini 2014).

The extracts from freshwater algae also contain many antioxidants and moisturizing substances, so that they are expected to be used as raw cosmetic products in the future. The antioxidative properties of algae extracts studied in vitro are described in Fabrowska et al. (2016), while the in vivo studies of cosmetic O/W emulsions containing extract from *C. glomerata* are reported in Fabrowska et al. (2017).

According to the results of Fabrowska et al. (2016), the algae species *C. fragilis* shows the greatest antioxidative potential, because it has the richest content of bioactive substances, including carotenoids and polyphenols. The extracts from *U. flexuosa* showed the greatest antioxidative potential from among freshwater green macroalgae.

There is evidence that after a 4-week application of a cream with the extract from *C. glomerata*, skin elasticity and moisturization significantly increased (Fabrowska et al. 2017). The extracts from the algae species described above can be used in cosmetic products of different types, because they contain a wide range of bioactive substances of different activities: moisturizing, nourishing, antiaging, or antiwrinkle. The moisturizing activity of algae extracts is related to the presence of humectants (sulfated polysaccharides) and emollients (fatty acids). The nourishing properties of the green algae studied are related to the presence of fatty acids, while their antiaging activity relates to the presence of antioxidants (carotenoids and polyphenols). Extracts from freshwater algae can be used in sunscreen body lotions, because of the relatively high content of carotenoids and phenolic compounds protecting against UV radiation. They can also be used in face masks, shampoos, and hair conditioners. Thanks to the content of a variety of minerals, e.g., calcium, magnesium, and zinc, along with ulvans and fatty acids, the extracts from *C. glomerata* can be active ingredients in face masks of moisturizing and nourishing activity. The content of foam-producing sulfated polysaccharides in this algae species means that its extracts can be used in shampoos. Moreover, the fatty acids and polysaccharides occurring in *C. glomerata* are commonly used as moisturizing agents and ingredients in hair conditioners and face masks (Molski 2010; Bedoux et al. 2014).

A number of bioactive compounds detected in freshwater green algae of antioxidative and moisturizing activity and capable of improving skin elasticity open up perspectives for the use of their extracts as an effective cosmetic raw product. An additional argument for using this group of algae in cosmetic products is their high availability – mass blooming in water bodies taking place every year – and thus the potentially low cost of harvesting the material.

All of the above arguments confirm that not only marine but also freshwater algae have high potential as natural raw cosmetic products.

References

Aguilar-Briseño JA, Cruz-Suarez LE, Sassi JF, Ricque-Marie D, Zapata-Benavides P, Mendoza-Gamboa E, Rodríguez-Padilla C, Trejo-Avila LM (2015) Sulphated polysaccharides from *Ulva clathrata* and *Cladosiphon okamuranus* seaweeds both inhibit viral attachment/entry and cell-cell fusion, in NDV infection. Mar Drugs 13:697–712

Ahmed OM, Ahmed RR (2014) Anti-proliferative and apoptotic efficacies of ulvan polysaccharides against different types of carcinoma cells in vitro and in vivo. J Cancer Ther 6:2002–2008

Akremi N, Cappoen D, Anthonissen R, Verschaeve L, Bouraoui A (2017) Phytochemical and in vitro antimicrobial and genotoxic activity in the brown algae *Dictyopteris membranacea*. S Afr J Bot 108:308–314

Allen VG, Pond KR, Saker KE, Fontenot JP, Bagley CP, Ivy RL, Evans RR, Schmidt RE, Fike JH, Zhang X, Ayad JY, Brown CP, Miller MF, Montgomery JL, Mahan J, Wester DB, Melton C (2001) Tasco: influence of a brown seaweed on antioxidants in forages and livestock – a review. J Anim Sci 79(Suppl. E):E21–E31

Alves A, Caridade SG, Mano JF, Sousa RA, Reis R (2010) Extraction and physico-chemical characterization of a versatile biodegradable polysaccharide obtained from green algae. Carbohydr Res 345:2194–2200

Alves A, Sousa RA, Reis RL (2013) A practical perspective on ulvan extracted from green algae. J Appl Phycol 25:407–424

Amornlerdpison D, Mengumphan K, Thumvijit S, Peerapornpisal Y (2011) Antioxidant and anti-inflammatory activities of freshwater macroalga, *Cladophora glomerata* Kützing. Thai J Agric Sci 44:283–291

Andrès E, Molinari J, Péterszegi G, Mariko B, Ruszova E, Velebny V, Faury G, Robert L (2006) Pharmacological properties of rhamnose-rich polysaccharides, potential interest in age dependent alterations of connectives tissues. Pathol Biol 54:420–425

Barbosa-Pereira L, Pocheville A, Angulo I, Paseiro-Losada P, Cruz JM (2013) Fractionation and purification of bioactive compounds obtained from a brewery waste stream. Biomed Res Int 2013:408491

Bedoux G, Hardouin K, Burlot A, Bourgougnon N (2014) Bioactive components from seaweeds: cosmetic applications and future development. Adv Bot Res 71:345–378

Bian Q, Gao S, Zhou J, Qin J, Taylor A, Johnson EJ, Tanq G, Sparrow JR, Gierhart D, Shanq F (2012) Lutein and zeaxanthin supplementation reduces photooxidative damage and modulates the expression of inflammation-related genes in retinal pigment epithelial cells. Free Radic Biol Med 53:1298–1307

Bliding C (1963) A critical survey of European taxa in Ulvales. Part 1. Casosiphonia, Blidingia, Enteromorpha. Opera Bot 8:1–160

Bliding C (1968) A critical survey of European taxa in Ulvales. Part 2. Ulva, Ulvaria, Monostoma, Kornmannia. Bot Notiser 121:534–629

Blomster J, Hoey EM, Maggs CA, Stanhope MJ (2000) Species-specific oligonucleotide probes for macroalgae: molecular discrimination of two marine fouling species of *Enteromorpha* (*Ulvophyceae*). Mol Ecol 2:177

Borowitzka MA (2013) High-value products from microalgae – their development and commercialisation. J Appl Phycol 25:743–756

Borowitzka MA (2013a) Energy from microalgae: a short history. In: Borowitzka MA, Moheimani NR (eds) Algae for biofuels and energy. Springer, Dordrecht, pp 1–15

Castelar B, Reis PR, Dos Santos Calheiros AC (2014) *Ulva lactuca* and *U. flexuosa* (*Chlorophyta, Ulvophyceae*) cultivation in Brazilian tropical waters: recruitment, growth, and ulvan yield. J Appl Phycol 26:1989–1919

Cefali LC, Ataide JA, Moriel P, Foglio MA, Mazzola PG (2016) Plant-based active photoprotectants for sunscreens. Int J Cosmet Sci 38:346–353

Chandika P, Ko SC, Jung WK (2015) Marine-derived biological macromolecule-based biomaterials for wound healing and skin tissue regeneration. Int J Biol Macromol 77:24–35

Cox S, Hamilton Turley G, Rajauria G, Abu-Ghannam N, Jaiswal AK (2014) Antioxidant potential and antimicrobial efficacy of seaweed (*Himanthalia elongata*) extract in model food systems. J Appl Phycol 26:1823–1831

Czerpak R, Jabłońska-Trypuć A (2008) Roślinne surowce kosmetyczne [Vegetable cosmetic raw materials] MedPharm

D'Orazio N, Gemello E, Gammone MA, de Girolamo M, Ficoneri C, Riccioni G (2012) Fucoxantin: a treasure from the sea. Mar Drugs 10:604–616

De Lima Yamaguchi KK, Ravazi Pereira LF, Lamarão CV, Silva Lima E, da Veiga-Junior VF (2015) Amazon acai: chemistry and biological activities: a review. Food Chem 179:137–151

Debacq-Chainiaux F, Borlon C, De Hertogh B, Remacle J, Morvan PY, Vallée R, Toussain O (2006) Identification of potential anti-photoageing algal compounds using an in-vitro model of photoageing. J Pharm Pharmacol 58:1577–1583

El-Baky HHA, Baz FKE, Baroty GSE (2009) Potential biological properties of sulphated polysaccharides extracted from the macroalgae *Ulva lactuca*. Am J Cancer Res 2:1–11

El-Baz FK, El-Senousy WM, El-Sayed AB, Kamel MM (2013) In vitro antiviral and antimicrobial activities of *Spirulina platensis* extract. J Appl Pharm Sci 3:52–56

Esquivel-Hernández DA, Ibarra-Garza IP, Rodríguez-Rodríguez J, Rostro-Alanis M, Alemán-Nava GS, García-Pérez JS, Parra-Saldívar R (2017) Green extraction technologies for high-value metabolites from algae: a review. Biofuels Bioprod Biorefin 11:215–231

Fabrowska J, Łęska B (2012) Algae and their chelating properties. In: Rybachenko VI (ed) From molecules to functional architecture, supramolecular interactions. East Publisher House, Donetsk, pp 495–511

Fabrowska J, Łęska B, Schroeder G (2015a) Freshwater *Cladophora glomerata* as a new potential cosmetic raw material. Chemik 69:491–497

Fabrowska J, Łęska B, Schroeder G, Messyasz B, Pikosz M (2015b) Biomass and extracts of algae as material for cosmetics. In: Kim S-K, Chojnacka K (ed) Marine algae extracts: processes, products, and applications. Wiley-VCH Verlag GmbH & Co. KGaA, Weinheim, pp 681–706

Fabrowska J, Ibañez E, Łęska B, Herrero M (2016) Supercritical fluid extraction as a tool to valorize underexploited freshwater green algae. Algal Res 19:237–245

Fabrowska J, Kapuścińska A, Łęska B, Feliksik-Skrobich K, Nowak I (2017) In vivo studies and stability study of *Cladophora glomerata* extract as a cosmetic active ingredient. Acta Pol Pharm 74:633–641

Fang JY, Chiu HC, Wu JT, Chiang YR, Hsu SH (2004) Fatty acids in *Botryococcus braunii* accelerate topical delivery of flurbiprofen into and across skin. Int J Pharm 276:163–173

Fleurence J (2004) Seaweed proteins. In: Yada RY (ed) Proteins in food processing. Woodhead Publishing Limited, Cambridge, pp 197–213

Gade R, Siva Tulasi M, Aruna Bhai V (2013) Seaweeds: a novel biomaterial. Int J Pharm Pharm Sci 5:40–44

Gadenne V, Lebrun L, Jouenne T, Thebault P (2013) Antiadhesive activity of ulvan polysaccharides covalently immobilized onto titanium surface. Colloids Surf B: Biointerfaces 112:229–236

Gellenbeck K (2011) Utilization of algal materials for nutraceutical and cosmeceutical applications – what do manufacturers need to know? J Appl Phycol 24(3):309–313

Gómez-Ordóñez E, Jiménez-Escrig A, Rupérez P (2012) Effect of the red seaweed Mastocarpus stellatus intake on lipid metabolism and antioxidant status in healthy Wistar rats. Food Chem 135:806–811

Guiry MD, Guiry GM, (2015) AlgaeBase. World-wide electronic publication. National University of Ireland, Galway, http://www.algaebase.org Accessed on 11.05.2017

Hardouin K, Bedoux G, Burlot AS, Nyvall-Collén P, Bourgougnon N (2014) Enzymatic recovery of metabolites from seaweeds: potential applications. Adv Bot Res 71:279–320

Heo SJ, Yoon WJ, Kim KN, Ahn GN, Kang SM, Kang DH, Affan A, Oh C, Jung WK, Jeon YJ (2010) Evaluation of anti-inflammatory effect of fucoxanthin isolated from brown algae in lipopolysaccharide-stimulated RAW 264.7 macrophages. Food Chem Toxicol 48:2045–2051

Hernández-Garibay E, Zertuche-González J, Pacheco-Ruíz I (2010) Isolation and chemical characterization of algal polysaccharides, from the green seaweed *Ulva clathrata* (Roth) C. Agardh. J Appl Phycol 23:537–542

Herrero M, Mendiola JA, Cifuentes A, Ibáñez E (2010) Supercritical fluid extraction: recent advances and applications. J Chromatogr A 1217:2495–2511

Igielska-Kalwat J, Wawrzyńczak A, Nowak I (2012) β-Carotene as an exemplary carotenoid and its application in cosmetic industry. Chemik 66:140–144

Ikawa M, Thomas VM, Buckley LJ, Uebel JJ (1973) Sulfur and the toxicity of the red alga *Ceramium rubrum* to *Bacillus subtilis*. J Phycol 9:302–304

Imhoff J, Labes A, Wiese J (2011) Bio-mining the microbial treasures of the ocean: New natural products. Biotechnol Adv 29:468–482

Kartal M, Orhan I, Abu-Asaker M, Senol FS, Atici T, Sener B (2009) Antioxidant and anticholinesterase assets and liquid chromatography-mass spectrometry preface of various fresh-water and marine macroalgae. Pharmacogn Mag 5:291–297

Komárek J, Kastovsky J, Mares J, Johansen JR (2014) Taxonomic classification of cyanoprokaryotes (cyanobacterial genera), using a polyphasic approach. Preslia 86:295–335

Lahaye M, Axelos MAV (1993) Gelling properties of water-soluble polysaccharides from proliferating marine green seaweeds (*Ulva* spp.). Carbohydr Polym 22:261–265

Lahaye M, Ray B (1996) Cell-wall polysaccharides from the marine green alga *Ulva "rigida" (Ulvales, Chlorophyta)* – NMR analysis of ulvan oligosaccharides. Carbohydr Res 283:161–173

Lahaye M, Alvarez-Cabal Cimadevilla E, Kuhlenkamp R, Quemener B, Lognone V, Dion P (1999) Chemical composition and ^{13}C NMR spectroscopic characterisation of ulvans from *Ulva (Ulvales, Chlorophyta)*. J Appl Phycol 11:1–7

Le Tutour B, Benslimane F, Gouleau MP, Gouygou JP, Saadan B, Quemeneur F (1998) Antioxidant and prooxidant activities of the brown algae, *Laminaria digitata, Himanthalia elongata, Fucus vesiculosus, Fucus serratus* and *Ascophyllum nodosum*. J Appl Phycol 10:121–129

Leiro M, Castro R, Arranz JA, Lamas J (2007) Immunomodulating activities of acidic sulphated polysaccharides obtained from the seaweed *Ulva rigida* C. Agardh. Int Immunopharmacol 7:879–888

Longo GO, Hay ME (2017) Seaweed allelopathy to corals: are active compounds on, or in, seaweeds? Coral Reefs 36:247–253

Ma L, Lin XM (2010) Effects of lutein and zeaxanthin on aspects of eye health. J Sci Food Agric 90:2–12

Machu L, Misurcova L, Ambrozova JV, Orsavova J, Mlcek J, Sochor J, Jurikova T (2015) Phenolic content and antioxidant capacity in algal food products. Molecules 20:1118–1133

Malinowska P (2011) Algae extracts as active cosmetic ingredients. In: Zeszyty Naukowe, Poznan University of Economic and Business, pp 123–129

Malinowska P, Gliszczynska-Świglo A, Szymusiak H (2014) Protective effect of commercial acerola, willow, and rose extracts against oxidation of cosmetic emulsions containing wheat germ oil. Eur J Lipid Sci Technol 116:1553–1562

Manoylov KM (2014) Taxonomic identification of algae (morphological and molecular) species concepts, methodologies, and their implications for ecological bioassessment. J Phycol 50:409–424

Maoka T, Tokuda H, Suzuki N, Kato H, Etoh H (2012) Anti-oxidative, anti-tumor-promoting, anti-carcinogenesis activities of nitroastaxanthin and nitrolutein, the reaction products of astaxanthin and lutein with peroxynitrite. Mar Drugs 10:1391–1399

Martini MC (2014) Kosmetologia i farmakologia skóry [Cosmetology and skin pharmacology] PZWL Warszawa

Messyasz B, Rybak A (2009) The distribution of green algae species from the *Ulva genera* (syn. Enteromorpha, Chlorophyta) in Polish inland waters. Oceanol Hydrobiol Stud 138:121–138

Messyasz B, Pikosz M, Schroeder G, Łęska B, Fabrowska J (2015a) Identification and ecology of macroalgae species existing in Poland. In: Kim S-K, Chojnacka K (ed) Marine algae extracts: processes, products, and applications. Wiley-VCH Verlag GmbH & Co. KGaA, Weinheim, pp 17–39

Messyasz B, Łęska B, Fabrowska J, Pikosz M, Rój E, Cieślak A, Schroeder G (2015b) Biomass of freshwater *Cladophora* as a raw material for agriculture and the cosmetic industry. Open Chem 13:1108–1118

Michalak I, Chojnacka K (2014) Algal extracts: technology and advances. Eng Life Sci 14:581–591

Michalak I, Chojnacka K (2015) Algae as production systems of bioactive compounds. Eng Life Sci 15:160–176

Michalak I, Witek-Krowiak A, Chojnacka K, Bhatnagar A (2015) Advances in biosorption of microelements – the starting point for the production of new agrochemicals. Rev Inorg Chem 35:115–133

Michalak I, Chojnacka K, Saeid A (2017) Plant growth biostimulants, dietary feed supplements and cosmetics formulated with supercritical CO_2 algal extracts. Molecules 22:66

Milledge JJ, Nielsen BV, Bailey D (2016) High-value products from macroalgae: the potential uses of the invasive brown seaweed, *Sargassum muticum*. Rev Environ Sci Biotechnol 15:67–88

Minchin D, Nunn J (2014) The invasive brown alga *Undaria pinnatifida* (Harvey) Suringar, 1873 (Laminariales: Alariaceae), spreads northwards in Europe. Bioinvasions Rec 3:57–63

Mišurcová L, Orsavová J, Ambrožová JV (2014) Algal polysaccharides and health. In: Ramawat KG, Mérillon JM (eds) Polysaccharides bioactivity and biotechnology. Springer International Publishing, Cham, pp 109–144

Molski M (2010) Chemia piękna [Chemistry of beauty] PWN, Warszawa

Moss B (1967) A Note on the estimation of chlorophyll A in freshwater algal communities. Limnol Oceanogr 340–342

Ngo DH, Kim SK (2013) Sulfated polysaccharides as bioactive agents from marine algae. Int J Biol Macromol 62:70–75

Ngo DH, Wijesekara I, Vo T, Ta QV, Kim SK (2011) Marine food-derived functional ingredients as potential antioxidants in the food industry: an overview. Food Res Int 44:523–529

O'Doherty JV, Dillon S, Figat S, Callan JJ, Sweeney T (2010) The effects of lactose inclusion and seaweed extract derived from *Laminaria* spp. on performance, digestibility of diet components and microbial populations in newly weaned pigs. Anim Feed Sci Technol 157:173–180

Onofrejová L, Vasícková J, Klejdus B, Stratil P, Misurcová L, Krácmar S, Kopecký J, Vacek J (2010) Bioactive phenols in algae: the application of pressurized-liquid and solid-phase extraction techniques. J Pharm Biomed Anal 51:464–470

Pádua D, Rocha E, Gargiulo D, Ramos AA (2016) Bioactive compounds from brown seaweeds: phloroglucinol, fucoxanthin and fucoidan as promising therapeutic agents against breast cancer. Phytochem Lett 14:91–98

Pankiewicz R, Łęska B, Messyasz B, Fabrowska J, Sołoducha M, Pikosz M (2016) First isolation of polysaccharidic ulvans from the cell walls of freshwater algae. Algal Res 19:348–354

Pavia H, Toth GB (2000) Influence of light and nitrogen on the phlorotannin content of the brown seaweeds *Ascophyllum nodosum* and *Fucus vesiculosus*. Hydrobiologia 440:299–305

Peña-Rodríguez A, Mawhinney TP, Ricque-Marie D, Cruz-Suárez LE (2011) Chemical composition of cultivated seaweed *Ulva clathrata* (Roth) C. Agardh. Food Chem 129:491–498

Pengzhan Y, Ning L, Xiguang L, Gefei Z, Quanbin Z, Pengcheng L (2003) Antihyperlipidemic effects of different molecular weight sulfated polysaccharides from *Ulva pertusa (Chlorophyta)*. Pharmacol Res 48:543–549

Pielesz A (2010) Algi i alginiany – leczenie, zdrowie i uroda [Algae and alginates – treatment, health and beauty], E-book

Qi H, Sun Y (2015) Antioxidant activity of high sulfate content derivative of ulvan in hyperlipidemic rats. Int J Biol Macromol 76:326–329

Rao AR, Reddy AH, Aradhya SM (2010) Antibacterial properties of *Spirulina platensis, Haematococcus pluvialis, Botryococcus braunii* micro algal extracts. Curr Trends Biotechnol Pharm 4:809–819

Raposo JMF, de Morais BAMM, de Morais SCRM (2015) Carotenoids from marine microalgae: a valuable natural source for the prevention of chronic diseases. Mar Drugs 13:5128–5155

Rastogi RP, Richa SRP, Singh SP, Häder DP (2010) Photoprotective compounds from marine organisms. J Ind Microbiol Biotechnol 37:537–558

Rivera SM, Canela-Garayoa R (2012) Analytical tools for the analysis of carotenoids in diverse materials. J Chromatogr A 1224:1–10

Robic A, Lahaye M (2007) Structure and functional properties of ulvan, a polysaccharide from green seaweeds. Biomacromolecules 6:1765–1774

Robic A, Sassi JF, Dion P, Lerat Y, Lahaye M (2009) Seasonal variability of physicochemical and rheological properties of ulvan in two *Ulva* species (*Chlorophyta*) from the Brittany Coast. J Phycol 45:962–973

Romera E, Gonzalez F, Ballester A, Blazquez ML, Munoz JA (2007) Comparative study of biosorption of heavy metals using different types of algae. Bioresour Technol 98:3344–3353

Samarkoon K, Jeon YJ (2012) Bio-functionalities of proteins derived from marine algae – a review. Food Res Int 48:948–960

Sánchez-Camargo AP, Mendiola JA, Ibáñez E, Herrero M (2014) Supercritical fluid extraction. In: Reedjik J (ed) Elsevier reference module in chemistry, Molecular Sciences and Chemical Engineering, MA. Elsevier, Waltham

Schagerl M, Pichler C (2000) Pigment composition of freshwater *charophyceae*. Aquat Bot 67:117–129

Schroeder G, Łęska B, Fabrowska J, Messyasz B, Pikosz M (2015) Analysis of green algae extract. In: Kim S-K, Chojnacka K (ed) Marine algae extracts: processes, products, and applications, Wiley-VCH Verlag GmbH & Co. KGaA, Weinheim, pp 81–99

Sebök S, Herppich WB, Hanelt D (2017) Development of an innovative ring-shaped cultivation system for a land-based cultivation of marine macroalgae. Aquac Eng 77:33–41

Serisawa Y, Yokohama Y, Aruga Y, Tanaka J (2004) Dark respiration of the stipe of *Ecklonia cava* (Laminariales, Phaeophyta) in relation to temperature. Phycol Res 52:174–179

Shah MMR, Liang Y, Cheng JJ, Daroch M (2016) Astaxanthin-producing green microalga Haematococcus pluvialis: from single cell to high value commercial products. Front Plant Sci 7:531–559

Shimoda H, Tanaka J, Shan SJ, Maoka T (2010) Anti-pigmentary activity of fucoxanthin and its influence on skin mRNA expression of melanogenic molecules. J Pharm Pharmacol 62:1137–1145

Siddhanta AK, Goswami AM, Ramavat BK, Mody KH, Mairh OP (2001) Water soluble polysaccharides of marine algal species of *Ulva (Ulvales, Chlorophyta)* of Indian waters. Indian J Mar Sci 30:166–172

Silberfeld T, Rousseau F, de Reviers B (2014) An updated classification of brown algae (Ochrophyta, Phaeophyceae). Cryptogam Algol 35:117–156

Soltani S, Saadatmand S, Khavarinejad R, Nejadsattari T (2011) Antioxidant and antibacterial activities of *Cladophora glomerata* (L.) Kütz. in Caspian Sea Coast. Iran Afr J Biotechnol 10:7684–7689

Starmach K (1972) Zielenice nitkowate [Green alga] (In: Flora słodkowodna Polski [Freshwater flora of Poland] Ed. K. Starmach) – PWN, Warszawa-Kraków

Stengel DB, Connan S, Popper ZA (2011) Algal chemodiversity and bioactivity: sources of natural variability and implications for commercial application. Biotechnol Adv 29:483–501

Stolz P, Obermayer B (2005) Manufacturing microalgae for skin care. Cosmet Toiletries 120:99–106

Sun Y, Chang R, Li Q, Li B (2016) Isolation and characterization of an antibacterial peptide from protein hydrolysates of Spirulina platensis. Eur Food Res Technol 242:685–692

Surget G, Roberto VP, Le Lann K, Mira S, Guerard F, Laize V, Poupart N, Cancela ML, Stiger-Pouvreau V (2017) Seasonal phenology and metabolomics of the introduced red macroalga *Gracilaria vermiculophylla*, monitored in the Bay of Brest (France). J Appl Phycol 29:575–584

Szopa A, Ekiert R, Ekiert H (2017) Current knowledge of *Schisandra chinensis* (Turcz.) Baill. (Chinese magnolia vine) as a medicinal plant species: a review on the bioactive components, pharmacological properties, analytical and biotechnological studies. Phytochem Rev 16:195–218

Takaichi S (2011) Carotenoids in algae: distributions, biosyntheses and functions. Mar Drugs 9:1101–1118

Tonnesen HH, Karlsen J (2002) Alginate in drug delivery system. Drug Dev Ind Pharm 28:621–630

Van den Hoek C, Mann DG, Jahns HM (1995) Algae. An introduction to phycology, 1st edn. Cambridge University Press, Cambridge

Verkleij FN (1992) Seaweed extracts in agriculture and horticulture – a review. Biol Agric Hortic 8:309–324

Vilchez C, Forjan E, Cuaresma M, Bedmar F, Garbayo I, Vega JM (2011) Marine carotenoids: biological functions and commercial applications. Mar Drugs 9:319–333

Wang J, Chen C (2009) Biosorbents for heavy metals removal and their future. Biotechnol Adv 27:195–226

Widomska J, Kostecka-Gugała A, Latowski D (2009) Calorimetric studies of the effect of cis-carotenoids on the thermotropic phase behavior of phosphatidylcholine bilayers. Biophys Chem 140:108–114

Yildiz G, Celikler S, Vatan O, Dere S (2012) Determination of the antioxidative capacity and bioactive compounds in green seaweed *Ulva rigida* C. Agardh. Int J Food Prop 15:1182–1189

Yu PZ, Zhang QB, Li N, Xu ZH, Wang YM, Li ZE (2003) Polysaccharides from *Ulva pertusa (Chlorophyta)* and preliminary studies on their antihyperlipidemia activity. J Appl Phycol 15:21–27

Zhou H, Ren J, Li Z (2017) Antibacterial activity and mechanism of pinoresinol from *Cinnamomum Camphora* leaves against food-related bacteria. Food Control 79:192–199

The Biomass of Algae and Algal Extracts in Agricultural Production

9

Bogusława Górka, Karolina Korzeniowska, Jacek Lipok, and Piotr P. Wieczorek

Abstract

Fertilizers and plant growth regulators (PGRs) based on seaweeds are commonly known, having been produced and utilized for dozens of years. Algae are rich in carbohydrates, proteins, minerals and chemical compounds that belong mainly to the following groups: polysaccharides, polyphenols, phlorotannins, plant pigments, unsaturated fatty acids, sterols and plant hormones.

Agricultural formulations containing algal extracts stimulate the growth and yielding of plants in a very efficient way because of their action at low concentrations. Concentrations of phytohormones depend mainly on the botanical origin of the obtained biomass, the time and place of its collection and the method of extraction of the active compounds. Plant growth regulators are obtained mainly from brown algae, due to the relatively high content of those substances and their consistent accessibility throughout the year. Algal extracts are often enriched with such substances as urea, humic acids, ammonium phosphate, potassium sulphate and additional doses of growth hormones. The composition of such products may influence not only the growth and development processes in plants but also indirect factors such as soil fertility or the presence of soil microorganisms. Bearing in mind the potential of algal extracts, it is worth paying attention to seaweeds other than brown algae, as well as to promising modern methods of extraction of those compounds of interest.

Keywords

Seaweed extracts · Biostimulants in agriculture · Biotic and abiotic stress in plants

B. Górka (✉) · K. Korzeniowska · J. Lipok · P. P. Wieczorek
Faculty of Chemistry, Department of Analytical and Ecological Chemistry, Opole University, Opole, Poland
e-mail: bgorka@uni.opole.pl

9.1 Seaweeds: Great Potential for Use in Agriculture

Seaweeds have been used in agriculture for thousands of years. Their history reaches back to ancient Roman times, when plant seedlings were wrapped with algae to create good conditions for their growth. In the coastal areas of Europe, farmers used the biomass of seaweed washed up onto the shore by the sea, mixing the algae with soil or making compost out of them. The agricultural usage of algae in the form of untreated biomass or compost has been tested in many field trials. However, since 1950, the use of algal biomass has been almost completely superseded by products containing extracts from various species of algae that have gained the approval of customers all over the world.

Plants, including algae, produce hundreds of biologically active compounds belonging to such chemical groups as polysaccharides, sterols, phenolic compounds and fatty acids. Commercially, not only is algal biomass used as biofertilizer, but so are the compounds isolated from algae in the form of pure chemical entities or extracts containing fractions of compounds, such as polysaccharides (laminarin, alginates and carrageenans) and their degradation products. Other plant growth-enhancing ingredients are micro- and macrominerals, betaines and, perhaps the most important, phytohormones (du Jardin 2015).

9.2 Active Ingredients in Seaweed Extracts

The concentration of active compounds depends on the species, the harvesting site, the season, environmental conditions and the extraction method used. These compounds not only play very important functions in plant cells and algae but also show interesting properties for other groups of organisms, including humans. In addition, plants and algae also synthesize compounds that act as regulators of growth and development that occur at very low concentrations. The best known phytohormones are auxins, cytokines and gibberellins; however, there are many other compounds that can

© Springer International Publishing AG, part of Springer Nature 2018

K. Chojnacka et al. (eds.), *Algae Biomass: Characteristics and Applications*, Developments in Applied Phycology 8,
https://doi.org/10.1007/978-3-319-74703-3_9

regulate the growth and development of photoautotrophic organisms.

Plant hormones are believed to be one of the most active components of seaweed extracts. These low-molecular-weight compounds, which are derived from groups with different chemical structures, can affect plant metabolism in very low concentrations. On the other hand, in higher doses, plant hormones can cause adverse effects and inhibit important processes occurring in plants (Tarakhovskaya et al. 2007).

Seaweeds are extremely rich in polysaccharides, which exhibit numerous biological activities. The total content of polysaccharides in algae varies from 4% to 76% of the dry weight of these organisms. These natural polymers serve in seaweeds as cell wall structural building blocks or as storage compounds. The basic polysaccharides of the cell wall of the green algae are ulvans, while red algae have agarans and carrageenans. In turn, brown algae contain alginates, fucoidans and laminarin as storage polymers (Jiao et al. 2011). Some polysaccharides, i.e. the carrageenans of red algae, are known plant stress response elicitors (Shukla et al. 2016).

Fatty acids are also a very important component of seaweed cells. Depending on the purpose and species of alga, both unsaturated and saturated fatty acids will be present. Usually, the total content of these compounds reaches 10% of dry weight, but in some cases, this value is much higher. For biofuel feedstock, a high content of saturated fatty acids has been established, for instance, in *Spatoglossum macrodontum*. On the other hand, species rich in polyunsaturated fatty acids may be used as nutraceuticals, e.g. *Derbesia tenuissima* (Gosch et al. 2012).

Phenolic compounds, as well as flavonoids, exhibit strong antioxidant properties; therefore, biosynthesis of these substances after exposure to stress factors is enhanced. A high total content of these compounds was determined in such species as *Fucus vesiculosus*, *Fucus serratus* and *Ascophyllum nodosum*. Brown algae contain the highest amount of phlorotannins, compared to green and red algae, up to 25–30% of dry matter, whereas in green and red algae, the content is maintained at a level of about 10% of dry matter (Singh and Sidana 2013).

Betaines serve as osmoregulators and osmoprotectants, which accumulate in cells during drought or temperature changes, thus playing an important role in the process of maintaining turgor and preventing dehydration (Mantri et al. 2012).

Besides the groups of compounds listed above, seaweed cells contain a diverse assortment of biologically active compounds, which may affect plant growth synergistically or antagonistically, as well as impose stimulatory or inhibitory effects.

9.3 Acquisition of Bioactive Compounds from Seaweeds

Many reports indicate that seaweeds contain relatively high concentrations of macro- and microelements, as well as plant hormones, amino acids, antibiotics and vitamins that are necessary for the development and growth of terrestrial plants. The chemical composition of algal extracts depends, to a large extent, on the extraction method and solvents used in the process. For this reason, the activity of extracts obtained from the same species, but using different extraction methods, may vary widely. Extraction of active ingredients from a plant matrix can be impeded by the presence of the cell wall. Frequently, in such cases, methods of breaking up the cells or crushing the biomass to be extracted at a later stage are used. There are several basic pretreatment methods for biomass, i.e. mechanical shredding or enzymatic hydrolysis (Nabti et al. 2017).

The choice of extraction method plays a very important role at the beginning of every experiment in which active ingredients are being isolated from complex matrixes. Economic considerations are essential in the course of obtaining substances from natural sources, so analysis of the costs associated with the extraction processes is often a crucial step in transferring the production process from the laboratory to the industrial scale. Nowadays, these methods are expected to be eco-friendly, less time- and money-consuming, easy and reproducible and, more than anything else, produce the highest possible yield. In many cases, researchers can reduce the number of experiments carried out by implementing statistical methods in experimental designing. After choosing the experimental design, which helps to estimate which experiments should be carried out in the area of study, optimization with the response surface methodology can be applied. These methods help to limit the number of tests and predict the best possible conditions for the experiment, depending on which variables need to be evaluated in the process (Bezerra et al. 2008). Independent variables like temperature, pH, time of extraction and impact of the solvent used are often considered.

Depending on the solubility of the desired compounds in water, or in organic solvents, and the specific pH of the solutions used, various methods of extraction may also be supported by several cell wall degradation techniques that can be applied in the isolation step of the research. Generally, higher yields can be obtained when the applied method of extraction has the ability for good matrix penetration and cell disruption. What is more, the different polarities of the solvents allow us to isolate the molecules of interest from seaweeds or remove unwanted substances (Zvyagintseva et al. 1999). Additives like urea, humic acids, ammonium phosphate, potassium sulphate and additional doses of growth hormones are often present in the composition of seaweed-based biostimulants.

So far, many methods of extraction have been utilized to obtain extracts containing carbohydrates at the highest possible yield, including alkaline and acidic extraction, enzyme-assisted extraction, reactive extrusion, pressurized liquid extraction, ultrasonic-assisted extraction, microwave-assisted extraction and supercritical fluid extraction (Kadam et al. 2013; Michalak and Chojnacka 2015).

9.4 Species Used for the Production of Biostimulants

The most widely utilized group of algae for agricultural purposes are brown seaweeds, among which *Ascophyllum nodosum* is the most important and most common. Out of 47 seaweed-based products, listed in the review by Sharma et al., more than 50% contain extracts made out of *Ascophyllum nodosum*. Other species that are used for the manufacturing of plant growth stimulants are *Fucus* spp., *Laminaria* spp., *Sargassum* spp. and *Turbinaria* spp. Brown algae are a good substrate for production, as they can reach notable sizes, are available throughout the year and can be cultivated if the natural resources are all spent. *Kappaphycus alvarezii* is an example of red algae that have been used in the production of biostimulants. In laboratory-scale experiments, *Ulva* spp. (green algae) are commonly used as the substrate biomass for preparation of extracts (Table 9.1).

Most of the products of this kind are derived from Asia, the United Kingdom, Brittany in France, near South Africa's western coast, the coastal regions of Norway, the United States or Canada, because of the great availability of the raw material and its good processing properties (Sharma et al. 2014).

Table 9.1 Some of the commercially available products containing seaweed extracts

Name of product	Species	Type of algae
Kelpak®	*Ecklonia maxima*	Brown algae
Goemar BM86®	*Ascophyllum nodosum*	
SeaTop®	*Ascophyllum nodosum*	
Maxicrop®	*Ascophyllum nodosum*	
Seamac®	*Ascophyllum nodosum*	
SuperFifty	*Ascophyllum nodosum*	
Chase SM6®	*Ascophyllum, Laminaria, Fucus*	
Tri-Kelp™	*Laminaria, Sargassum, Ascophyllum nodosum*	
Seasol®	*Durvillaea potatorum*	
Aquasap™	*Kappaphycus alvarezii*	Red algae
Vegegrow™	*Kappaphycus alvarezii, Eucheuma denticulatum, Sargassum polycystum*	Red and brown algae

9.5 Direct or Indirect Stimulation of Plant Growth Conditioned by the Method of Algal Extracts Application

Due to the diversity of active compounds present in seaweed cells, the growth and development of cultivated plants can be affected in two ways. Seaweed extracts or products containing such ingredients have a positive impact on crop plants, not only because of the direct stimulation of their growth and metabolisms, e.g. nutrient uptake, but also indirectly, through improvement of the soil condition (hydration, structure) or enhancement of the number of soil microorganisms resulting from the creation of favourable conditions for their development. Products containing algal extracts not only stimulate the growth and development of crops, thus increasing the yield, but also improve their quality and nutritional values (Battacharyya et al. 2015).

For laboratory tests evaluating the effect of new agricultural products, several species of plant possessing certain features are usually chosen. The preliminary tests are usually carried out on model plants like *Arabidopsis thaliana* or other plants with a very fast growth rate. Sometimes, plants of high economic importance, or those that are the basic source of nutrients for big populations and that grow in environments with significant climatic changes, are chosen.

In most studies describing tests on plants, the first step consists of assessment of the application method for a new biostimulating product, e.g. algal extract. Several methods are usually utilized, such as seed treatment, foliar spraying, soil enrichment, soil drenching (before or after sowing) or a combination of these. Also, the concentration, volume and number of doses must be adjusted before supplementing the plants in the fields. In many cases, the field experiments do not reflect the laboratory tests, since under field conditions, the combination of all of these factors may mitigate the beneficial effects of the applied biostimulator, and thus the significant results are not obtained in some cases.

The impact is determined by comparing the following parameters in relation to the control series: number of germinated seeds, weight and length of the shoots and roots, number and surface of leaves, photosynthetic dye content, total content of phenolic compounds, flavonoids, proteins, lipids, nutrient uptake or antioxidant potential. Also, the activity of germination and antioxidative enzymes may be one of the factors evaluated. In further stages of plant development, the number of flowers, fruits and total yield is also evaluated (Povero et al. 2016).

9.6 Seaweed Composts in Plant Nutrition

The earliest written references concerning preparation of composts were dated approx. 2300 BC and were discovered in the Akkadian Empire. However, the great bloom of popularity of compost application was observed in the twentieth century, as the detailed studies relating to compost performance became available. The technology of compost production subsequently improved, as the numerous experiments of the time provided confirmation about the usefulness of compost as a soil additive. To date, several contemporary composting systems have become common, such as windrow composting, silo systems and both reactor and nonreactor composting (open air) (Fitzpatrick et al. 2005). In many cases, the composted biomass consists of seaweed, fish waste (fish bones, shrimp shells, etc.), straw, peat or other organic wastes, in order to obtain a final, mature compost filled with highly nutritional compounds for the cultivation of crops.

Mixes of green algae *Ulva ohnoi* and sugarcane bagasse were used in the composting process, and the different ratio of each biomass was tested in order to obtain the proportion of C:N. The composting process lasted 16 weeks, and the C:N ratio was measured before and after this period of time. This was an important parameter, because plants sown in compost with a C:N ratio higher than 20:1 have limited access to nitrogen. Several mature composts were then mixed with soil in which sugarcane seedlings had been placed. Compost rich in seaweed biomass resulted in a sevenfold higher total above-ground biomass of sugarcane than low seaweed composts and a fourfold higher total above-ground biomass than sugarcane grown in commercial compost that did not contain seaweed. Researchers concluded that the high proportion of nitrogen (C:N; 22:1) in the initial biomass for composting containing 82% of seaweed is highly beneficial for the tested plants and that the *Ulva* biomass is able to retain these nitrogen levels throughout the composting process. Due to the fact that, during the composting process, carbon losses are observed, the initial biomass should usually contain a slightly higher C:N ratio than 20:1 (Cole et al. 2016).

In the greenhouse trials, the activities of fish-seaweed-pine bark waste compost in several doses, as well as application of one dose of mineral fertilizer and a certified organic fertilizer made from dehydrated broiler litter on a tomato crop and its residual effects on the succeeding lettuce crop, were compared. The control treatment was without fertilization. The main species of seaweed present in the compost were *Laminaria* spp. and *Cystoseira* spp. All of the treatments were performed as superficial ploughing. The activity of fish-seaweed compost at a dose of 66 t ha^{-1} was found to be the most effective in fertilization of tomato plants, with increased fruit weight and larger fruit diameter compared to crops receiving mineral fertilization or no fertilization. Also, the yield in terms of fresh weight of lettuce, which was sown in the soil previously occupied by tomato plants, was greater in comparison with the control in the trials with fish-seaweed compost. The best results in the case of the fresh weight of lettuce were obtained in the treatment with broiler litter. As tomato plants demand more nutrients than lettuce, it was possible to obtain good outcomes in both experiments, as they were correctly organized. Also, the residual effect of organic fertilizer application may lead to increased production for up to 4 years after the initial application, which makes such treatments superior to traditional mineral fertilization (Illera-Vives et al. 2015).

A similar set of organic and mineral fertilizers (fish-seaweed-pine bark compost, dehydrated broiler litter, mineral fertilizer) was used by Illera-Vives et al. in trials with potato plants (*Solanum tuberosum*). The yield obtained by the highest dose of marine-based compost was the greatest (53% over the control), but all of the treatments gave superior yield in comparison with the control plants, although in some cases, the differences were not significant (Illera-Vives et al. 2017).

9.7 Seaweed Extracts Stimulate the Growth of Plants

Biostimulants prepared out of macro- and microalgae can positively influence plant metabolism, enhancing respiration, chlorophyll production, photosynthesis, nucleic acid synthesis and ion uptake. Such products also enhance water-holding capacity and antioxidant formation (Sharma et al. 2014). Seaweed and seaweed extracts affect plant growth, and greater responses might be obtained, especially when plants grow under optimal environmental conditions, in terms of temperature, water, nutrient access and lack of harmful pathogens (Calvo et al. 2014).

Application of liquid extracts of the brown marine alga *Stoechospermum marginatum* was evaluated on the growth, biochemical changes and yield of brinjal (*Solanum melongena*), a vegetable crop. A promoting effect was noticed with the use of extract concentration up to 1.5%. The assessed parameters of shoot and root length, total fresh and dry weight, leaf area and content of moisture, photosynthetic pigments, protein, amino acids, reducing sugars, ascorbic acid and nitrate reductase activity were found to be enhanced in the leaves of the brinjal plants. Higher concentrations of the seaweed extract (above 1.5%) were found to show a retarding effect on all of the growth parameters of the tested plant. Algal extracts impact quite a number of parameters related to the growth and development of the studied plants, suggesting that they are the right products for self-use (Ramya et al. 2015).

In the studies by Safinaz and Ragaa, the effect of biomass of three species of red algae on the content of macroelements, shoot length and root length and their mass, as well as the number of maize leaves in potted tests, was investigated. The dried and powdered algae *Laurencia obtusa*, *Jania rubens* and *Corallina elongata* in the amount of 3 g kg^{-1} of soil applied singly and in combination over a 60-day cycle were used for the study. Regardless of the biomass used, all shoots exhibited a significant increase in length compared to the controls, and the largest, an almost 50% increase, occurred with the mixture of *Laurencia obtusa* and *Jania rubens*. The same mixture significantly affected the potassium content (+ 61.84%), while the use of *Jania rubens* only resulted in a significant increase in nitrogen (+ 129.23%) compared to the control. These studies confirm the usefulness and appropriateness of using unprocessed red algal biomass as a source of nutrients and for stimulating the growth of crops (Safinaz and Ragaa 2013).

Foliar application of microalgal extract stimulated the growth of *Solanum lycopersium* and *Capsicum annum* plants. Extract of *Spirulina platensis* containing mainly polysaccharides was used in the different growth stages of test plants. Plant growth was enhanced after treatment with a polysaccharides solution, 30% over the control for tomatoes and 20% for peppers. What is more, the weight of the tomato roots increased by 230%, while in case of peppers, the improvement was 67%. The number of nodes per plant was also higher in both test plants, as well as leaf area. These findings suggest that total polysaccharide extract is a promising plant growth enhancer (Elarroussia et al. 2016).

Presoaking of seeds of *Zea mays* L. and *Helianthus annuus* L. in a seaweed extract solution was performed in order to evaluate the activity of compounds isolated from *Gracilaria corticata* and *Enteromorpha flexuosa*. Seaweeds were also distributed for mixing with soil and as a solution for irrigation. Greenhouse trials were performed at room temperatures. In further stages of experiment, after 60 days, it was observed that treatments with both species of seaweed gave superior results in comparison with the control; however, better results were obtained after the use of *Enteromorpha flexuosa* extract. Application of seaweed extracts enhanced shoot length, root length, shoot and root dry weights, photosynthetic pigments, carbohydrate and protein contents and nutrient uptake. Also, the presence of phytohormones in the extracts was evaluated with gas chromatography. This study revealed that *Enteromorpha flexuosa* extract was richer in plant hormones, especially from the cytokinins class, which may be attributed to better growth results (Omar et al. 2015).

An interesting example of the stimulation of oil plant metabolism by algae preparation may be the use of a product available under the trade name Aquasap™. Karthikeyan and his collaborators used a biostimulator containing *Kappaphycus alvarezii* extract in field studies involving peanut and sunflower seedlings. Better crop yield (31.69% for nuts and 51.30% for sunflower compared to the control) and increased oil content in plant seeds treated with 5% solution (15.77% for sunflower and 14.27% for walnut) were observed. Application of the seaweed-based product greatly influenced crop yield and quality in terms of nutritional values (Karthikeyan and Shanmugam 2015).

9.8 Effect of Seaweed Extracts on Plants Affected by Biotic and Abiotic Stress

Environmental pollution, global warming and environmental anomalies, often caused by human activity, cause plants and algae to be permanently exposed to various environmental stressors. These factors can be divided into two groups, i.e. biotic (related to plant diseases or the presence of pests) and abiotic factors. The latter include temperature fluctuations, droughts or floods, excessive salinity, heavy metal pollution, excessive radiation and the presence of organic pollutants such as herbicides, pesticides and insecticides (Vardhini 2015). A changing climate and greater demand for food due to an increasing population number have caused an ongoing search for products that mitigate the effects of these changes on crops. Stress tolerance mechanisms have been widely studied, but many of the explanations remain unknown (Sangha et al. 2014). As the seaweed extracts are rich in bioactive compounds, products containing such extracts have been tested on plants growing under stress conditions, e.g. drought, temperature changes, salinity or pathogen attack (mostly fungal or bacterial) (Van Oosten et al. 2017). Signals of environmental stress can lead to severe changes in plant metabolism and morphology. However, similar morphogenic responses such as inhibition or elongation, changes in cell differentiation or regulation of cell division are found to be induced by different stressors (Patakas 2012).

9.8.1 Drought

Plant response to drought corresponds with many changes on the molecular, biochemical and physiological levels. Water deficiency affects plant growth through limited photosynthesis and aperture of stomata. Under such conditions, plants are also vulnerable due to possible oxidative stress triggered by reactive oxygen species. Plants can deal with drought by following a few mechanisms: (a) drought avoidance, (b) drought tolerance, (c) drought escape and (d) drought recovery. The first two mechanisms are the main ones that plants exhibit. Under conditions of water shortage, plants can sustain a sufficient water level in their tissues by limiting water transpiration connected with stomatal closure and enhanced wax production or by increased water uptake through the net of roots (Fang and Xiong 2015). Roots, as place of nutrient and water uptake, are very sensitive to changes in the physicochemical parameters of the soil. Nowadays, insight into plants' mechanisms for coping with drought and rewatering, together with investigation of possible compounds that may minimize water stress, is very important. This could be very useful in the cultivation of plants exposed to severe climatic changes (Xu et al. 2010).

For the purpose of increasing drought tolerance, numerous studies have been conducted, in which extracts of the brown seaweed *Ascophyllum nodosum* have been widely exploited. Besides that, a few reports are available in the literature concerning the application of green algal extracts on plants in order to verify their activity during water shortages.

One example of the utilization of an *Ascophyllum nodosum*-based product (Acadian®) was the research by Martynenko and co-workers explaining the effect on soybeans during induced drought stress. In this experiment, thermal imaging was used for evaluation of leaf temperature changes, with leaf turgor and angle also being measured. During the first 2 days of drought, leaf surface temperature increased, while on the third day, a decrease was observed, for both the treated and the untreated plants. However, after this period of time, it was concluded that treatment with the seaweed-based product gave positive results, due to the stomatal closure connected with the action of abscisic acid and maintaining of leaf turgor, as the treated plants adapted to drought conditions (Martynenko et al. 2016).

A greenhouse experiment was conducted to evaluate the effects of foliar application of an *Ascophyllum nodosum* seaweed extract (AZAL5) on the growth, nutrient uptake and yield of winter wheat in a surface soil. The experiments were performed in soil with 75% and 45% water field capacity, with and without additional nitrogen sources, and the plants were treated with two concentrations of AZAL5. Results revealed a beneficial effect of seaweed extracts on all tested parameters of the wheat, but only if the N-fertilizer was present. This fact may suggest that seaweed extracts stimulated the wheat for better soil nutrient uptake. In the case when the water field capacity decreased, twice the higher concentration of seaweed extract was required to obtain similar positive effects as with 75% water field capacity (Stamatiadis et al. 2014).

In the experiments by Elansary, *Ascophyllum nodosum* extracts at concentrations of 5 and 7 mL L^{-1} were applied as a soil drench or foliar spray on *Spiraea nipponica* "Snowmound" and *Pittosporum eugenioides* "Variegatum". Drought stress was indicated by watering with the volume of water reduced by 50%. There were no significant differences between tested parameters when plants obtained the usual dose of water, and results were only slightly better in the case of a content of phenolics and flavonoids. Treatment with a higher dose of seaweed extract during moderate drought gave a better outcome in the case of increasing the phenolic and flavonoid content and reducing proline accumulation, as well as enhancing plant water status and gas exchange parameters. Also, the leaf number and area, dry weights and plant heights recorded for seaweed extract treatments were superior to those in the control (Elansary et al. 2016).

Extracts of *Ulva rigida* and *Fucus spiralis* were used as a foliar treatment on bean plants (*Phaseolus vulgaris* L.) under severe and moderate drought conditions. Growth parameters, chlorophyll, glycine, betaine and phenolic compound content and antioxidant enzyme activity were all assessed. Foliar applications of two seaweed extracts with a concentration of 25% were effective and helped treated plants to overcome stress and improved their vegetative growth. Increased activity of antioxidant enzymes, like ascorbate peroxidase, superoxide dismutase and catalase, and polyphenol content may indicate that seaweed extract treatments contributed to the protection mechanisms against peroxidation and enhanced tolerance against water shortages on bean plants (Mansori et al. 2015).

A similar experiment using *Ulva rigida* extract was also performed by Mansori et al., but in this case, the plants were of the common herb *Salvia officinalis* L. Induced drought represented three levels of water field capacity (82, 18 and 7%). Vegetative growth parameters such as shoot length and number and area of leaves during water shortage were enhanced after the application of seaweed extract. Highly beneficial results were obtained after spraying with 25% seaweed extract during moderate water stress; this treatment led to growth parameters increased by 40% in comparison with the control plants. Also, the phenolic compound content and activity of antioxidant enzymes were superior in treated plants, an outcome that suggests that seaweed extract protects plants against lipid peroxidation induced by water stress (Mansori et al. 2016).

Container-grown "Hamlin" sweet oranges, trees of "Carrizo" citrange and "Swingle" citrumelo rootstocks were treated weekly with a commercial extract of the brown seaweed *Ascophyllum nodosum* at doses of 5 and 10 mL L^{-1} as foliar sprays or a soil drench. Drought stress was induced by watering the plants with an amount of water decreased by 50%. It was noticed that trees that were treated with seaweed extracts and subjected to drought had better water relations, e.g. higher stem water potential, but only in the case of soil drenches. Also, stress vulnerability was reduced in plants treated with extracts in comparison with the control. Growth of untreated plants was reduced by 30% during the drought period. Application of seaweed extracts turned out to be a viable approach for maintaining the growth of citrus nursery trees grown under greenhouse conditions in non-uniform irrigation systems (Spann and Little 2011).

However, in some cases, tolerance towards drought was not being enhanced, even when the seaweed extracts were applied, although the growth parameters increased. This may be caused by the fact that specific species of plant may be vulnerable to such abiotic stress factors, no matter what kind of protecting agent is used. Crucial in such situations was the method of application, e.g. substrate drench or foliar application, as in the research by Li and Mattson. Treatments with *Ascophyllum nodosum* extract were performed on *Petunia hybrida* and *Solanum lycopersicum* in order to evaluate the effect on plant growth and tolerance during drought. Foliar application of seaweed extract resulted in enhanced growth and postharvest life of petunia and tomato transplants, but stress tolerance was not enhanced in this case. Drenches with seaweed extract improved the stress tolerance measured in terms of leaf angle changes (Li and Mattson 2015).

The effect of *Ascophyllum nodosum* extract application through foliar application, soil drenching and a combination of both on spinach in a growth chamber experiment was evaluated. Drought stress was induced by irrigating the plants with doses of water reduced by 50%. Under full irrigation, seaweed extract had no effect on leaf growth, physiology or nutrition value, but under drought stress, it improved plant growth. The phenolic, flavonoid and carotenoid content, and antioxidant capacity, was not affected by treatment with seaweed extract. The consequences caused by mild drought stress in *Spinacia oleracea* L. might be mitigated by application of *Ascophyllum nodosum* extract (Xu and Leskovar 2015).

Excessive use of mineral fertilizers in dry soil, providing loads of nitrogen, may cause root degradation and therefore degradation of the whole plant, or it may lead to certain diseases. During field plot experiments on creeping bentgrass (*Agrostis palustris* Huds.) grown in soil enriched with nitrogen (in urea form), the application of fortified *Ascophyllum nodosum* extract and a solution of propiconazole was examined. In addition, each of the treatments was enriched with a chemical compound containing chelated iron ions. Drought stress was induced after transplanting the plant into metal containers, by placing them under a rain shelter for 3 weeks. After that, the plants were irrigated to water field capacity three times a week. Their leaf colour, water status and cumulative evapotranspiration Se-Kwon Kim were measured during the period when water was withheld. Foliar application of plant growth stimulators, regardless of the addition of iron ions, positively affected the leaf water status of the treated plants. Creeping bentgrass grown in soil with high nitrogen fertility exhibited greater cumulative evapotranspiration during water shortages. Researchers noted that plants that obtained an additional source of cytokinin-like substances during drought were able to maintain integrity of their tonoplast membrane and repress lipoxygenase (the enzyme responsible for leaf senescence). After 3 weeks of drought, the application of plant growth regulators had a beneficial impact on root growth, and a lower amount of wilting of leaves was observed (Nabati et al. 2008).

9.8.2 Temperature

Changes in temperature during the night and day force plants to adapt and sustain optimal conditions inside their cells in order to carry out the necessary physiological processes. Several complex responses are performed, depending on the directions of temperature changes, duration and rate. During heat stress, plants may induce evaporative cooling by opening their stomata, reduce the surface area exposed to sun by decreasing leaf turgor and change leaf orientation or even cause them to fall off. Reduced temperatures can result in plant dormancy or excretion of antifreezing compounds into the cytoplasm, so as to lower the freezing temperature of plant fluids (Zinn et al. 2010; Walbot 2011; Bita and Gerats 2013). Fluctuations in temperature strongly affect plant development and productivity; thus, products alleviating the effect of such changes are strongly desired. Optimal temperatures for the growth of plants strongly depend on the considered species. For each species, a defined range of maximum and minimum temperatures form the boundaries of observable growth. For most plant species, vegetative development usually has a higher optimum temperature than reproductive development (Hatfield and Prueger 2015).

Arabidopsis thaliana was used as the model plant in the temperature stress tests evaluating the activity of *Ascophyllum nodosum* extracts and their lipophilic fractions. Application of extracts significantly increased tolerance to freezing temperatures during in vitro and in vivo assays. Untreated plants exhibited severe chlorosis and tissue damage and failed to recover from freezing treatments, while the extract-treated plants recovered from freezing temperatures of 7.5 °C in vitro and 5.5 °C in vivo assays. What is more, during the recovery stage after the period with decreased temperature, plants treated with seaweed extracts exhibited 70% less chlorophyll damage in comparison with untreated plants. That suggests that *Ascophyllum nodosum* extracts are able to induce specific systemic physiological responses in plants (Rayirath et al. 2009).

Two cucumber (*Cucumis sativus* L.) cultivars were exposed to low temperature conditions (7 ± 1 °C) and the effect of the application of seaweed extracts (seaforce and seamino). Decreased temperature was set after the expansion of cotyledon leaves. Such developmental parameters as number of female flowers and fruit setting percentage were enhanced after treatment with seaweed extracts. Also, the yield in terms of number of fruit per plant, fruit weight, yield per plant, yield per square metre and total yield were significantly increased and was greatest in treatments that combined both seaweed products. Low temperature did not strongly affect the yield of cucumber, but the tests carried out confirmed the beneficial effect of the foliar sprays with seamino and seaforce (Sarhan and Ismael 2014).

9.8.3 Salinity

Salinity causes water stress, which limits the growth of the plant; affects photosynthesis, cell elongation and differentiation and stomatal closure; reduces leaf expansion; and causes leaf senescence, deterioration of energy production and biosynthesis of many important bioactive compounds (Carillo et al. 2011). Plants have developed a collection of responses at the molecular, cellular, metabolic, physiological and whole-plant levels that ease the occurrence of salt stress. Among them, ion homeostasis, ion transport and uptake, biosynthesis of osmoprotectants, activation of antioxidant enzymes and production of antioxidant compounds, synthesis of polyamines, generation of nitric oxide and hormonal regulation can all be enumerated (Gupta and Huang 2014). Salinity of soil or water may influence plant growth and cause adverse effects in the metabolism, yield or quality of vegetables and crops. Some plants exhibit higher salt tolerance and are called halophytes, while species that cannot cope with salt stress as well are called glycophytes; more importantly, most crops belong to the second group. That is why it is so relevant to enhance the resistance of these plants and search for products able to maintain good crop health, regardless of the negative environmental conditions.

One approach being investigated is the application of *Ascophyllum nodosum* seaweed extracts. In the research by Aziz et al., the growth, flowering and chemical constituent performance of *Amaranthus tricolor* plants were investigated, influenced by application of seaweed *Ascophyllum nodosum* extract under salt stress conditions. Four concentrations of saline water were used for irrigation. Growth parameters decreased in the series where the concentration of salt was at the level of 2000 ppm and 3000 ppm. Treatment with seaweed extract alleviated the salt stress, and what is more, scientists noted that watering with 1000 ppm saline water, together with foliar application of 2.5 mL L^{-1} of seaweed extract, produced superior results for vegetative growth, flowering and chemical constituents of *Amaranthus tricolor* plants (Aziz et al. 2011).

Guinan and co-workers have described the resistance of the lettuce "Little Gem" towards salinity stress in the presence of seaweed extracts. Two *Ascophyllum nodosum* extracts were applied, one prepared in the elevated temperature media, over 125 °C, and the second in a temperature below 75 °C. A significant difference in growth between the treated lettuce and the control was observed, when the salinity conditions occurred. The high-temperature extract increased lettuce weight by about 42%, while the low-temperature extract resulted in about 28%. Under nonsaline conditions, no noteworthy effects of the extracts were observed (Guinan et al. 2012).

Another example of treatment with *Ascophyllum nodosum* extract during salt stress is the experiment by Hegazi et al. Three levels of salt concentration combined with three concentrations of seaweed extract were tested on eggplants in pot experiments. Seaweed extract alleviated the negative effects of induced salinity on eggplants up to certain, moderate concentrations of NaCl. Salt stress enhanced the activity of antioxidant enzymes – superoxide dismutase and ascorbic peroxidase. Treatments with *Ascophyllum nodosum* extract positively affected the total content of tannins and phenolic compounds and chlorophyll, as well as vegetative growth (Hegazi et al. 2015).

Seeds of the radish *Raphanus sativus* L. were primed with the extract of either *Codium taylorii* or *Pterocladia capillacea* and then sown in clay-sandy soil and left to grow for 35 days. During that period, seedlings were under the effect of either 150 or 200 mM NaCl, and the control was pure water. Adverse effects of salinity were visible in decreased water content, shoot length, photosynthetic pigments and total lipid content. Researchers also carried out a molecular analysis of the leaf protein profile of the radishes treated with seaweed extracts, which produced results about newly synthesized stress proteins. This suggests that seeds primed with such extracts might serve as a great tool in procedures that alleviate salinity stress (El-Aziz Kasim et al. 2016).

Water extract from the green algae *Ulva lactuca* was used in order to evaluate the growth of salinized wheat plants. Seeds of *Triticum aestivum* L. were presoaked in algal extracts for 12 h prior to sowing on sterilized Petri plates. Concentrations of NaCl water solution within the range of 50–250 mM were added after germination of the plants. The experimental period lasted 7 days, and after that, several parameters were measured. Enhancement of seed germination in the salinized plants was observed, together with the seedlings' dry-fresh matter. What is more, the enzyme activity measurements revealed increased activity of catalase and superoxide dismutase with the increasing concentration of seaweed extract used for seed treatment. On the other hand, ascorbate peroxidase and glutathione reductase activity decreased. Presoaking of grains with *Ulva lactuca* extracts alleviated salinity stress and induced the formation of 12 new protein bands in the protein pattern of the treated plants (Ibrahim et al. 2014).

Foliar application of seaweed extract on *Foeniculum vulgare* during salt stress was then tried. A dose of 4.5 mL

L^{-1} significantly increased plant height, stem diameter, dry weight of the vegetative growth and the roots' fresh weight in the second season, compared to the control, although the application of 25 g L^{-1} of active dry yeast gave the best results. Other treatments were performed with humic acid solutions and amino acids. All of the examined growth enhancers positively affected the growth of fennel. Superior results were obtained in the analysis of essential oil content in the series in which seaweed extract was used (Mostafa 2015).

Rouphael et al. measured the physiological and anatomical effects, as well as the changes in mineral composition, of greenhouse zucchini squash (*Cucurbita pepo* L.) treated with *Ecklonia maxima* extract (Kelpak®) during induced salinity stress on three levels. Foliar application of seaweed extract increased yield and shoot biomass by 12.0% and 17.4%, respectively, as well as the fruits' dry matter and total soluble solid contents in comparison with untreated plants. This research also indicated changes in leaf morphology, as the sizes of stomata, increased lamina thickness and volume of intercellular spaces in the mesophyll were influenced by the application of saline water. Treatment with seaweed extract enhanced the marketable yield and shoot biomass production of zucchini (Rouphael et al. 2016).

The total carotenoid, tocopherol, phenolic and protein contents in whole grains of wheat plants irrigated with 10% and 20% (v/v) seawater, in response to water extracts of the microalgae *Spirulina maxima* and *Chlorella ellipsoidea*, were measured. Additionally, some exogenous plant growth enhancers, such as ascorbic acid and benzyladenine treatments, were investigated to compare the effectiveness of the algal extracts. Results showed that application of algal extracts triggered the production of biologically active substances in wheat. The accumulation of higher amounts of antioxidant compounds in the yielded grains was observed in the plants treated with 20% saline water and the extract of *Spirulina maxima*, in comparison with other treatments. Overall, algal extracts helped wheat plants to overcome the salt stress, also improving the content of antioxidative components and proteins, which make such whole grains more attractive to consumers (El-Baky et al. 2010).

9.8.4 Pathogen Attack

Plants, especially those growing outdoors under field conditions, are constantly exposed to pathogen attack. The defence mechanism against such invasions is based on the combination of two elements: physical barriers (e.g. cuticle), which inhibit the pathogens from gaining access to the internal tissues of plants, and biochemical processes, which take place in the cells and lead to the production of toxic compounds or substances that inhibit the growth of pathogens inside the plant (Jibril et al. 2016).

As one of the primary vegetables in many countries, the tomato and its cultivation deserve special attention. That is why in many researches, information concerning good practices and protective agents are described. The work of Jiménez et al. might serve as the most exemplary example of research concerning the application of seaweed extracts as agents mitigating pathogen infection. Aqueous and ethanolic extracts obtained from nine species of seaweed, *Macrocystis pyrifera*, *Macrocystis integrifolia*, *Lessonia nigrescens*, *Lessonia trabeculata*, *Durvillaea antarctica*, *Gracilaria chilensis*, *Porphyra columbina*, *Gigartina skottsbergii* and *Ulva costata*, were tested on tobacco leaves (*Nicotiana tabacum* L.) infected with tobacco mosaic virus and tomato plants (*Solanum esculentum*, cv. Patron) infected with the fungus *Botrytis cinerea*. Additionally, microplate tests evaluating the antibacterial potential of seaweed extracts on *Erwinia carotovora* and *Pseudomonas syringae* and antifungal activity against *Phytophthora cinnamomi* and *Botrytis cinerea* were performed on proper media. Extracts were prepared from the biomass collected in four subsequent seasons during 1 year. Results indicated aqueous and ethanolic extracts of *Durvillaea antarctica* as the most effective among others against the tobacco mosaic virus. *Lessonia trabeculata* organic extract inhibited bacterial growth of *Botrytis cinerea* and reduced necrotic lesions on tomato leaves. Also, the extracts of *Gracilaria chilensis* were found to be able to reduce the growth of *Phytophthora cinnamomi*. What is more, this research also assessed the activity of extracts prepared from the biomass collected in different seasons, and in some cases, dissimilarities were revealed. Such information indicates the best time to collect the seaweeds from the Chilean coast that possess antipathogenic activity (Jiménez et al. 2011).

In the work of Esserti et al., the effectiveness of *Cystoseira myriophylloides*, *Laminaria digitata* and *Fucus spiralis* methanolic extracts against the tomato pathogens *Verticillium dahliae* and *Agrobacterium tumefaciens* has been evaluated. Greenhouse experiments resulted in the inhibition of mycelial growth, no matter which seaweed was used to prepare the extract, but seed treatment was found to be a superior method of extract application to foliar spraying. What is more interesting, the in vitro test of seaweed extract on the fungal pathogen did not show any inhibiting activity; thus, the researchers proposed the mechanism of induced resistance as the best course of action. In the case of the bacterial pathogen, a significant reduction of crown gall disease was obtained after application of extracts made from *Cystoseira myriophylloides* and *Fucus spiralis* (Esserti et al. 2017).

Begum et al. evaluated the activity of *Turbinaria conoides* extract against the root rot pathogen *Fusarium oxysporum*.

The greatest inhibition of mycelial growth was measured on the plates on which 15% and 20% seaweed extracts were applied, even up to 6 days after incubation. As *Fusarium* fungi are one of the main pathogens for tomatoes, this study revealed great potential of *Turbinaria conoides* alcoholic extracts as protective agents in the cultivation of *Lycopersicon esculentum* Mill (Begum et al. 2016).

Another research in which tomatoes (*Solanum lycopersicum*) served as the test plant is the work of Hernández-Herrera et al. Water-soluble compounds extracted from the green algae *Ulva lactuca* and *Caulerpa sertularioides* and the brown algae *Padina gymnospora* and *Sargassum liebmannii* were assessed as potential protection agents against the necrotrophic fungus *Alternaria solani*. Tomato seedlings were sprayed with seaweed extract solution 30 days after sowing. After that, plants were infected with fungus, and the severity of the disease was evaluated after another 15 days. Tomato plants treated with algal extracts exhibited significantly infected leaf areas in comparison with the control plants. Among the main compounds of the extracts, ulvans, alginates, fucans and laminarin (β-1,3 glucan) were also present; all of the named polysaccharides are known to trigger defence responses. Algal extracts increased activity of defence enzymes such as polyphenol oxidase, guaiacol peroxidase and proteinase inhibitors, suggesting that application of such extracts may alleviate the biotic stress of plants infected with *Alternaria solani* (Hernández-Herrera et al. 2014).

One more research concerned tomatoes, in which alkaline extracts of *Ulva lactuca*, *Sargassum filipendula* and *Gelidium serrulatum* were used on tomato plants infected with *Alternaria solani* and *Xanthomonas campestris* pv. *vesicatoria*. During the trials, the extracts applied reduced disease severity, and elevated rates of activities of defence enzymes were observed. Treatments with such phytoelicitors caused activation of jasmonate defence systems, as well as salicylic acid signalling pathways, thus alleviating a pathogen attack (Ramkissoon et al. 2017).

Foliar spray and soil drenching with Stimplex™, a formulation made with the extract of *Ascophyllum nodosum*, were used on cucumbers during biotic stress in the greenhouse experiment. Treated plants were inoculated with four fungal pathogens, including *Alternaria cucumerinum*, *Didymella applanata*, *Fusarium oxysporum* and *Botrytis cinerea*. A combination of both methods of extract application was found to be the most effective one in reducing the infections. Also, the activity of defence-related enzymes, including chitinase, β-1,3-glucanase, peroxidase, polyphenol oxidase, phenylalanine ammonia lyase and lipoxygenase, was enhanced in the treated plants, as well as the concentration of phenolic compounds (Jayaraman et al. 2011).

The effect of seaweed extract application and its activity towards the necrotrophic pathogens *Sclerotinia sclerotiorum* and *Alternaria brassicae* was examined in the greenhouse experiment on lettuce seedlings in the controlled environment. *Ascophyllum nodosum* extracts prepared with low- and high-temperature extractions were assessed as agents that alleviate effects of the infection. Local and systemic reductions of lesions were observed in plants treated with low-temperature extract, while on the other hand, high-temperature extracts were more effective in mitigating salinity stress (as in the example described earlier). Such experiments lead to the conclusion that seaweeds possess numerous biological activities and that the method of final product preparation is extremely important. Knowledge about preparation with proper parameters is essential in order to obtain a product that will fit certain requirements (Guinan et al. 2012).

9.8.5 Postharvest Treatment

Postharvest treatments are widely used in order to maintain and extend the shelf life of fresh products, such as fruits and vegetables. Currently, several physical and chemical methods of saving such fresh product properties as appearance, texture, flavour and nutritional value are being performed. Among them, heat treatment, irradiation, edible coatings, nitric oxide and sulphur dioxide and gaseous treatments (e.g. ethylene, ozone) are widely used to keep horticultural products fresh during storage, transport and distribution (Mahajan et al. 2014).

There is also some information concerning the application of seaweed extracts as protective agents. Among the available data, the research by Augusto and co-workers might serve as a source of valuable knowledge. In this study, the application of *Fucus spiralis*, *Bifurcaria bifurcata*, *Codium tomentosum* and *Codium vermilara* extracts as postharvest treatments in minimally processed Fuji apples was investigated. Slices of apples were stored at low temperature, and after 20 days, after application of the seaweed extracts, the moisture content, soluble solid concentration, firmness and browning index were evaluated, as well as the activity of enzymes responsible for the response during stress connected to cutting. Based on the obtained results, researchers concluded that 0.5% *Codium tomentosum* extract may serve as a potential natural additive for postharvest treatment, due to a reduced browning index, as the extract imposed an inhibitory effect on polyphenol oxidase and peroxidase (Augusto et al. 2016).

Another example of the use of a seaweed-based product as a protective agent, prolonging the shelf life of fruits, was described by Omar. Orange fruits *Citrus sinensis* were dipped in the seaweed extract Alga600® and in solutions of calcium chloride in order to compare the effect of natural and mineral postharvest treatments. After that, the oranges were stored at decreased, or even room, temperature. The evaluation parameters measured every 15 days were as follows: weight loss, percentage of fruit rot, percentage of soluble solids, titratable acidity, vitamin C and total and percentage reducing sugars. Treatment with 4% seaweed extract increased storage life in terms of fruit quality, weight loss and fruit decay, as compared with calcium chloride and controls (Omar 2014).

9.9 Conclusions

Even the most resistant crops cultivated all over the world need continuous care and maintenance of the conditions for their growth and development. The examples of the use of algal biomass, as well as extracts and commercial products described in this chapter, clearly support advantageous use of those materials as stimulators of plant growth and development. The positive effect on plant metabolism and stimulation of defence responses to stress factors seem to be the most important benefits. Most studies concern the use of natural biostimulants in countries with large populations, where large plant production is thus required, e.g. India and Thailand. These countries often have access to naturally occurring algae along their coastal areas, and hence this type of fertilization is preferred because of the availability of algae and the reduction of soil degradation, as well as the possibility of avoiding the costs associated with the use of mineral fertilizers.

Due to the fact that algal extracts can be used in tests determining their effect on plant growth and development without providing detailed composition, qualitative and quantitative analyses may be omitted. However, thanks to development of the techniques of instrumental analysis, the composition of extracts, especially with respect to plant hormones, is often determined, because too high a dose may adversely affect the treated plants.

Numerous examples of treatments with seaweed extracts have proven their effectiveness in many ways. Small amounts of seaweed extract expended on the treatment of plants may lead to great beneficial outcomes, not only in regard to optimal growth conditions but also under different biotic and abiotic stress conditions.

References

Augusto A, Simões T, Pedrosa R, Silva SFJ (2016) Evaluation of seaweed extracts functionality as post-harvest treatment for minimally processed Fuji apples. Innov Food Sci Emerg 33:589–595

Aziz NGA, Mahgoub MH, Siam HS (2011) Growth, flowering and chemical constituents performance of *Amaranthus tricolor* plants as influenced by seaweed (*Ascophyllum nodosum*) extract application under salt stress conditions. J Appl Sci 7:1472–1484

Battacharyya D, Babgohari MZ, Rathor P, Prithiviraj B (2015) Seaweed extracts as biostimulants in horticulture. Sci Hortic 196:39-48

Begum AJM, Selvaraju P, Vijayakumar A (2016) Evaluation of antifungal activity of seaweed extract (*Turbinaria conoides*) against *Fusarium oxysporum*. J Appl Nat Sci 8:60–62

Bezerra MA, Santelli RE, Oliveira EP, Villar LS, Escaleira LA (2008) Response surface methodology (RSM) as a tool for optimization in analytical chemistry. Talanta 76:965–977

Bita CE, Gerats T (2013) Plant tolerance to high temperature in a changing environment: scientific fundamentals and production of heat stress-tolerant crops. Front Plant Sci 4:1–18

Calvo P, Nelson L, Kloepper JW (2014) Agricultural uses of plant biostimulants. Plant Soil 383:3–41

Carillo P, Annunziata MG, Pontecorvo G, Fuggi A, Woodrow P (2011) Salinity stress and salt tolerance. In: Shanker A, Venkateswarlu B (eds) Abiotic stress in plants – mechanisms and adaptations. InTech, Rijeka. https://doi.org/10.5772/22331

Cole AJ, Roberts DA, Garside AL, de Nys R, Paul NA (2016) Seaweed compost for agricultural crop production. J Appl Phycol 28:629–642

du Jardin P (2015) Plant biostimulants: definition, concept, main categories and regulation. Sci Hortic 196:3–14

Elansary HO, Skalicka-Woźniak K, King IW (2016) Enhancing stress growth traits as well as phytochemical and antioxidant contents of *Spiraea* and *Pittosporum* under seaweed extract treatments. Plant Physiol Biochem 105:310–320

Elarroussia H, Elmernissia N, Benhimaa R, El Kadmiria IM, Bendaou N, Smouni A, Wahbya I (2016) Microalgae polysaccharides a promising plant growth biostimulant. J Algal Biomass Utln 7:55–63

El-Aziz Kasim WA, Saad-Allah KM, Hamouda M (2016) Seed priming with extracts of two seaweeds alleviates the physiological and molecular impacts of salinity stress on radish (*Raphanus sativus*). Int J Agric Biol 18:653–660

El-Baky HHA, El-Baz FK, Baroty GS (2010) Enhancing antioxidant availability in wheat grains from plants grown under seawater stress in response to microalgae extract treatments. J Sci Food Agric 90:299–303

Esserti S et al (2017) Protective effect of three brown seaweed extracts against fungal and bacterial diseases of tomato. J Appl Phycol 29:1081–1093

Fang Y, Xiong L (2015) General mechanisms of drought response and their application in drought resistance improvement in plants. Cell Mol Life Sci 72:673–689

Fitzpatrick GE, Worden EC, Vendrame WA (2005) Historical development of composting technology during the 20th century. HortTechnology 15:48–51

Gosch BJ, Magnusson M, Paul NA, de Nys R (2012) Total lipid and fatty acid composition of seaweeds for the selection of species for oil-based biofuel and bioproducts. Glob Chang Biol Bioenergy 4:919–930

Guinan KJ et al (2012) Discrete roles for extracts of *Ascophyllum nodosum* in enhancing plant growth and tolerance to abiotic and biotic stresses. Acta Hortic 1009:127–136

Gupta B, Huang B (2014) Mechanism of salinity tolerance in plants: physiological, biochemical, and molecular characterization. Int J Genomics 2014:1–18

Hatfield JL, Prueger JH (2015) Temperature extremes: effect on plant growth and development. Weather Clim Extrem 10(Part A):4–10

Hegazi AM, El-Shraiy AM, Ghoname AA (2015) Alleviation of salt stress adverse effect and enhancing phenolic anti-oxidant content of eggplant by seaweed extract. Gesunde Pflanz 67:21–31

Hernández-Herrera RM, Virgen Calleros G, Ruiz López M, Zañudo Hernández J, Délano-Frier JP, Sánchez-Hernández C (2014) Extracts from green and brown seaweeds protect tomato (*Solanum lycopersicum*) against the necrotrophic fungus *Alternaria solani*. J Appl Phycol 26:1607–1614

Ibrahim WM, Ali RM, Hemida KA, Sayed MA (2014) Role of *Ulva lactuca* extract in alleviation of salinity stress on wheat seedlings. Sci World J. https://doi.org/10.1155/2014/847290

Illera-Vives M, Seoane Labandeira S, Brito LM, López-Fabal A, López-Mosquera ME (2015) Evaluation of compost from seaweed and fish waste as a fertilizer for horticultural use. Sci Hortic 186:101–107

Illera-Vives M, Seoane Labandeira S, Iglesias Loureiro L, López-Mosquera ME (2017) Agronomic assessment of a compost consisting of seaweed and fish waste as an organic fertilizer for organic potato crops. J Appl Phycol 29:1663–1671

Jayaraman J, Norrie J, Punja ZK (2011) Commercial extract from the brown seaweed *Ascophyllum nodosum* reduces fungal diseases in greenhouse cucumber. J Appl Phycol 23:353–361

Jiao G, Yu G, Zhang J, Ewart HS (2011) Chemical structures and bioactivities of sulfated polysaccharides from marine algae. Mar Drugs 9:196–233

Jibril SM, Jakada BH, Kutama AS, Umar HY (2016) Plant and pathogens: pathogen recognision, invasion and plant defense mechanism. Int J Curr Microbiol App Sci 5:247–257

Jiménez E, Dorta F, Medina C, Ramírez A, Ramírez I, Peña-Cortés H (2011) Anti-phytopathogenic activities of macro-algae extracts. Mar Drugs 9:739–756

Kadam SU, Tiwari BK, O'Donnell CP (2013) Application of novel extraction technologies for bioactives from marine algae. J Agric Food Chem 61:4667–4675

Karthikeyan K, Shanmugam M (2015) Yield and oil content of peanut (var. TMV-7) and sunflower (var. Co-2) applied with biostimulant AQUASAP manufactured from seaweed. Afr J Agric Res 10:2537–2543

Li Y, Mattson NS (2015) Effects of seaweed extract application rate and method on post-production life of petunia and tomato transplants. HortTechnology 25:505–510

Mahajan PV, Caleb OJ, Singh Z, Watkins CB, Geyer M (2014) Postharvest treatments of fresh produce. Philos Trans A Math Phys Eng Sci. https://doi.org/10.1098/rsta.2013.0309

Mansori M, Chernane H, Latique S, Benaliat A, Hsissou D, El Kaoua M (2015) Seaweed extract effect on water deficit and antioxidative mechanisms in bean plants (*Phaseolus vulgaris* L.). J Appl Phycol 27:1689–1698

Mansori M, Chernane H, Latique S, Benaliat A, Hsissou D, El Kaoua M (2016) Effect of seaweed extract (*Ulva rigida*) on the water deficit tolerance of *Salvia officinalis* L. J Appl Phycol 28:1363–1370

Mantri N, Patade V, Penna S, Ford R, Pang E (2012) Abiotic stress responses in plants: present and future. In: Ahmad P, Prasad MNV (eds) Abiotic stress responses in plants: metabolism, productivity and sustainability. Springer, New York, pp 1–19

Martynenko A, Shotton K, Astatkie T, Petrash G, Fowler C, Neily W, Critchley AT (2016) Thermal imaging of soybean response to drought stress: the effect of *Ascophyllum nodosum* seaweed extract. Springerplus. https://doi.org/10.1186/s40064-016-3019-2

Michalak I, Chojnacka K (2015) Production of seaweed extracts by biological and chemical methods. In: Kim S-K, Chojnacka K (eds) Marine algae extracts. Wiley-VCH Verlag GmbH & Co. KGaA, Weinheim, pp 121–144

Mostafa GG (2015) Improving the growth of fennel plant grown under salinity stress using some biostimulants. Am J Plant Physiol 10:77–83

Nabati DA, Schmidt RE, Khaleghi ES, Parrish DJ (2008) Assessment of drought stress on physiology growth of *Agrostis palustris* Huds. as affected by plant bioregulators and nutrients. Asian J Plant Sci 7:717–723

Nabti E, Jha B, Hartmann A (2017) Impact of seaweeds on agricultural crop production as biofertilizer. Int J Environ Sci Technol 14:1119–1134

Omar AEDK (2014) Use of seaweed extract as a promising post-harvest treatment on Washington Navel orange (*Citrus sinensis* Osbeck). Biol Agric Hortic 30:198–210

Omar HH, Abdullatif BM, Al-Kazan MM, El-Gendy AM (2015) Various applications of seaweed improves growth and biochemical constituents of *Zea Mays* L. and *Helianthus Annuus* L. J Plant Nutr 38:28–40

Patakas A (2012) Abiotic stress-induced morphological and anatomical changes in plants. In: Ahmad P, Prasad M (eds) Abiotic stress responses in plants: metabolism, productivity and sustainability. Springer, New York, pp 21–39

Povero G, Mejia JF, Di Tommaso D, Piaggesi A, Warrior P (2016) A systematic approach to discover and characterize natural plant biostimulants. Front Plant Sci. https://doi.org/10.3389/fpls.2016.00435

Ramkissoon A, Ramsubhag A, Jayaraman J (2017) Phytoelicitor activity of three Caribbean seaweed species on suppression of pathogenic infections in tomato plants. J Appl Phycol. https://doi.org/10.1007/s10811-017-1160-0

Ramya SS, Vijayanand N, Rathinavel S (2015) Foliar application of liquid biofertilizer of brown alga *Stoechospermum marginatum* on growth, biochemical and yield of *Solanum melongena*. Int J Recycl Org Waste Agric 4:167–173

Rayirath P, Benkel B, Mark Hodges D, Allan-Wojtas P, MacKinnon S, Critchley AT, Prithiviraj B (2009) Lipophilic components of the brown seaweed, *Ascophyllum nodosum*, enhance freezing tolerance in *Arabidopsis thaliana*. Planta 230:135–147

Rouphael Y, de Micco V, Arena C, Raimondi G, Colla G, de Pascale S (2016) Effect of *Ecklonia maxima* seaweed extract on yield, mineral composition, gas exchange, and leaf anatomy of zucchini squash grown under saline conditions. J Appl Phycol. https://doi.org/10.1007/s10811-016-0937-x

Safinaz AF, Ragaa AH (2013) Effect of some red marine algae as biofertilizers on growth of maize (*Zea mays* L.) plants. Int Food Res J 20:1629–1632

Sangha JS, Kelloway S, Critchley AT, Prithiviraj B (2014) Seaweeds (macroalgae) and their extracts as contributors of plant productivity and quality. Adv Bot Res 71:189–219

Sarhan TZ, Ismael SF (2014) Effect of low temperature and seaweed extracts on flowering and yield of two cucumber cultivars (*Cucumis sativus* L.). Int J Agri Food Res 3:41–45

Sharma HSS, Fleming C, Selby C, Rao JR, Martin T (2014) Plant biostimulants: a review on the processing of macroalgae and use of extracts for crop management to reduce abiotic and biotic stresses. J Appl Phycol 26:465–490

Shukla PS, Borza T, Critchley AT, Prithiviraj B (2016) Carrageenans from red seaweeds as promoters of growth and elicitors of defense response in plants. Front Mar Sci 3:1–9

Singh IP, Sidana J (2013) Phlorotannins. In: Dominguez H (ed) Functional ingredients from algae for foods and nutraceuticals. Woodhead Publishing, Cambridge, pp 181–204

Spann TM, Little HA (2011) Applications of a commercial extract of the brown seaweed *Ascophyllum nodosum* increases drought tolerance in container-grown 'hamlin' sweet orange nursery trees. Hortic Sci 46:577–582

Stamatiadis S, Evangelou L, Yvin JC, Tsadilas C, Mina JMG, Cruz F (2014) Responses of winter wheat to *Ascophyllum nodosum* (L.) Le Jol. extract application under the effect of N fertilization and water supply. J Appl Phycol 27:589–600

Tarakhovskaya ER, Maslov YI, Shishova MF (2007) Phytohormones in algae. Russ J Plant Physiol 54:163–170

Van Oosten MJ, Pepe O, De Pascale S, Silletti S, Maggio A (2017) The role of biostimulants and bioeffectors as alleviators of abiotic stress in crop plants. Chem Biol Technol Agric. https://doi.org/10.1186/s40538-017-0089-5

Vardhini BV (2015) Brassinosteroids are potential ameliorators of heavy metal stresses in plants. In: Parvaiz A (ed) Plant metal interaction: emerging remediation techniques. Elsevier, Amsterdam, pp 209–237

Walbot V (2011) How plants cope with temperature stress. BCM Biol 9:79–79

Xu C, Leskovar DI (2015) Effects of *A. nodosum* seaweed extracts on spinach growth, physiology and nutrition value under drought stress. Sci Hortic 183:39–47

Xu Z, Zhou G, Shimizu H (2010) Plant responses to drought and rewatering. Plant Signal Behav 5:649–654

Zinn KE, Tunc-Ozdemir M, Harper JF (2010) Temperature stress and plant sexual reproduction: uncovering the weakest links. J Exp Bot 61:1959–1968

Zvyagintseva TN, Shevchenko NM, Popivnich IB, Isakov VV, Scobun AS, Sundukova EV, Elyakova LA (1999) A new procedure for the separation of water-soluble polysaccharides from brown seaweeds. Carbohydr Res 322:32–39

Algae As Fertilizers, Biostimulants, and Regulators of Plant Growth

10

Agnieszka Dmytryk and Katarzyna Chojnacka

Abstract

Currently, legislation restricts the use of mineral fertilizers and pesticides and thus forces a new approach to reducing the use of chemical products through either parallel application or partial replacement with formulations capable of enhancing the efficiency of conventional treatment. Among the natural materials of such capability are algae, which contain a variety of biologically active compounds verified to have a beneficial influence on plants. Algal derivatives have been confirmed to provide crops with nutrients, increase biomass production, and activate the natural ability of plants to respond properly to stress agents. Depending on the formulation, algae-based products might show the functionality of organic fertilizers or components of organo-mineral fertilizers, soil amendments (improvers), (bio)stimulants, and pesticides. However, current European rules are not harmonized at the union level. Until a single market is established, algae's potential for plant growth enhancement will not be sufficiently developed. There are, however, new strategies for elaborating EU-wide standards and regulations governing products obtained within value chains based on secondary feedstock.

Keywords

Sustainable agriculture · Plant stress · Signaling pathways · Phytohormones · Phlorotannins · Algal polysaccharides

10.1 Introduction

A discussion about modern agriculture and associated strategies for the future is of current relevance. On one hand, increasing population growth necessitates the development of activities that will enhance crop yields and prevent their loss as the result of abiotic[1] and biotic[2] stress conditions, while on the other hand, forthcoming regulations restrict the use of mineral fertilizers and chemical plant protection products to limits that make it difficult to cope with the relevant issues (Pimentel 1997; Van Velthuizen 2007; Balconi et al. 2012). According to recent trends, improvement of plant productivity – meaning enhancement of both efficiency of nutrient use and stress resistance – would preferably be effected through solutions that develop plants' natural capability for adaptation over the multiplication of agrochemical input. The reduction of chemical plant protection products is of particular concern, as they accumulate in plants and soil, posing a risk to humans, along with the environment, and lead to pathogen drug resistance. Hence, partial replacement of chemicals with naturally based substitutes that do not generate residues is expected to provide cultivation that is free from pathogens and toxic constituents. As treating plants with active compounds of biological origin has been proven to influence the growth and quality of crop yield beneficially, various natural sources have been investigated (Cutler and Cutler 2000; du Jardin 2012), with algae included among the leading choices.

The applicability of algae to agriculture has been known since plant breeding became a practice (Milton 1964; Caliceti et al. 2002; Craigie 2011). In the 1960s and 1970s, the beneficial effect of algae on plants became a subject of research (Booth 1969; Blunden 1971; Lyn 1972; Stephenson 1974). At that time, the value of algae related to primary nutrient content was suggested as being complemented by the activity of micronutrients and metabolites (Booth 1969), shedding new light on the material (Beckett and Van Staden 1989; Crouch et al. 1990; Hankins and Hockey 1990).

A. Dmytryk (✉) · K. Chojnacka
Faculty of Chemistry, Department of Advanced Material Technologies, Wrocław University of Science and Technology, Wrocław, Poland
e-mail: agnieszka.dmytryk@pwr.edu.pl

[1] Abiotic stress – negative effect or condition suffered by a living organism as a result of exposure to nonliving factors, including high and low temperatures, drought, humidity, light intensity, supply of water and minerals, CO_2, salinity, heavy metals, and ozone (Schulze et al. 2005; Ferrero 2014).

[2] Biotic stress – negative effect or condition suffered by a living organism as a result of interactions with other organisms, leading to infection (pathogen attack), mechanical damage (including trampling), symbiosis, or parasitism (Schulze et al. 2005).

© Springer International Publishing AG, part of Springer Nature 2018

K. Chojnacka et al. (eds.), *Algae Biomass: Characteristics and Applications*, Developments in Applied Phycology 8,
https://doi.org/10.1007/978-3-319-74703-3_10

As subsequent algae-derived enhancers of plant growth were identified, their potential to regulate plant reaction pathways appeared (Sanderson et al. 1987; Crouch and Van Staden 1993; Stirk et al. 2004; Rayorath et al. 2007; Khan et al. 2009; Sharma et al. 2012a). The algae-based products capable of enhancing plant resistance and mitigating damage caused by external agents are still used worldwide as emergency kits in cases of both abiotic and biotic stress (Demir et al. 2006; Spinelli et al. 2010; Zhang and Ervin 2004). The current chapter focuses on algae derivatives suitable for plant treatment and their market potential through the current and forthcoming European regulations.

10.2 Algal Bioactive Compounds for Use in Agriculture

Algae, as bio-based material, are a good source of macro- and micronutrients (Ito and Hori 1989; Ruperez et al. 2002; Rao et al. 2007; Sharma et al. 2012b, 2014). Besides, they are well known for their high content of biologically active compounds. However, there are several derivatives of particular interest when taking plant treatment into account: plant hormones (phytohormones) and hormone-like compounds, poly- and oligosaccharides, and phenolic compounds (Craigie 2011).

Phytohormones and hormone-like compounds naturally occur in terrestrial plants, playing a crucial role in the regulation of growth, development, and reaction to stimuli. Such constituents have also been identified in algae. Since they meet the expectation of beneficial effects when applied to plants externally, phytohormones and derivatives have been the best described algae-based compounds for plant treatment (Tarakhovskaya et al. 2007; Craigie 2011; Sharma et al. 2014). In Table 10.1, the biological activity of major plant hormones and hormone-like constituents is reviewed.

Signaling pathways mediated by hormones and hormone-like compounds might be induced by poly- and oligosaccharides, which comply with one of the most frequently reported metabolites derived from algae (Kraan 2012). Besides their wide use in the food industry (see Chap. 4), poly- and oligosaccharides show potential for agricultural application, which is summed up in Table 10.2.

Similar to phytohormones, stress and defense response in both terrestrial plants and algae also involves phenolic compounds (Bhattacharya et al. 2010; Onofrejová et al. 2010). The compounds differ in their available structure as a given source is considered. Among algae, brown seaweeds, including *Ascophyllum* sp., *Ecklonia* sp., *Fucus* sp., *Sargassum fusiforme*, *Ishige okamurae*, *Saccharina japonica*, *Sargassum thunbergii*, and *Undaria pinnatifida*, were proven to contain phenolic constituents defined under the common name phlorotannins. Phlorotannins – subclassed into fuhalols and phlorethols, fucols, fucophloroethols, and eckols – are not found in terrestrial plants (Nagayama et al. 2003; Parys et al. 2007; Yoon 2008; Yoon et al. 2008; Li et al. 2009, 2011). At the same time, typical plant polyphenols, flavonoids, have not so far been isolated from algal biomass (Onofrejová et al. 2010). Despite their origin and chemical differences, both algae and plant phenolic compounds are produced during the organism's exposure to nonliving stress agents and pathogens. Thus, they show a great biological effect – meaning antioxidant, radioprotective, antiviral, and antimicrobial activity, along with the capability to regulate enzyme-mediated reactions (Schmitz-Hoerner and Weissenbock 2003; Clé et al. 2008; Li et al. 2011). Such an effect is expected from plants treated with algal phenol metabolites. As high-value products, fractionated phlorotannins are components of nutraceuticals, pharmaceuticals, and cosmeceuticals (Li et al. 2011), while for agricultural purposes, phenolic-containing extracts are used instead (Craigie 2011).

Among high-value compounds extracted from algae, there are two groups with especially high potential for application when taking into account plant growth and development – carotenoids and fatty acids. The former participates in key physiological processes, i.e., respiration and photosynthesis, as well as influencing plant functioning at the level of enzyme- and hormone-dependent regulation, including mechanisms against stress and pathogen attack (Cazzonelli et al. 2010). The latter, usually associated with storage material, comprises essential components of the lipid bilayer in biological membranes. As such, fatty acids are involved in coordination of the majority of vital cellular functions and responses affecting the proper work of the plant body (Du Granrut and Cacas 2016; Hou et al. 2016). Because of the market value, the practice of using carotenoids and fatty acids in agriculture is weak. Instead, application of supercritical algal extract rich in the compounds of interest seems to be the promising approach (Michalak et al. 2016).

Both the composition diversity of biomass and the multitude of its processing configuration enable various algae-based products to be applied in plant treatment. A brief description of the legal conditions for launching such products in the European market is given in the next section of the chapter.

10.3 Applicability of Algae-Based Products According to the Current and Forthcoming Legislation

Although fertilizers are considered low-value products, agriculture itself has remained one of the major industrial sectors. Thus, great attention is paid to implementing the sustainable growth concept in activities related to plant

Table 10.1 The activity of phytohormones and hormone-like compounds when applied to plants

Compound		Effect on plants[a]
Abscisic acid	Phaeophyceae: species of *Ascophyllum*, *Saccharina*	Promotion of drought resistance by induction of protein synthesis and stomatal closure
	Chlorophyta: species of *Ulva*	Induction of protein storage in seeds
	Microalgae: species of *Chlorella*, *Dunaliella*, *Haematococcus*	Induction of seed dormancy
		Induction of gene transcription for proteinase inhibitors
		Inhibition of shoot growth
Auxins	Phaeophyceae: species of *Ascophyllum*, *Ecklonia*, *Fucus*, *Laminaria*, *Macrocystis*, *Nereocystis*, *Sargassum*, *Undaria*	Induction of elongation growth
	Chlorophyta: species of *Caulerpa*, *Cladophora*, *Ulva*	Promotion of root formation and growth
	Rhodophyta: species of *Botryocladia*, *Pyropia*	Involvement in phloem differentiation
	Microalgae: species of *Chlorella*	Induction of apical dominance
	Cyanophyta: species of *Chlorogloeopsis*, *Oscillatoria*	Induction of tropisms
Betaines	Phaeophyceae: species of *Ascophyllum*, *Fucus*, *Laminaria*	Role of methyl donor
		Role of nitrogen source when provided in low concentrations
		Role of osmolyte at higher concentrations (osmoregulation)
		Induction of drought, high salinity, high temperature, and frost resistance
		Induction of resistance to osmotic stress induced by drought and salinity stress
		Induction of defense response in pathogenesis
		Involvement in slowdown of leaf chlorophyll degradation
Brassinosteroids	Chlorophyta: species of *Hydrodictyon*	Involvement in cell division, growth by elongation, vascular differentiation
		Promotion of ethylene production
		Inhibition of root growth
Cytokinins	Phaeophyceae: species of *Ascophyllum*, *Cystoseira*, *Ecklonia*, *Fucus*, *Macrocystis*, *Sargassum*, *Undaria*	Most crop responses to seaweed-based products
	Rhodophyta: species of *Pyropia*	Promotion of cell division and morphogenesis
	Charophyta: species of *Chara*	Promotion of protein synthesis
	Microalgae: species of *Chlorella*, *Chlamydomonas*, *Desmococcus*, *Mychonastes*, *Scenedesmus*	Induction of lateral bud release
	Cyanophyta: species of *Arthronema*, *Calotrix*	Promotion of leaf expansion (cell enlargement)
	Euglenophyta: species of *Euglena*	Stimulation of chlorophyll synthesis, promotion of chloroplast maturation, and delay of leaf senescence (cell cycle control)
		Enhancement of stomatal opening
Ethylene	Phaeophyceae: species of *Ecklonia*	Induction of senescence
		Induction of defense response
Gibberellins	Phaeophyceae: species of *Cystoseira*, *Ecklonia*, *Fucus*, *Petalonia*, *Sargassum*	Induction of seed germination – activation of α-amylase
	Chlorophyta: species of *Caulerpa*, *Ulva*	Enhancement of plant stem elongation
	Rhodophyta: species of *Pyropia*	
Jasmonates (oxylipins)	Phaeophyceae: species of *Fucus*	Induction of defense and stress responses
	Rhodophyta: species of *Gelidium*	Synthesis of proteinase inhibitors
	Microalgae: species of *Chlorella*	Promotion of tuber formation and senescence
	Dunaliella, *Gelidium*	Inhibition of growth and seed germination
	Cyanophyta: species of *Arthrospira*	
	Euglenophyta: species of *Euglena*	
Lunaric acid	Chlorophyta: species of *Ulva*	Similar functions to abscisic acid
Polyamines	Rhodophyta: species of *Gelidium*, *Grateloupia*	Influence on stability of various conformational states of RNA and DNA
	Phaeophyceae: species of *Dictyota*	Involvement in growth, cell division, and normal development
	Chlorophyta: species of *Ulva*	Imparting stability to different cellular membranes
	Microalgae: species of *Chlorella*, *Cyanidium*	
	Euglenophyta: species of Euglena	

(continued)

Table 10.1 (continued)

Compound		Effect on plants[a]
Rhodomorphin	Rhodophyta: species of *Griffithsia*	Involvement in morphogenesis after removing one of the intercalary cells, induction of fusion of fragments, and restoration of filament integrity
		Induction of gamete adhesion and fusion – pheromone effect
Signal peptides	Rhodophyta: species of *Ceratodictyon, Ellisolandia, Pyropia*	Induction of defense response
	Phaeophyceae: species of *Undaria*	Involvement in self-incompatibility recognition
	Microalgae: species of *Chlorella, Navicula, Pavlova*	
	Cyanophyta: species of *Arthrospira*	

[a] Effect on Plants relates to general activity of the compound (not the compound extracted from the species in the same raw as specific effect)
Rossano et al. (2003), Aneiros and Garateix (2004), Tarakhovskaya et al. (2007), Khan et al. (2009), Craigie (2011), Panda et al. (2012), Samarakoon and Jeon (2012), and Sharma et al. (2014)

breeding. Finding solutions to limit dependency on critical (primary) raw materials and, in parallel, increasing the use of second-generation sources are of global concern. Yet, the European Union seems to have taken the lead in elaboration of the relevant strategies (Star-COLIBRI 2011; PPP BBI 2012, 2013, 2017). Such a prominent role for the EU follows the need to harmonize the legislation and actions in all member countries.

Currently, there are three key valid regulations supervising the European market of products for plant treatment:

1. Regulation (EC) No 2003/2003 of October 13, 2003, relating to fertilizers
2. Directive 2009/128/EC of October 21, 2009, establishing a framework for community action to achieve the sustainable use of pesticides
3. Regulation (EC) No 1107/2009 of October 21, 2009, concerning the placing of plant protection products on the market

It is assumed that up to 30% of nonorganic fertilizers could already have been replaced by products recycled from bio-based feedstock (Star-COLIBRI 2011; PPP BBI 2012; European Commission 2016a). There are, however, no adequate rules governing this field, since Regulation No 2003/2003 entirely focuses on well-characterized mineral (inorganic) products, addressing synthetic organic compounds only within the context of their micronutrient-chelating properties. Due to a lack of reference to organo-mineral and organic fertilizers, as well as plant growth enhancing formulations (soil amendments, stimulants/biostimulants), such products are traded on the market of a given member country under legislation at the national level. The regulation particularly does not include control mechanisms and safeguards for the dependable use of organic or secondary materials while taking the inherent variability of the feedstock into account. Hence, no relation between the fertilizer market and waste management has been established – though some categories of organic waste and by-products are legally allowed to be processed into crop-applicable formulations (Commission Decision of May 3, 2000, replacing Decision 94/3/EC, Directive 2008/98/EC of November 19, 2008, Regulation (EC) No 1069/2009 of October 21, 2009). Until the valorization of organic waste and its by-products, including nutrient recovery, is defined with the harmonized rules, there will be no possibility for free movement of all CE-marked fertilizing products across the European Union (European Commission 2016b, 2017a).

Legislation on pesticides, i.e., formulations showing their controlling effect on harmful organisms, covers both plant protection and biocidal products (Directive 2009/128/EC of October 21, 2009; European Commission 2017b). The latter are not applicable for crops (Directive 2009/128/EC of October 21, 2009, Directive 98/8/EC of February 16, 1998, no longer in force) and thus will not be further discussed in the current chapter. Therefore, the term "pesticides" hereinafter corresponds to "plant protection products" only.

Regulation (EC) No 1107/2009 of October 21, 2009, provides quite a clear scope for the functionality of plant protection products, as follows:

- Protection of plants or plant products against all harmful organisms or prevention of the action of such organisms
- Influence on the life processes of plants, in the manner of substances other than nutrients
- Preservation of plant products
- Destruction of undesired plants or parts of plants
- Check on or prevention of undesired growth of plants

Yet, these statements seem to be too general, since they lead to confusion in terms of distinguishing between pesticide and biostimulant activity (see the EBIC definition later in this section). Pesticides have also been covered by the strategy promoting innovation in products obtained within bio-based value chains (PPP BBI 2013, 2017).

In order to establish more precise guidelines that meet the market demand and opportunities, the European Commission started a dialogue with organizational structures formed by

Table 10.2 The activity of poly- and oligosaccharides when applied to plants

Compound		Effect on plants
Alginates and oligoalginates	Phaeophyceae: species of *Ascophyllum*, *Fucus*, *Laminaria*, *Lessonia*, *Sargassum*	Promotion of plant stress response and defense response in pathogenesis
		Induction of seed germination
		Enhancement of plant growth, of both above- and underground parts, and seed yield
		Improvement of plant yield quality (e.g., content of essential oils)
		Enhancement of nitrogen assimilation
Carrageenans (including κ-carrageenans, λ-carrageenans, κ- ι-, and λ-oligo-carrageenans)	Rhodophyta: species of *Eucheuma*, *Chondracanthus*, *Gigartina*, *Hypnea*, *Kappaphycus*	Involvement in cell cycle and cell division
		Enhancement of efficiency of both photosynthesis and basal metabolism
		Involvement in CO_2 and ammonium fixation and Krebs cycle
		Enhancement of plant growth and development, including biomass production (shoots, stem, leaves, roots) and flowering
		Improvement and induction of both plant resistance and immune response, including hormone and hormone-like compound-mediated pathways
		Improvement of plant yield quality (e.g., content of essential oils)
Fucans	Phaeophyceae: species of *Ascophyllum*, *Fucus*, *Laminaria*, *Pelvetia*, *Sargassum*	Improvement of plant resistance to stress, including induction of early and late defense response
		Induction of defense response in pathogenesis
		Involvement in red pigment formation
Laminarans (laminarins)	Phaeophyceae: species of *Ascophyllum*, *Fucus*, *Laminaria*, *Sargassum*	Promotion of plant resistance to stress, including hormone and hormone-like compound-mediated pathways
		Mimicking pathogen attack
		Involvement in cellular recognizing mechanisms
		Induction defense response in pathogenesis
Ulvans	*Chlorophyta*: species of *Ulva*	Induction of defense response in pathogenesis and mitigation of disease effects, including hormone and hormone-like compound-mediated pathways
		Promotion of fast and efficient immune response through the priming effect
		Induction of more frequent hypersensitive responses in epidermal cells
		Control of biotrophic and hemibiotrophic fungi

Mercier et al. (2001), Rai (2002), Klarzynski et al. (2003), Briand et al. (2007), Lahaye and Robic (2007), Laporte et al. (2007), Forcat et al. (2008), Valluru and Van den Ende (2008), Livingston et al. (2009), Jaulneau et al. (2010), Bi et al. (2011), Sharma et al. (2012b), González et al. (2013), and Stadnik and de Freitas (2014)

the entrepreneurs of member countries. Besides Fertilizers Europe, which has an almost 30-year-long tradition, three representative entrepreneurial organizations are especially worth mentioning: the European Biostimulant Industry Council (EBIC), the European Crop Protection Association (EPCA), and the Bio-based Industries (BBI) Consortium.

The EBIC facilitates interaction between key stakeholders – including regulators, investors, farmers, consumers, and scientists, to implement the approach of a single market for biostimulants to improve agricultural sustainability. For that purpose, standard harmonization is required, which has already begun with development of the consistent definition of biostimulants as containing "substance(s) and/or microorganisms whose function when applied to plants or the rhizosphere is to stimulate natural processes to enhance/benefit nutrient uptake, nutrient efficiency, tolerance to abiotic stress, and crop quality" (EBIC 2017). Since no relevant rules are in force, the definition is the reference point for biostimulant industry members.

The EPCA plays a similar role to the EBIC, but for the pesticide industry. The organization focuses on raising awareness and stewardship activities within member countries to assure the safe and sustainable use of pesticides in Europe (EPCA 2017).

As a representative of the private sector, the BBI Consortium is included in the Bio-based Industries Joint Undertaking – a public-private partnership with the European Commission, established to constitute one of the pillars of the European Commission Bioeconomy Strategy (BBI-C 2017).

Aiming at the objectives of the European strategy for 2030 to increase bio-based chemical production and enhance

Europe's reindustrialization by valorizing second-generation materials into marketable products (Star-COLIBRI 2011; PPP BBI 2012), in 2016, the European Commission published the proposal for a new regulation concerning a genuine single market for fertilizing products (European Commission 2016b, 2017a). The proposal did not provide unambiguous definitions of organic and organo-mineral fertilizers but rather emphasized the use of bio-based feedstock for fertilizer production. The rules have not yet been established, as negotiations between the European Commission and Fertilizers Europe are still being tackled.

On the other hand, the proposal clearly distinguished between the activities of plant protection products and those of biostimulants. For that purpose, the proposed emendation to the Regulation (EC) No 1107/2009 was to separate the definition of "plant biostimulant" (Article 3, point 34) as "a product stimulating plant nutrition processes independently of the product's nutrient content with the sole aim of improving one or more of the following characteristics of the plant: (a) nutrient use efficiency; (b) tolerance to abiotic stress; (c) crop quality traits." The European legislative authorities noted the need to exclude biostimulants from pesticides while including their marking requirements under the regulation on fertilizers (European Commission 2016b).

Taking into account the forthcoming rules, there is a great opportunity to launch algae-based products, especially as biostimulants and plant protection products – due to their confirmed antistressor activity (see Sect. 10.2). This is also in accordance with the newest European strategy, which points the potential development of algae in the direction of extracting mid- or high-value compounds, rather than using raw biomass for low-value products (PPP BBI 2017). The current market for algae-based formulations of interest is dominated by brands: Göemar (now in Arysta LifeScience offer, arystalifescience.com), Valagro (valagro.com), and Kelpak (kelpak.com). The creation of new bio-based value chains promoted by the new regulation would enhance the diversity of and access to such products and hence positively influence the sustainability of European agriculture.

10.4 Conclusions

Algae have been known for their beneficial effects on plants for centuries, but their market potential still seems to be underestimated. As biomass, excluded from primary raw materials, and a rich natural source of compounds with high biological activity against stress factors, algae fit very well into recent European strategies for the chemical industry, including production of fertilizers and pesticides. Accordingly, meeting the forthcoming regulation, algae-based biostimulants and plant protection products are a great solution for the future for assuring sustainable agriculture.

References

Aneiros A, Garateix A (2004) Bioactive peptides from marine sources: pharmacological properties and isolation procedures. J Chromatogr B 803(1):41–53

Arysta LifeScience. https://www.arystalifescience.com/. Accessed on 4 Sept 2017

Balconi C, Stevanato P, Motto M, Biancardi E (2012) Breeding for biotic stress resistance/tolerance in plants. In: Ashraf M, Öztürk M, Ahman MSA, Aksoy A (eds) Crop production for agricultural improvement. Springer, Dordrecht, pp 57–114

Beckett RP, van Staden J (1989) The effect of seaweed concentrate on the growth and yield of potassium stressed wheat. Plant Soil 116:29–36

Bhattacharya A, Sood P, Citovsky V (2010) The roles of plant phenolics in defence and communication during *Agrobacterium* and *Rhizobium* infection. Mol Plant Pathol 11(5):705–719

Bi F, Iqbal S, Arman M, Ali A, Hassan MU (2011) Carrageenan as an elicitor of induced secondary metabolites and its effects on various growth characters of chickpea and maize plants. J Saudi Chem Soc 15:269–273

Bio-Based Industry Council (BBI-C) (2017). http://biconsortium.eu/. Accessed on 4 Sept 2017

Blunden G (1971) The effect of aqueous seaweed extract as fertilizer additives. Proc Int Seaweed Symp Tokyo 7:584–589

Booth E (1969) The manufacture and properties of liquid seaweed extracts. Proc Int Seaweed Symp 6:655–662

Briand X, Cluzet S, Dumas B, Esquerre-Tugaye MT, Salamagne S (2007) Use of ulvans as activators of plant defence and resistance reactions against biotic or abiotic stresses. US Patent 2007/0232494 A1

Caliceti M, Argese E, Sfriso A, Pavoni B (2002) Heavy metal contamination in the seaweeds of the Venice lagoon. Chemosphere 47:443–454

Cazzonelli CI, Nisar N, Hussain D, Carmody ME, Pogson BJ (2010) Biosynthesis and regulation of carotenoids in plants – micronutrients, vitamins and health benefits. In: Pua EC, Davey MR (eds) Plant developmental biology-Biotechnological perspectives, vol 2. Springer, Berlin, pp 117–137

Clé C, Hill LM, Niggeweg R, Martin CR, Guisez Y, Prinsen E, Jansen MA (2008) Modulation of chlorogenic acid biosynthesis in *Solanum lycopersicum*; consequences for phenolic accumulation and UV-tolerance. Phytochemistry 69:2149–2156

Commission Decision of 3 May 2000 replacing Decision 94/3/EC establishing a list of wastes pursuant to Article 1(a) of Council Directive 75/442/EEC on waste and Council Decision 94/904/EC establishing a list of hazardous waste pursuant to Article 1(4) of Council Directive 91/689/EEC on hazardous waste

Craigie JS (2011) Seaweed extract stimuli in plant science and agriculture. J Appl Phycol 23:371–393

Crouch IJ, van Staden J (1993) Evidence for the presence of plant growth regulators in commercial seaweed products. Plant Growth Regul 13:21–29

Crouch IJ, Beckett RP, van Staden J (1990) Effect of seaweed concentrate on growth and mineral nutrition of nutrient stressed lettuce. J Appl Phycol 2:269–271

Cutler HG, Cutler SJ (2000) Growth regulators, plant. In: Kirk-Othmer encyclopedia of chemical technology (published online 15 October 2004). http://onlinelibrary.wiley.com/doi/10.1002/0471238961.161 2011403212012.a01.pub2/full. Accessed on 4 Sept 2017

Demir N, Dural B, Yildirim K (2006) Effect of seaweed suspensions on seed germination of tomato, pepper and aubergine. J Biol Sci 6:1130–1133

Directive 2008/98/EC of the European Parliament and of the Council of 19 November 2008 on waste and repealing certain Directives (Text with EEA relevance)

Directive 2009/128/EC of the European Parliament and of the Council of 21 October 2009 establishing a framework for Community action to achieve the sustainable use of pesticides (Text with EEA relevance)

Directive 98/8/EC of the European Parliament and of the Council of 16 February 1998 concerning the placing of biocidal products on the market, no longer in force

Du Granrut ADB, Cacas JL (2016) How very-long-chain fatty acids could signal stressful conditions in plants? Front Plant Sci 7:1–13. https://doi.org/10.3389/fpls.2016.01490

Du Jardin P (2012) The science of plant biostimulants–a bibliographic analysis, ad hoc study on biostimulants products. European Commission, Contract 30-CE0455515/00-96

European Biostimulant Industry Council (EBIC) (2017). http://www.biostimulants.eu/. Accessed on 4 Sept 2017

European Commission (2016a) Circular economy: new regulation to boost the use of organic and waste-based fertilisers. Press release, Brussels. 17 Mar 2016

European Commission (2016b) Circular economy package. Proposal for a regulation of the European Parliament and of the Council laying down rules on the making available on the market of CE marked fertilising products and amending regulations (EC) No 1069/2009 and (EC) No 1107/2009 (text with EEA relevance). Brussels. 17 Mar 2016

European Commission (2017a) Report from the commission to the European Parliament, the council, the European economic and social committee and the committee of the regions on the implementation of the circular economy action plan. Brussels. 26 Jan 2017

European Commission (2017b) Pesticides. https://ec.europa.eu/food/plant/pesticides_en. Accessed on 3 Nov 2017

European Crop Protection Association (EPCA) (2017). http://www.ecpa.eu/. Accessed on 4 Sept 2017

Ferrero A (ed) (2014) Role in sustainable agriculture, detrimental effects and management strategies. Nova Science Publishers, New York

Forcat S, Bennett MH, Mansfield JW, Grant MR (2008) A rapid and robust method for simultaneously measuring changes in the phytohormones ABA, JA and SA in plants following biotic and abiotic stress. Plant Methods 4:16–23

González A, Castro J, Vera J, Moenne A (2013) Seaweed oligosaccharides stimulate plant growth by enhancing carbon and nitrogen assimilation, basal metabolism, and cell division. J Plant Growth Regul 32:443–448

Hankins SD, Hockey HP (1990) The effect of a liquid seaweed extract from *Ascophyllum nodosum* on the 2 spotted red spider mite *Tetranychus urticae*. Hydrobiologia 204:555–559

Hou Q, Ufer G, Bartels D (2016) Lipid signalling in plant responses to abiotic stress. Plant Cell Environ 39(5):1029–1048

Ito K, Hori K (1989) Seaweed: chemical composition and potential foods uses. Food Rev Int 5:101–144

Jaulneau V, Lafitte C, Jacquet C, Fournier S, Salamagne S, Briand X, Esquerré-Tugayé MT, Dumas B (2010) Ulvan, a sulfated polysaccharide from green algae, activates plant immunity through the jasmonic acid signaling pathway. J Biomed Biotechnol 2010:1–11. https://doi.org/10.1155/2010/525291

Kelpak. http://www.kelpak.com/. Accessed on 4 Sept 2017

Khan W, Rayirath UP, Subramanian S, Jithesh MN, Rayorath P, Hodges DM, Critchley AT, Craigie JS, Norrie J, Prithiviraj B (2009) Seaweed extracts as biostimulants of plant growth and development. J Plant Growth Regul 28:386–399

Klarzynski O, Descamps V, Plesse B, Yvin JC, Kloareg B, Fritig B (2003) Sulfated fucan oligosaccharides elicit defense responses in tobacco and local and systemic resistance against tobacco mosaic virus. Mol Plant-Microbe Interact 16:115–122

Kraan S (2012) Algal polysaccharides, novel applications and outlook. In: Chang CF (ed) Carbohydrates – comprehensive studies on glycobiology and glycotechnology. InTech, Rijeka. https://doi.org/10.5772/51572

Lahaye M, Robic A (2007) Structure and functional properties of ulvan, a polysaccharide from green seaweeds. Biomacromolecules 8:1765–1774

Laporte D, Vera J, Chandía NP (2007) Structurally unrelated algal oligosaccharides differentially stimulate growth and defense against tobacco mosaic virus in tobacco plants. J Appl Phycol 19:79–88

Li Y, Qian ZJ, Ryu BM, Lee SH, Kim MM, Kim SK (2009) Chemical components and its antioxidant properties in vitro: an edible marine brown alga, *Ecklonia cava*. Bioorg Med Chem 17:1963–1973

Li YX, Wijesekara I, Li Y, Kim SK (2011) Phlorotannins as bioactive agents from brown algae. Process Biochem 46:2219–2224

Livingston DP, Dirk KH, Heyer AG (2009) Fructan and its relationship to abiotic stress tolerance in plants. Cell Mol Life Sci 66:2007–2023

Lyn LB (1972) The chelating properties of seaweed extract *Ascophyllum nodosum* vs *Macrocystis pyrifera* on the mineral nutrition of sweet peppers, *Capsicum annum*. MSc thesis, Clemson University, Clemson

Mercier L, Lafitte C, Borderies G, Briand X, Esquerré-Tugayé MT, Fournier J (2001) The algal polysaccharide carrageenans can act as an elicitor of plant defense. New Phytol 149:43–51

Michalak I, Chojnacka K, Dmytryk A, Wilk R, Gramza M, Rój E (2016) Evaluation of supercritical extracts of algae as biostimulants of plant growth in field trials. Front Plant Sci 7:1–11. https://doi.org/10.3389/fpls.2016.01591

Milton RF (1964) Liquid seaweed as a fertilizer. In: Proceedings of the 4th international seaweed symposium, vol 4. The Macmillan Co, London, pp 428–431

Nagayama K, Shibata T, Fujimoto K, Honjo T, Nakamura T (2003) Algicidal effect of phlorotannins from the brown alga *Ecklonia kurome* on red tide microalgae. Aquaculture 218:601–611

Onofrejová L, Vašíčková J, Klejdus B, Stratil P, Mišurcová L, Kráčmar S, Kopecký J, Vacek J (2010) Bioactive phenols in algae: the application of pressurized-liquid and solid-phase extraction techniques. J Pharm Biomed Anal 51(2):464–470

Panda D, Pramanik K, Nayak BR (2012) Use of sea weed extracts as plant growth regulators for sustainable agriculture. Int J Biores Stress Manag 3(3):404–411

Parys S, Rosenbaum A, Kehraus S, Reher G, Glombitza KW, König GM (2007) Evaluation of quantitative methods for the determination of polyphenols in algal extracts. J Nat Prod 70(12):1865–1870

Pimentel D (1997) Techniques for reducing pesticide use: economic and environmental benefits. Wiley, Hoboken

Public-Private Partnership on Bio-Based Industries (PPP BBI) (2012) The bio-based industries vision. Accelerating innovation and market uptake of bio-based products

Public-Private Partnership on Bio-Based Industries (PPP BBI) (2013) Strategic innovation and research agenda. Bio based and renewable industries for development and growth in Europe

Public-Private Partnership on Bio-Based Industries (PPP BBI) (2017) Strategic innovation and research agenda. Bio-based industries for development and growth in Europe

Rai VK (2002) Role of amino acids in plant responses to stresses. Biol Plant 45:481–487

Rao PVS, Mantri VA, Ganesan K (2007) Mineral composition of edible seaweed *Porphyra vietnamensis*. Food Chem 102:215–218

Rayorath P, Narayanam JM, Farid A, Khan W, Palanisamy R, Hankins SD, Critchley AT, Prithiviraj B (2007) Rapid bioassays to evaluate the plant growth promoting activity of *Ascophyllum nodosum* (L.) Le Jol. using a model plant, *Arabidopsis thaliana* (L) Heynh. J Appl Phycol 20:423–429

Regulation (EC) No 1069/2009 of the European Parliament and of the Council of 21 October 2009 laying down health rules as regards animal by-products and derived products not intended for human consumption and repealing Regulation (EC) No 1774/2002 (Animal by-products Regulation)

Regulation (EC) No 1107/2009 of the European Parliament and of the Council of 21 October 2009 concerning the placing of plant protection products on the market and repealing Council Directives 79/117/EEC and 91/414/EEC

Regulation (EC) No 2003/2003 of the European Parliament and of the Council of 13 October 2003 relating to fertilisers (Text with EEA relevance)

Rossano R, Ungaro N, D'Ambrosio A, Liuzzi GM, Riccio P (2003) Extracting and purifying R-phycoerythrin from Mediterranean red algae *Corallina elongata* Ellis & Solander. J Biotechnol 101(3):289–293

Ruperez P, Ahrazem O, Leal JA (2002) Potential antioxidant capacity of sulphated polysaccharides from the edible brown seaweed *Fucus vesiculosus*. J Agric Food Chem 50:840–845

Samarakoon K, Jeon YJ (2012) Bio-functionalities of proteins derived from marine algae – a review. Food Res Int 48(2):948–960

Sanderson KJ, Jameson PE, Zabkiewiez JA (1987) Auxin in a seaweed extract: identification and quantification of indole-3-acetic acid by gas chromatography-mass spectrometry. J Plant Physiol 129:363–367

Schmitz-Hoerner R, Weissenbock G (2003) Contribution of phenolic compounds to the UV-B screening capacity of developing barley primary leaves in relation to DNA damage and repair under elevated UV-B levels. Phytochemistry 64:243–255

Schulze ED, Beck E, Hohenstein KM (2005) Environment as stress factor: stress physiology of plants. In: Plant ecology. Springer, Berlin/Heidelberg, pp 7–11

Sharma SHS, Lyons G, McRoberts C, McCall D, Carmichael E, Andrews F, McCormack R (2012a) Brown seaweed species from Strangford Lough: compositional analyses of seaweed species and biostimulant formulations by rapid instrumental methods. J Appl Phycol 24:1141–1157

Sharma SHS, Lyons G, McRoberts C, McCall D, Carmichael E, Andrews F, Swan R, McCormack R, Mellon R (2012b) Biostimulant activity of brown seaweed species from Strangford Lough: compositional analyses of polysaccharides and bioassay of extracts using mung bean (*Vigna mungo* L.) and pak choi (*Brassica rapa chinensis* L.). J Appl Phycol 24:1081–1091

Sharma SHS, Fleming C, Selby C, Rao JR, Martin T (2014) Plant biostimulants: a review on the processing of macroalgae and use of extracts for crop management to reduce abiotic and biotic stresses. J Appl Phycol 26(1):465–490

Spinelli F, Fiori G, Noferini M, Sprocatti M, Costa G (2010) A novel type of seaweed extract as a natural alternative to the use of iron chelates in strawberry production. Sci Hortic 125:263–269

Stadnik MJ, de Freitas MB (2014) Algal polysaccharides as source of plant resistance inducers. Trop Plant Pathol 39:111–118

Star-COLIBRI (2011) Joint European biorefinery vision for 2030. Strategic targets for 2020 – Collaboration Initiative on Biorefineries

Stephenson WA (1974) Seaweed in agriculture and horticulture, 3rd edn. Barglya & Gylver Rateaver, San Diego

Stirk WA, Arthur GD, Lourens AF, Novák O, Strnad M, Van Staden J (2004) Changes in cytokinin and auxin concentrations in seaweed concentrates when stored at an elevated temperature. J Appl Phycol 16:31–39

Tarakhovskaya ER, Maslov YI, Shishova MF (2007) Phytohormones in algae. Russ J Plant Phys 54(2):163–170

Valagro. https://www.valagro.com/en/corporate/#. Accessed on 4 Sept 2017

Valluru R, Van den Ende W (2008) Plant fructans in stress environments: emerging concepts and future prospects. J Exp Bot 59:2905–2916

Van Velthuizen H (2007) Mapping biophysical factors that influence agricultural production and rural vulnerability. Food and Agriculture Organization of the United Nations and the International Institute of Applied Systems Analysis, Rome, pp 1–84. http://www.fao.org/docrep/010/a1075e/a1075e00.htm. ISBN:978-92-5-105689-9. Accessed on 4 Sept 2017

Yoon NY (2008) Cholinesterase and lens aldose reductase inhibitory activities of phlorotannins from *Ecklonia stolonifera* and their protective effects on tacrine induced hepatotoxicity and hyperlipidemic rat models. Ph.D. thesis, Pukyong National University, Republic of Korea

Yoon NY, Chung HY, Kim HR, Choi JS (2008) Acetyl- and butyrylcholinesterase inhibitory activities of sterols and phlorotannins from *Ecklonia stolonifera*. Fish Sci 74:200–207

Zhang X, Ervin EH (2004) Cytokinin-containing seaweed and humic acid extracts associated with creeping bentgrass leaf cytokinins and drought resistance. Crop Sci 44:1737–1745

Algae Biomass in Animal Production

11

Damian Konkol, Wanda Górniak, Marita Świniarska, and Mariusz Korczyński

Abstract

Constantly increasing populations have forced the producers of food into increasing the scale of their production. Consumers expect that, with the increase in food products, their quality will also improve. This is especially evident in the livestock sector, as the demand for animal protein is systematically increasing. This situation led to the search for innovative products of natural origin that could be used in animal husbandry and breeding. This product could be an alga containing ingredients in its biomass that have a positive impact on animal and human organisms. Not only can algae-based feed additives improve production parameters and animal health, they can also affect the quality of animal products. Several studies have been conducted to develop algae in feed for poultry, pigs, cattle and horses. These studies have shown that the use of algae as feed additives can bring many benefits, due to their unique properties.

Keywords

Algae · Feed additives · Animal nutrition · Biologically active compounds

11.1 Introduction

The food industry runs along two main lines, those being crop production and animal production. Animal production is developing at an increasingly rapid pace, which is primarily attributable to an increase in the demand for animal protein, most noticeable in developing countries. This trend has led to a stabilization of the food market for products of animal origin and consequently has also affected the fodder market. The global feed market is based on five main lines of production; however, the most important for the market is feed production for pigs, poultry and cattle. Feed production is increasingly leaning on materials of plant origin, determined by the physiology of farm animals. The main items of strategic importance for feed manufacturers are cereals, animal fats and food waste (Korczyński et al. 2015). In addition to those materials, feed additives are of great importance for the value of the feed. Depending on the type of additives applied, they can increase the digestibility of nutrients, compensate for nutrient deficiencies (vitamins, minerals, exogenous amino acids) and improve the taste, aroma and quality of the feed. Feed additives can also be used to improve the taste and composition of animal products. The most controversial feed additives are those containing antibiotics and hormones in their composition. The use of these feed additives has been banned in 27 EU countries. Since 2006, only coccidiostats and histomonostats have been authorized (Korczyński and Opaliński 2012; Opaliński et al. 2012). The use of antibiotics as feed additives has been banned, among other practices, due to the increasing drug resistance of many pathogens. In some cases, the industry has managed to replace antibiotics with substances of natural origin (Kupczyński et al. 2013). The high quality of animal products, along with ensuring the proper welfare of the animals, has economic benefits; thus new feed additives are being sought. This search is mainly concentrated on the better availability of natural components containing macroelements in their composition (Dobrzański et al. 2008; Dolińska et al. 2011; Michalak et al. 2011; Opaliński et al. 2012; Saeid et al. 2013a, b), animal products containing a lot of polyunsaturated fatty acids (PUFAs) (Gładkowski et al. 2011; Kupczyński et al. 2012b) and mechanisms for improvement of the conditions of animal health (Adamski et al. 2011; Kupczyński et al. 2012b, 2013). Materials of natural origin that are able to combine these dietary functions in the context of breeding and animal husbandry include both marine and freshwater algae. These materials are used in many regions of the world as both animal feed and human food (Spolaore et al. 2006). The European Union allows the use of algae as raw feed materials (Korol 2002). Algae, due to their unique composition, can be used as functional additives that are applied in order to enrich animal products. Algae, apart from

D. Konkol · W. Górniak · M. Świniarska · M. Korczyński (✉)
Department of Environment Hygiene and Animal Welfare, Wrocław University of Environmental and Life Sciences, Wrocław, Poland
e-mail: mariusz.korczynski@upwr.edu.pl

© Springer International Publishing AG, part of Springer Nature 2018
K. Chojnacka et al. (eds.), *Algae Biomass: Characteristics and Applications*, Developments in Applied Phycology 8,
https://doi.org/10.1007/978-3-319-74703-3_11

their nutritional value, are also rich in biologically active ingredients, natural antioxidants and antibacterial compounds (Gupta and Abu-Ghannam 2011). Algae can be applied as feed additives and also directly distributed to animals. Algae biomass is digested by cattle, swine and sheep (Michalak and Chojnacka 2008). Biologically active ingredients included in the algae biomass include, among others, polysaccharides, proteins, polyunsaturated fatty acids (PUFAs), polyphenols, dyes and mineral elements.

Polysaccharides are biopolymers composed of monosaccharides bound by glycosidic bonds. The content of polysaccharides in the dry substance of algae can reach as high as 76% and is dependent on the algal group (Holdt and Kraan 2011). The polysaccharides found in green algae are primarily galactans and xylans; those in brown algae, alginates, fucoidan and laminarin; and those in red – carrageenan (Senthilkumar et al. 2013). These compounds, thanks to their antibacterial properties, can be an alternative to antibiotics. The supplementation of polysaccharide-rich algae improves the stabilization of alimentary tract microflora and increases the absorption of nutrients (Leonard et al. 2010; O'Doherty et al. 2010).

The content of proteins in the dry substance of algae can reach 44% but usually doesn't exceed 5% (Leonard et al. 2010). Brown algae are characterized by the lowest level of protein, e.g. Phaeophyceae (*Fucus* spp. 3–17% in dry mass). Red algae (Rhodophyta), e.g. *Porphyra tenera* and *Palmaria palmata*, are characterized by a protein content on the level of appropriately 30–44% (Galland-Irmouli et al. 1999). The highest protein level was recorded in *A. platensis*, which was 63% (Tokuşoglu and Üunal 2003). The vast majority of known species of alga contain all essential amino acids, although their number and mutual proportions are conditional on the season of occurrence. The amino acid profile of proteins contained in such algae as *Arthrospira platensis*, *Chlorella vulgaris* and *Dunaliella bardawil* is similar to the profile of the white of the chicken egg (Becker 2007). The peptides contained in the algae biomass that deserve special attention are lectins (Boyd et al. 1966). They bind carbohydrates (Kole et al. 2010; Holdt and Kraan 2011), regulate the activity of many enzymes, participate in interactions between cells and exhibit antibacterial, anti-inflammatory and antiviral effects. Lectins have been detected in such algae as *Ulva* sp., *Hypnea japonica*, *Galaxaura marginata* and *Eucheuma serra* (Holdt and Kraan 2011; Chojnacka et al. 2012; Zhang et al. 2012).

The major groups of lipids appearing in algae biomass are phospholipids and glycolipids. The content of these connections usually doesn't exceed 5%. The amount of fatty acids is dependent on the species of alga, the temperature and the salinity (Holdt and Kraan 2011). Brown algae such as *Dictyota acutiloba* and *Dictyota sandvicensis* are characterized by the highest content of these connections, amounting appropriately to 16.2% and 20.2%, respectively (McDermid and Stuercke 2003). During a fall in their environmental temperature, algae accumulate polyunsaturated fatty acids (PUFAs); therefore algae biomass found in cool climatic zones contains more PUFAs than the algae biomass that come from warm zones (Holdt and Kraan 2011). Particularly noteworthy are the omega-3 fatty acids (n-3 PUFAs). The most important functions in both human and animal organisms are fulfilled by docosahexaenoic acid (DHA), eicosapentaenoic acid (EPA) and α-linolenic acid (ALA) (Yates et al. 2014). Due to the fact that essential unsaturated fatty acids are not synthesized by the human body (Kalogeropoulos et al. 2010) and that their pro-health effects have been proven, demand for products containing these substances has increased (Adarme-Vega et al. 2014). It has been shown that deficiencies of essential unsaturated fatty acids and the inappropriate ratio of omega-6 acids to omega-3 acids constitute the cause of many diseases. Diseases caused by the aforementioned factors include atherosclerosis, hypertension, rheumatoid arthritis, obesity, diabetes and some types of cancer (Lewis et al. 2000; Wongcharoen and Chattipakorn 2005; Fredriksson et al. 2006; Kolanowski 2007; Kassis et al. 2010; Khozin-Goldberg et al. 2011; Fraeye et al. 2012; Kupczyński et al. 2012a; Ryckebosch et al. 2012). This capability allows for the use of algae as a raw material or feed additive, which at the same time eliminates the problem of the fishy aftertaste and the scent accompanying the use of fish meals and fish oil (Barclay et al. 1994; Harun et al. 2010; Kassis et al. 2010; Khozin-Goldberg et al. 2011).

A very important group of compounds that can be extracted from algae is polyphenols (Holdt and Kraan 2011), which counteract the formation of free radicals (Gupta and Abu-Ghannam 2011). The dry substance of algae can contain anywhere from 1% to 10% of these compounds (Craigie 2011). In cases of food enrichment with polyunsaturated fatty acids, the addition of polyphenols is extremely important, since they contribute to reducing the degree of oxidizing lipids (Kassis et al. 2010). Thanks to their high content of polyphenols, algae can successfully replace synthetic antioxidants such as BHT and BHA, which have carcinogenic effects. Thanks to their antioxidant properties, polyphenols contained in algae can be used in the prevention and treatment of many diseases. They show protective qualities towards the nervous system, have anti-inflammatory and anti-cancer properties and support the fight against obesity (Bravo 1998; O'Sullivan et al. 2011; Pangestuti and Kim 2011). The properties of algae such as *Laminaria* spp., *Saccharina* spp., *Pyropia* spp. and *Porphyra* spp. reduce the risk of falling ill with cancers of the mammary gland and the intestine (Yuan and Walsh 2006). It has also been shown that the polyphenols contained in brown algae protect organisms against cardiovascular diseases (Kang et al. 2003). Nwosu et al. (2011) showed that polyphenols could limit the appearance of diabetes.

In algae, it is possible to single out three major groups of dyes, that is, chlorophylls, carotenoids and phycobiliproteins (Holdt and Kraan 2011; Stengel et al. 2011). Similarly to polyphenols, it is possible to use pigments in the prevention and treatment of many diseases, since they suppress the formation of free radicals (Pangestuti and Kim 2011; Bruneel et al. 2013). Algal dyes can also be used to improve the coloration of animal products, such as egg yolks. These products are more attractive and desirable to consumers (Pangestuti and Kim 2011).

Algae biomass contains a lot of mineral elements in its composition, such as iodine, zinc, iron, copper, calcium, magnesium, sodium and potassium. The content of these elements is much higher in algae than in edible land plants, even exceeding 40% (Holdt and Kraan 2011). In cases of mineral deficiency, algae can successfully replace traditional feed additives (Rupérez 2002; Rao et al. 2007; Kumar et al. 2011; Nwosu et al. 2011; Chojnacka et al. 2012; Tuhy et al. 2012). He et al. (2002) proved that the bioavailability of micro- and macroelements from algae biomass is significantly higher than that of inorganic compounds such as salt.

As mentioned above, algae contain a number of other biologically active compounds. Among them, it is possible to distinguish indoles and terpenes, being secondary metabolites of red and brown algae that have antibacterial, antifungal and antiviral properties (Jiménez et al. 2010; Raj 2016). Algae, in their composition, also contain enzymes such as superoxide dismutase, peroxidase, catalase or glutathione reductase, which have a protective effect on organisms (Ganesan et al. 2011). Additionally, the presence of saponins with anti-inflammatory and antifungal properties was demonstrated in red algae (Raj 2016).

11.2 Algae in Animal Nutrition

11.2.1 Algae in Poultry Feed

Algae can be successfully used in the feeding of poultry, pigs, ruminants and horses. The relatively low price and high dietary value of poultry products has caused them to enjoy great interest from consumers. This situation has led to the rapid advancement of poultry farming worldwide. Today's lines of laying hens and chickens for fattening are characterized by high production parameters. At the present scale of production, not only are solutions sought to improve the health and welfare of the birds, but those solutions must also improve the quality parameters and pro-health properties of the finished products. Such a solution may be algae-based feed additives. Use of algae in poultry feed is aimed at improving production parameters of laying hens and broilers, improving the quality parameters of eggs, making poultry products rich in biologically active compounds and improving the health of the birds. Sirakov et al. (2012) explored the possibility of the addition of algae in amounts not exceeding 5–10% of the dry substance, perhaps to be used as a substitute for traditional protein sources and the fishmeal in poultry feed. Research (Bruneel et al. 2013) showed an increase in ALA and DHA fatty acids by 7% and 53%, respectively, in chicken eggs fed with *Nannochloropsis gaditana* (5% dry substance) after 14 days of experimentation. Additionally, the EPA content in the eggs grew from 0 up to 2.1 mg. The same studies also showed an increased colouring of the yolks on the La Roche scale from 10 in the control group to 14 and 15 in those groups receiving 5 and 10% addition of the *Nannochloropsis gaditana*. Supplementation of the commercial feed mixtures for laying hens with marine algae (in the amount of 20%), as detailed by Fredriksson et al. (2006), caused an increased concentration of PUFAs in the phospholipids fraction of the yolk. Additionally, the authors showed that supplementation of algae with rapeseed oil increases omega-3 fatty acids by about 15%, reducing the content of omega-6 acids by about 8% for chickens only receiving the addition of rapeseed oil. Studies by the same authors have shown that the 4-week supplementation of *Nannochloropsis oculata* in the amount of 20% in feed for laying hens increased the overall quantity of carotenoids (from 9.7 up to 37 mg kg^{-1}), luteins (from 8 up to 22 mg kg^{-1}), β-carotene (from 0 up to 1.3 mg kg^{-1}) and canthaxanthin (from 0 up to 7.7 mg kg^{-1}) in eggs, compared with the birds from the control group. Research (Cachaldora et al. 2005) was aimed at comparing the supplementation of oil of microalgae and fish oil in the amount of 1.7%. Based on the results, the increase in EPA and DHA content in chicken eggs supplemented with microalgal oil was 46% and 17%, respectively. Kalogeropoulos et al. (2010) showed a fivefold increase in DHA content, an improvement in the ratio of omega-3 to omega-6 fatty acids and increased carotenoids in the muscles of broiler chickens receiving the addition of the *Schizochytrium* sp., compared with chickens that were fed the conventional feed mixtures. Lemahieu et al. (2014) found an increased content of omega-3 fatty acids in the egg yolks of chickens receiving microalgal supplementation. Kostik et al. (2015) demonstrated that the addition of *Schizochytrium* spp. (1.27% and 1.77%) increased the DHA content in egg yolks. Improvement of the fatty acid profiles in eggs after application of *Schizochytrium* was also demonstrated by Park et al. (2015). Rizzi et al. (2009) have confirmed the possibility of producing eggs fortified with DHA through supplementation of microalga in the diet of laying hens. An experiment conducted by Michalak et al. (2016) showed that the use of supercritical extracts of *Spirulina* sp. as an addition to drinking water in an amount of 2% resulted in a slight increase in omega-3 and omega-6 fatty acids in the yolk. After 120 days of experimentation, there was an increase in GLA, ETE and DPA content in egg yolks of 12%, 13% and 11%, respectively, compared to the control group. The use of *Arthrospira* microalga as a feed additive in 15 g/kg^{-1} resulted in an increase in EPA and DPA of 12 and 20%. Nitsan et al. (1999) showed the intensification of the colour of yolks by increasing the contents of carotenoids in the eggs of hens being fed the algae supplement from the *Nannochloropsis* group. Ševčíková et al. (2006) found that the meat of the broilers whose feed was supplemented with yeast and *Chlorella* algae enriched with selenium was characterized by a higher content of this element, compared with the meat of the birds from the control group. Other studies (Michalak et al. 2011) have shown that the addition of algae enriched with micronutrients increased egg weight and shell thickness, thus reducing the number of broken eggs. On the basis of these studies, it was also found that the body weight of the chickens receiving the tested supplements was greater than that of the control chickens. The experimental groups were also characterized by more intense egg yolk coloration. These studies also showed a higher concentration of copper and manganese in the blood of chickens receiving the macroalgae additive. The concentration was higher by 14% and 31%, respectively. Additionally, the whites and yolks of eggs supplemented with macroalgae were shown to have higher chromium and copper content. The white was also characterized by higher zinc concentration and lower manganese concentration. In yolks, manganese concentration also increased, while cobalt and zinc concentrations decreased. Skrivan et al. (2006) obtained similar results, stating that the weight of the eggs from hens fed with selenium-enriched *Chlorella* algae was higher than that of the eggs from hens not receiving such a supplement. These studies also showed that the whites and yolks of hens from the experimental group had been characterized by a higher content of selenium, compared with the control group. Zahid et al. (1995) showed that chicks given feed with 10, 20 and 30% addition of green algae (*Ulva intestinalis*, *Ulva lactuca*, *Ulva taeniata*, *Caulerpa taxifolia*, *Codium flabellatum*, *Codium iyengarii*, *Halimeda tuna*, *Bryopsis pennata*, *Caulerpa scalpelliformis*) were characterized by a higher body weight compared with chicks from the control group. Ginzberg et al. (2000), by adding *Porphyridium* sp. to the feed, reduced the concentration of the total cholesterol in the blood serum of chickens of the White Leghorn race.

These studies clearly show that the use of algae as an additive in poultry feed brings many benefits. Apart from the better use of the bird's genetic potential, improved health and consequent well-being, a high-quality product also has a positive impact on the health of consumers.

11.2.2 Algae in Swine Feed

Besides poultry, pigs are also very important for the food industry. Pigs, apart from large-commercial breeding, are also being raised in many little family households. The distinct flavour and nutritional value of pork have led to these products being chosen by the consumer in numbers often equal to those of poultry products. Today's breeds and breeding lines of pigs are characterized by rapid weight gain and low carcass fat. Research associated with the use of algae in feeding pigs has focused, above all, on concentrating on pro-health elements of the pork from the point of view of the consumer, increasing production parameters of pigs, improvement in the health of the animals and better use of nutrients. Research (Saeid et al. 2013a) on the use of *Arthrospira maxima* algae supplementation in pigs did not show any effect on production parameters (average daily feed intake, feed conversion ratio). However, these studies showed that faeces of pigs receiving the additive were characterized by a 60% lower content of copper, compared to faeces of pigs from the control group. Other studies (Saeid et al. 2013b) on the use of *Arthrospira maxima* algae enriched in copper through a biosorption process showed no statistically significant differences in growth, feed intake or feed conversion ratio between the control and experimental groups. There were also no statistically significant differences in the excretion of faeces and urine nitrogen. It was found that animals in the experimental group had a lower LDL level and a lower total cholesterol level of 17.05% and 9.43%, respectively, relative to the control group. Additionally, the content of copper, manganese and selenium in the meat was increased by 18%, 25% and 23%, respectively, while the zinc and selenium contents decreased in the livers of the pigs tested. Sardi et al. (2006) conducted an experiment in which they applied algae supplement to feeders. The study did not show any effect of the additive used on the production parameters of the fattening pigs nor on the quality parameters of the carcasses. Statistically significant differences in pH and meat colour, fat composition and iodine content in subcutaneous fat were also not found. However, a significant increase in docosahexaenoic acid was observed in both the loin and subcutaneous fat. Baňoch et al. (2010) noticed an increase in the iodine level in the meat of pigs fed with *Laminaria* spp. algae, compared with pigs receiving iodine supplementation in an inorganic form. Similar conclusions have been made by Smet (2012) using pigs fed with *Laminaria digitata* and *Ascophyllum nodosum*.

Studies on the use of algae-based feed additives in pigs did not show their effect on the production parameters of the animals. However, they did show that, with addition of the supplements, pork products could be made rich in elements essential for the correct functioning of human organisms.

11.2.3 Algae in Cattle Feed

The algae-based diet for dairy cattle is widely described in the modern literature, in great detail. The impact of the algae on the milk has been proven to be beneficial by a vast number of researchers. In dairy cattle feeding, the addition of algae improves the microbial protein synthesis in the rumen, increases lactation efficiency and improves milk quality, as well as reduces somatic cell counts, prevents postpartum mortality and prepares the cow for re-calving and increases concentration of certain nutrients in the milk (Holman and Malau-Aduli 2013). Kulpys et al. (2009) found that milk production increased by 21% when cows were fed with a *Spirulina* diet. Simkus et al. (2007, 2008) showed an increase in the fat, protein and lactose in the milk of cows receiving algae, compared to cows that did not receive them. Moreover, a beneficial effect of *Spirulina* on the fatty acid profile of milk has also been shown to be beneficial, as the share of unsaturated acids (Papadopoulos et al. 2002) including DHA was found to have increased (Stamey et al. 2012). This research was reinforced by Christaki et al. (2012), who found that the content of saturated fatty acids in milk was decreasing and the amount of unsaturated fatty acids and PUFAs in cows receiving *Spirulina* was increasing. López-Alonso et al. (2016) aimed to assess the influence of supplementation of a standard diet with marine algae on the quality of milk and the effectiveness of its production. Under investigation were 32 randomly selected Holstein Friesian lactating cows. It appeared that algae supplementation and rumen boluses, or a combination of those two additives, caused significant iodine growth and a tendency to increase selenium concentrations in milk.

Furthermore, studies have shown the usefulness of algae with a high proportion of EFAs in reducing the amount of methane produced, which can be very desirable in greenhouse gas reduction strategies. Fievez et al. (2007) showed that in vitro incubation with 25 ml of buffered almond and oil (sunflower or linseed) fluid inhibited methane production up to 80% within 24 h.

However, the usability of algae as an additive in the standard diet of dairy cattle is not always beneficial. In some cases, algae supplementation caused a decrease in feed intake. Moate et al. (2013) reported that such a decrease can reach as much as 16%. Furthermore, milk fat, and therefore milk yield, was reduced by almost 24%. This was also proven by Kupczyński et al. (2011), who investigated the addition of fish oil and algae to the standard feed. It appeared that the feed consulting and quality and yield of milk were also reduced. Furthermore, the additives used contributed to the reduction of fat and protein in the milk. Apart from the milk quality and yield, the usability of algae can affect such processes as, for example, digestibility. This was proven by Drewery et al. (2014), who conducted experiment design in a Latin square implementing varied doses of algae with respect to the body weight of the steer. The designed experiment allowed them to indicate the optimal value for improved digestibility of low-quality straw forage. It was found that the optimal share of additives is 100 mgN with respect to the body weight of the steer. Inasmuch as different results have been reported in the studies, it should be mentioned that algae can vary considerably depending on the area of harvest and method of production, as well as the medium used.

11.2.4 Algae in Horse Feed

Nowadays, there is a high availability of dietary and algae-rich supplements for horses. Fodder companies have given consistent emphasis to the manufacture of formulas enriched with novel ingredients. Generally, the use of dietary supplements combined with algae filtrates brings the expected effects. The animals stimulated in this way show significantly better health, fitness and higher productivity. Algae for horses are presented as an excellent addition to improving the growth and quality of the hoof, the condition of the skin and coat, the general health of the horse and the stimulation of the immune system (Ememe and Ememe 2017). In the veterinary literature, there are many reports on the beneficial effects of algae on BGA illnesses, downregulating inflammation in colitis, liver disease, joint disease and neuropathic pain in horses.

It has been found that gastric ulceration is quite common among racehorses and sport horses (Murray et al. 1996; Bezděková et al. 2005). Moir et al. (2016) have proven that calcium-enriched algae can be an effective method of neutralizing gastric juices and supporting the prevention of peptic ulcer disease. The research considers a trial of horses with the diagnosed disease, which was further rated, in accordance with the Equine Gastric Ulcer Council (EGUC), at the score of 2.2 pts. (+/−0.75). Those horses were fed with their standard diet supplemented with 40 g of calcium-enriched red marine algae. After 30 days of treatment, the horses underwent re-gastroscopy. It appeared that the ulcer was significantly smaller and was rated, on average, at 0.3 pts. (+/−0.48), according to the EGUC. Furthermore, significant improvement was observed in all of the considered horses. What was found was that seven horses were completely cured of the ulcer disease, and in the remaining ones, gastric mucosal remnants and healed ulcers were observed. The above study confirms the validity of earlier experiments, according to which fodder containing a higher calcium share (Andrews et al. 2005) and algae products tested in vitro represent similar effects on the health of the animal (Moore-Colyer et al. 2014).

The fodders' algae enrichment can also result in relief of chronic lower airway diseases, such as recurrent airway obstruction (RAO) and inflammatory airway disease (IAD). Reducing the amount of dust in the diet and the environment can have a beneficial effect in alleviating symptoms. Nogradi et al. (2015) have undertaken studies that show that supplementing the diet with omega-3 fatty acids reduces disease symptoms even more. In this case, researchers have used an algae-based omega-3 supplement that also contains vitamin C, methylsulfonylmethane (MSM) and a mushroom complex, in order to evaluate the influence of the algae on horses suffering from RAO or IAD, based on changes in Visual Analogue Scale (VAS) for cough, lung function and bronchoalveolar lavage fluid (BALF). The investigation lasted 9 months and involved examination of 32 horses (14 suffering from RAO and 18 diagnosed with IAD). After the results of the analysis came in, it appeared that VAS for cough was significantly improved in the case of all of the considered horses. Moreover, it appeared that all of the horses had healthier BALF and improved lung function. Hence, omega-3 PUFA supplementation may be an additional option for symptom relief, treatment and prophylaxis of chronic airway disease.

Degenerative joint disease (DJD) and osteoarthritis are painful afflictions for horses, resulting in their exclusion from sport and economic losses, due to the fact that there are no remedies for such illnesses nowadays (Oke and McIlwraith 2010). Moreover, some medications used to counter those illnesses cause serious side effects. For this reason, veterinarians and scientists are looking for natural products that could aid in the horse's health and slow down the progression of the disease. One such product is blue-green algae extract. The research done by Taintor et al. (2014) has shown that supplementing the diet of horses with mild forms of DJD and osteoarthritis with C-phycocyanin for 12 months resulted in less intra-articular corticosteroids. C-phycocyanin and biliprotein from *Arthrospira platensis* have an anti-inflammatory and antioxidant effect and are available as a dietary supplement for humans, dogs and horses (Reddy et al. 2000). However, the usage of those additives may also be hazardous to health, due to the fact that collection and processing of the blue-green algae is not regulated and produces toxic microcystins related to algal blooms. Mittelman et al. (2016) described the case of an 8-year-old Holsteiner gelding, the diet of which was supplemented with a commercial powdered blue-green algae for 2 months for purported hoof health benefits. Unexpectedly, after this period, liver failure and toxicity-related brain disease were diagnosed. The remaining horses in the stable did not show signs of illness. In order to find the source of the sickness, three of the purchased C-phycocyanin supplements were examined, and it appeared that one of them (open and added to the feed) was found to be the toxin that caused the disorder.

Nevertheless, the possibilities of using algae seem to be endless. A vast number of reported studies and practice have proven that marine algae additives in the diet are an innovative tool for stimulating and activating the genetic potential of breeding animals to a high extent. Their utilization results in a positive impact on the quality of raw materials, animal welfare and productivity, while maintaining the total safety of the food produced.

References

Adamski M, Kupczyński R, Chladek G, Falta D (2011) Influence of propylene glycol and glycerin in Simmental cows in periparturient period on milk yield and metabolic changes. Arch Tierz 54(3):238–248

Adarme-Vega TC, Thomas-Hall SR, Schenk PM (2014) Towards sustainable sources for omega-3 fatty acids production. Curr Opin Biotechnol 26:14–18

Andrews F, Buchanan B, Elliot S, Clariday N, Edwards L (2005) Gastric ulcers in horses. J Anim Sci 83(13_suppl):E18–E21

Baňoch T, Fajt Z, Drabek J, Svoboda M (2010) Iodine and its importance in human and pigs. Veterinarstvi 60(12):690–694

Barclay W, Meager K, Abril J (1994) Heterotrophic production of long chain omega-3 fatty acids utilizing algae and algae-like microorganisms. J Appl Phycol 6(2):123–129

Becker E (2007) Micro-algae as a source of protein. Biotechnol Adv 25(2):207–210

Bezděková B, Jahn P, Vyskočil M, Plachý J (2005) Prevalence of equine gastric ulceration in Standardbred racehorses in Czech Republic. Acta Vet Brno 74(1):59–65

Boyd WC, Almodóvar LR, Boyd LG (1966) Agglutinins in marine algae for human erythrocytes. Transfusion 6(1):82–83

Bravo L (1998) Polyphenols: chemistry, dietary sources, metabolism, and nutritional significance. Nutr Rev 56(11):317–333

Bruneel C, Lemahieu C, Fraeye I, Ryckebosch E, Muylaert K, Buyse J, Foubert I (2013) Impact of microalgal feed supplementation on omega-3 fatty acid enrichment of hen eggs. J Funct Foods 5(2):897–904

Cachaldora P, De Blas J, de Blas J, García-Rebollar P, Álvarez C, Méndez J (2005) Short communication. Effects of type and level of supplementation with dietary n-3 fatty acids on yolk fat composition and n-3 fatty acid retention in hen eggs. Span J Agric Res 3(2):209–212

Chojnacka K, Saeid A, Witkowska Z, Tuhy L (2012) Biologically active compounds in seaweed extracts—the prospects for the application. Open Conf Proc J 1:20–28

Christaki E, Karatzia M, Bonos E, Florou-Paneri P, Karatzias C (2012) Effect of dietary *Spirulina platensis* on milk fatty acid profile of dairy cows. Asian J Anim Vet Adv 7:597–604

Craigie JS (2011) Seaweed extract stimuli in plant science and agriculture. J Appl Phycol 23(3):371–393

Dobrzański Z, Korczyński M, Chojnacka K, Górecki H, Opaliński S (2008) Influence of organic forms of copper, manganese and iron on bioaccumulation of these metals and zinc in laying hens. J Elem 13(3):309–319

Dolińska B, Opaliński S, Zieliński M, Chojnacka K, Dobrzański Z, Ryszka F (2011) Iodine concentration in fodder influences the dynamics of iodine levels in hen's egg components. Biol Trace Elem Res 144(1–3):747–752

Drewery M, Sawyer J, Pinchak W, Wickersham T (2014) Effect of increasing amounts of postextraction algal residue on straw utilization in steers. J Anim Sci 92(10):4642–4649

Ememe M, Ememe C (2017) Benefits of super food and functional food for companion animals. In: Superfood and functional food-an overview of their processing and utilization. InTech, London, pp 309–322

Fievez V, Boeckaert C, Vlaeminck B, Mestdagh J, Demeyer D (2007) In vitro examination of DHA-edible micro-algae: 2. Effect on rumen methane production and apparent degradability of hay. Anim Feed Sci Technol 136(1):80–95

Fraeye I, Bruneel C, Lemahieu C, Buyse J, Muylaert K, Foubert I (2012) Dietary enrichment of eggs with omega-3 fatty acids: a review. Food Res Int 48(2):961–969

Fredriksson S, Elwinger K, Pickova J (2006) Fatty acid and carotenoid composition of egg yolk as an effect of microalgae addition to feed formula for laying hens. Food Chem 99(3):530–537

Galland-Irmouli A-V, Fleurence J, Lamghari R, Luçon M, Rouxel C, Barbaroux O, Bronowicki J-P, Villaume C, Guéant J-L (1999) Nutritional value of proteins from edible seaweed *Palmaria palmata* (Dulse). J Nutr Biochem 10(6):353–359

Ganesan K, Kumar KS, Rao PS (2011) Comparative assessment of antioxidant activity in three edible species of green seaweed, Enteromorpha from Okha, Northwest coast of India. Innov Food Sci Emerg 12(1):73–78

Ginzberg A, Cohen M, Sod-Moriah UA, Shany S, Rosenshtrauch A, Arad S (2000) Chickens fed with biomass of the red microalga *Porphyridium* sp. have reduced blood cholesterol level and modified fatty acid composition in egg yolk. J Appl Phycol 12(3):325–330

Gładkowski W, Kiełbowicz G, Chojnacka A, Gil M, Trziszka T, Dobrzański Z, Wawrzeńczyk C (2011) Fatty acid composition of egg yolk phospholipid fractions following feed supplementation of Lohmann Brown hens with humic-fat preparations. Food Chem 126(3):1013–1018

Gupta S, Abu-Ghannam N (2011) Recent developments in the application of seaweeds or seaweed extracts as a means for enhancing the safety and quality attributes of foods. Innov Food Sci Emerg 12(4):600–609

Harun R, Singh M, Forde GM, Danquah MK (2010) Bioprocess engineering of microalgae to produce a variety of consumer products. Renew Sust Energ Rev 14(3):1037–1047

He M, Hollwich W, Rambeck W (2002) Supplementation of algae to the diet of pigs: a new possibility to improve the iodine content in the meat. J Anim Physiol Anim Nutr 86(3–4):97–104

Holdt SL, Kraan S (2011) Bioactive compounds in seaweed: functional food applications and legislation. J Appl Phycol 23(3):543–597

Holman B, Malau-Aduli A (2013) *Spirulina* as a livestock supplement and animal feed. J Anim Physiol Anim Nutr 97(4):615–623

Jiménez JT, O'Connell S, Lyons H, Bradley B, Hall M (2010) Antioxidant, antimicrobial, and tyrosinase inhibition activities of acetone extract of *Ascophyllum nodosum*. Chem Pap 64(4):434–442

Kalogeropoulos N, Chiou A, Gavala E, Christea M, Andrikopoulos NK (2010) Nutritional evaluation and bioactive microconstituents (carotenoids, tocopherols, sterols and squalene) of raw and roasted chicken fed on DHA-rich microalgae. Food Res Int 43(8):2006–2013

Kang K, Park Y, Hwang HJ, Kim SH, Lee JG, Shin H-C (2003) Antioxidative properties of brown algae polyphenolics and their perspectives as chemopreventive agents against vascular risk factors. Arch Pharm Res 26(4):286–293

Kassis N, Drake SR, Beamer SK, Matak KE, Jaczynski J (2010) Development of nutraceutical egg products with omega 3 rich oils. LWT-Food Sci Technol 43(5):777–783

Khozin-Goldberg I, Iskandarov U, Cohen Z (2011) LC-PUFA from photosynthetic microalgae: occurrence, biosynthesis, and prospects in biotechnology. Appl Microbiol Biotechnol 91(4):905

Kolanowski W (2007) Długołańcuchowe wielonienasycone kwasy tłuszczowe omega-3 – znaczenie zdrowotne w obniżeniu ryzyka chorób cywilizacyjnych. Bromatol Chem Toksykol 3:229–237

Kole C, Michler CH, Abbott AG, Hall TC (2010) Transgenic crop plants. Springer, Berlin

Korczyński M, Opaliński S (2012) Dodatki paszowe jako potencjalna perspektywa dla przemysłu chemicznego. Przem Chem 5:792–795

Korczyński M, Witkowska Z, Opaliński S, Świniarska M, Dobrzański Z (2015) Algae extract as a potential feed additive. In: Marine algae extracts: processes, products, and applications. Wiley-VCH Verlag GmbH & Co. KGaA, Weinheim, pp 603–626

Korol W (2002) Mineralne dodatki paszowe-regulacje prawne. Pasze Przem 4:18–19

Kostik V, Gjorgjeska B, Bauer B, Filev K (2015) Production of shell eggs enriched with n-3 fatty acids. ISORPHR 5(8):48–51

Kulpys J, Paulauskas E, Pilipavicius V, Stankevicius R (2009) Influence of cyanobacteria *Arthrospira (Spirulina) platensis* biomass additive towards the body condition of lactation cows and biochemical milk indexes. Agron Res 7:823–835

Kumar M, Kumari P, Trivedi N, Shukla MK, Gupta V, Reddy C, Jha B (2011) Minerals, PUFAs and antioxidant properties of some tropical seaweeds from Saurashtra coast of India. J Appl Phycol 23(5):797–810

Kupczyński R, Janeczek W, Pogoda-Sewerniak K, Dzieciol M, Szoltysik M, Zawadzki W (2011) Wpływ zastosowania alg morskich i oleju rybnego na wydajność, skład mleka oraz parametry biochemiczne krwi krów. Acta Sci Pol, Medicina Veterinaria 10(4):35–46

Kupczyński R, Adamski M, Falta D, Roman A (2012a) The efficiency of propolis in post-colostral dairy calves. Arch Tierzucht 55:315–324

Kupczyński R, Kuczaj M, Szołtysik M, Stefaniak T (2012b) Influence of fish oil, palm oil and glycerol on milk fatty acid composition and metabolism in cows during early lactation. Archiv für Tierzucht 55(6):540–551

Kupczyński R, Piasecki T, Bednarski M (2013) Zioła i propolis jako dodatki funkcjonalne dla przemysłu paszowego. Przem Chem 8(92):1549–1553

Lemahieu C, Bruneel C, Termote-Verhalle R, Muylaert K, Buyse J, Foubert I (2014) Effect of different microalgal n− 3 PUFA supplementation doses on yolk color and n− 3 LC-PUFA enrichment in the egg. Algal Res 6:119–123

Leonard S, Sweeney T, Pierce K, Bahar B, Lynch B, O'Doherty J (2010) The effects of supplementing the diet of the sow with seaweed extracts and fish oil on aspects of gastrointestinal health and performance of the weaned piglet. Livest Sci 134(1):135–138

Lewis NM, Seburg S, Flanagan N (2000) Enriched eggs as a source of n-3 polyunsaturated fatty acids for humans. Poult Sci 79(7):971–974

López-Alonso M, Rey-Crespo F, Orjales I, Rodríguez-Bermúdez R, Miranda M (2016) Effects of different strategies of mineral supplementation (marine algae alone or combined with rumen boluses) in organic dairy systems. J Anim Physiol Anim Nutr 100(5):836–843

McDermid KJ, Stuercke B (2003) Nutritional composition of edible Hawaiian seaweeds. J Appl Phycol 15(6):513–524

Michalak I, Chojnacka K (2008) The application of macroalga *Pithophora varia* Wille enriched with microelements by biosorption as biological feed supplement for livestock. J Sci Food Agric 88(7):1178–1186

Michalak I, Chojnacka K, Dobrzański Z, Górecki H, Zielińska A, Korczyński M, Opaliński S (2011) Effect of macroalgae enriched with microelements on egg quality parameters and mineral content of eggs, eggshell, blood, feathers and droppings. J Anim Physiol Anim Nutr 95(3):374–387

Michalak I, Świniarska M, Korczyński M, Opaliński S, Rój E, Chojnacka K (2016) Supercritical algal extracts as the component of pro-health preparations for animals. In: BIT's 9th annual world congress of industrial biotechnology, Seoul, South Korea

Mittelman N, Engiles J, Murphy L, Vudathala D, Johnson A (2016) Presumptive iatrogenic microcystin-associated liver failure and encephalopathy in a Holsteiner gelding. J Vet Intern Med 30(5):1747–1751

Moate P, Williams S, Hannah M, Eckard R, Auldist M, Ribaux B, Jacobs J, Wales W (2013) Effects of feeding algal meal high in docosahexaenoic acid on feed intake, milk production, and methane emissions in dairy cows. J Dairy Sci 96(5):3177–3188

Moir T, O'Brien J, Hill S, Waldron L (2016) The influence of feeding a high calcium, algae supplement on gastric ulceration in adult horses. J Appl Anim Nutr 4. https://doi.org/10.1017/jan.2016.7

Moore-Colyer M, O'gorman DM, Wakefield K (2014) An in vitro investigation into the effects of a marine-derived, multimineral supplement in simulated equine stomach and hindgut environments. J Equine Vet Sci 34(3):391–397

Murray M, Schusser G, Pipers F, Gross SJ (1996) Factors associated with gastric lesions in Thoroughbred racehorses. Equine Vet J 28(5):368–374

Nitsan Z, Mokady S, Sukenik A (1999) Enrichment of poultry products with ω3 fatty acids by dietary supplementation with the alga *Nannochloropsis* and mantur oil. J Agric Food Chem 47(12):5127–5132

Nogradi N, Couetil L, Messick J, Stochelski M, Burgess J (2015) Omega-3 fatty acid supplementation provides an additional benefit to a low-dust diet in the management of horses with chronic lower airway inflammatory disease. J Vet Intern Med 29(1):299–306

Nwosu F, Morris J, Lund VA, Stewart D, Ross HA, McDougall GJ (2011) Anti-proliferative and potential anti-diabetic effects of phenolic-rich extracts from edible marine algae. Food Chem 126(3):1006–1012

O'Doherty J, Dillon S, Figat S, Callan J, Sweeney T (2010) The effects of lactose inclusion and seaweed extract derived from *Laminaria* spp. on performance, digestibility of diet components and microbial populations in newly weaned pigs. Anim Feed Sci Technol 157(3):173–180

O'sullivan A, O'Callaghan Y, O'Grady M, Queguineur B, Hanniffy D, Troy D, Kerry J, O'Brien N (2011) In vitro and cellular antioxidant activities of seaweed extracts prepared from five brown seaweeds harvested in spring from the west coast of Ireland. Food Chem 126(3):1064–1070

Oke S, McIlwraith C (2010) Review of the economic impact of osteoarthritis and oral joint-health supplements in horses. In: Proceedings of the 56th Annual Convention of the American Association of Equine Practitioners, Baltimore, Maryland, USA, 4–8 December 2010. American Association of Equine Practitioners (AAEP), pp 12–16

Opaliński S, Dolińska B, Korczyński M, Chojnacka K, Dobrzański Z, Ryszka F (2012) Effect of iodine-enriched yeast supplementation of diet on performance of laying hens, egg traits, and egg iodine content. Poult Sci 91(7):1627–1632

Pangestuti R, Kim S-K (2011) Biological activities and health benefit effects of natural pigments derived from marine algae. J Funct Foods 3(4):255–266

Papadopoulos G, Goulas C, Apostolaki E, Abril R (2002) Effects of dietary supplements of algae, containing polyunsaturated fatty acids, on milk yield and the composition of milk products in dairy ewes. J Dairy Res 69(03):357–365

Park J, Upadhaya S, Kim I (2015) Effect of dietary marine microalgae (*Schizochytrium*) powder on egg production, blood lipid profiles, egg quality, and fatty acid composition of egg yolk in layers. Asian-Australas J Anim Sci 28(3):391

Raj EDS (2016) UV–VIS and HPLC studies on *Amphiroa anceps* (Lamarck) Decaisne. Arab J Chem 9:S907–S913

Rao PS, Mantri VA, Ganesan K (2007) Mineral composition of edible seaweed *Porphyra vietnamensis*. Food Chem 102(1):215–218

Reddy CM, Bhat VB, Kiranmai G, Reddy MN, Reddanna P, Madyastha K (2000) Selective inhibition of cyclooxygenase-2 by C-phycocyanin, a biliprotein from *Spirulina platensis*. Biochem Biophys Res Commun 277(3):599–603

Rizzi L, Bochicchio D, Bargellini A, Parazza P, Simioli M (2009) Effects of dietary microalgae, other lipid sources, inorganic selenium and iodine on yolk n-3 fatty acid composition, selenium content and quality of eggs in laying hens. J Sci Food Agric 89(10):1775–1781

Rupérez P (2002) Mineral content of edible marine seaweeds. Food Chem 79(1):23–26

Ryckebosch E, Bruneel C, Muylaert K, Foubert I (2012) Microalgae as an alternative source of omega-3 long chain polyunsaturated fatty acids. Lipid Technol 24(6):128–130

Saeid A, Chojnacka K, Korczyński M, Korniewicz D, Dobrzański Z (2013a) Biomass of *Spirulina maxima* enriched by biosorption process as a new feed supplement for swine. J Appl Phycol 25(2):667–675

Saeid A, Chojnacka K, Korczyński M, Korniewicz D, Dobrzański Z (2013b) Effect on supplementation of *Spirulina maxima* enriched with Cu on production performance, metabolical and physiological parameters in fattening pigs. J Appl Phycol 25(5):1607–1617

Sardi L, Martelli G, Lambertini L, Parisini P, Mordenti A (2006) Effects of a dietary supplement of DHA-rich marine algae on Italian heavy pig production parameters. Livest Sci 103(1):95–103

Senthilkumar K, Manivasagan P, Venkatesan J, Kim S-K (2013) Brown seaweed fucoidan: biological activity and apoptosis, growth signaling mechanism in cancer. Int J Biol Macromol 60:366–374

Ševčíková S, Skřivan M, Dlouhá G, Koucký M (2006) The effect of selenium source on the performance and meat quality of broiler chickens. Czech J Anim Sci 51(10):449–457

Simkus A, Oberauskas V, Laugalis J, Želvytė R, Monkevičienė I, Sederevičius A, Šimkienė A, Pauliukas K (2007) The effect of weed *Spirulina platensis* on the milk production in cows. Vet Zootec 38(60):74–77

Simkus A, Oberauskas V, Zelvyte R, Monkeviciene I, Laugalis J, Sederevicius A, Simkiene A, Juozaitiene V, Juozaitis A, Bartkeviciute Z (2008) The effect of the microalga *Spirulina platensis* on milk production and some microbiological and biochemical parameters in dairy cows. Paper presented at the Zhivotnov'dni Nauki

Sirakov I, Velichkova K, Nikolov G (2012) The effect of algae meal (*Spirulina*) on the growth performance and carcass parameters of rainbow trout (*Oncorhynchus mykiss*). J BioSci Biotech:151–156

Skrivan M, Simane J, Dlouhá G, Doucha J (2006) Effect of dietary sodium selenite, Se-enriched yeast and Se-enriched Chlorella on egg Se concentration, physical parameters of eggs and laying hen production. Czech J Anim Sci 51(4):163

Smet SD (2012) Meat, poultry, and fish composition: strategies for optimizing human intake of essential nutrients. Anim Front 2(4):10–16

Spolaore P, Joannis-Cassan C, Duran E, Isambert A (2006) Commercial applications of microalgae. J Biosci Bioeng 101(2):87–96

Stamey J, Shepherd D, de Veth M, Corl B (2012) Use of algae or algal oil rich in n-3 fatty acids as a feed supplement for dairy cattle. J Dairy Sci 95(9):5269–5275

Stengel DB, Connan S, Popper ZA (2011) Algal chemodiversity and bioactivity: sources of natural variability and implications for commercial application. Biotechnol Adv 29(5):483–501

Taintor JS, Wright J, Caldwell F, Dymond B, Schumacher J (2014) Efficacy of an extract of blue-green algae in amelioration of lameness caused by degenerative joint disease in the horse. J Equine Vet Sci 34(10):1197–1200

Tokuşoglu Ö, Üunal M (2003) Biomass nutrient profiles of three microalgae: *Spirulina platensis*, *Chlorella vulgaris*, and *Isochrisis galbana*. J Food Sci 68(4):1144–1148

Tuhy Ł, Witkowska Z, Saeid A, Chojnacka K (2012) Zastosowanie ekstraktów glonowych w wytwarzaniu nawozów, pasz, żywności i kosmetyków. Przem Chem 5(91):1031–1034

Wongcharoen W, Chattipakorn N (2005) Antiarrhythmic effects of n-3 polyunsaturated fatty acids. Asia Pac J Clin Nutr 14(4):307

Yates CM, Philip CC, Rainger GE (2014) Pharmacology and therapeutics of omega-3 polyunsaturated fatty acids in chronic inflammatory disease. Pharmacol Ther 141(3):272–282

Yuan YV, Walsh NA (2006) Antioxidant and antiproliferative activities of extracts from a variety of edible seaweeds. Food Chem Toxicol 44(7):1144–1150

Zahid PB, Aisha K, Ali A (1995) Green seaweed as component of poultry feed. Bangladesh J Bot 2(24):153–156

Zhang C, Li X, S-k K (2012) Application of marine biomaterials for nutraceuticals and functional foods. Food Sci Biotechnol 21(3):625–631

Economic Aspects of Algae Biomass Harvesting for Industrial Purposes. The Life-Cycle Assessment of the Product

12

Grzegorz Schroeder, Beata Messyasz, and Bogusława Łęska

Abstract

Biomass derived from algae is a valuable raw product for agriculture and the chemical industry. The chemical composition of the algae biomass obtained from the natural environment, culture under natural conditions, and culture in bioreactors determines its application to the energy, fuel, and cosmetic industries. The use of biomass, as well as the extracts derived from biomass, is discussed in regard to the economic aspect and life-cycle assessment. The economic aspects of obtaining biomass algae in the product life cycle are discussed for the bioproducts industry.

Keywords

Algae biomass · Algal bioproducts · Economic aspects · Application of algae · Life-cycle assessment

12.1 Introduction

In the countries of Central and Eastern Europe – located in a temperate zone with a warm and transitory climate between the marine and the continental, while characterized by high frequency of weather changes both in the annual and multiyear cycles – a significant interest in the use of algae biomass as a raw product for industry has recently arisen. The inclusion of countries in this region to the European Union in 2004 brought about a substantial increase in general ecological awareness. The emphasis on pro-ecological solutions in Europe stems not only from a comparison of the conditions and quality of life in different European countries but primarily from the implementation of legal regulations imposing pro-ecological activity in all of the countries in the EU. The idea is that the development of civilization must be based on principles of sustainable development (Atkinson et al. 2009; United Nations 2015; United Nations Development Program 2015; Transforming our Word 2015; United Nations Official Document 2016; Sustainable development goals 2016). The principles of sustainable development concern a number of solutions leading to the protection and preservation of the three areas of development: culture, nature, and economy.

Agenda 21 is an action plan document that presents methods for working out and implementing sustainable development rules at local levels (Report 1992; Agenda21 1992). This document was approved at the "United Nations Conference on Environment and Development," Rio de Janeiro, 1992. In Agenda 21, the notion of "sustainable development" was defined as the right to satisfy the developmental aspirations of present generations without restriction of the rights of future generations to satisfy their own developmental needs. According to this definition, the economic and civilizational development of the present generation should not lead to depletion of nonrenewable resources and damage to the natural environment, so that we might ensure that future generations will be able to realize their right to satisfy their aspirations. In the documents of Agenda 21, the term "eco-development" was introduced at the national level, being defined as "the development at regional and local levels, consistent with the potentials of the area involved, with attention given to the adequate and rational use of natural resources, technological styles and organizational forms that respect the natural ecosystems and local social and cultural patterns." This definition of eco-development resulted, in an indirect but definite way, in the introduction into political and economic practice of the notion of "ecological space," referring to the yield of renewable and nonrenewable resources and the capacity of the environment to support the life of mankind (on the global scale), races (on the continental scale), nations

G. Schroeder (✉) · B. Łęska
Faculty of Chemistry, Adam Mickiewicz University in Poznan, Poznan, Poland
e-mail: schroede@amu.edu.pl

B. Messyasz
Faculty of Biology, Department of Hydrobiology, Adam Mickiewicz University in Poznan, Poznan, Poland

© Springer International Publishing AG, part of Springer Nature 2018
K. Chojnacka et al. (eds.), *Algae Biomass: Characteristics and Applications*, Developments in Applied Phycology 8,
https://doi.org/10.1007/978-3-319-74703-3_12

(on the country scale), and local communities. According to the assumed pro-ecological policy, we are obliged to invest in the ecological space in order to stop the ongoing degradation of the natural environment, rationalize the use of natural resources, improve efficiency of production, improve the comfort of living, and provide future generations with the best possible conditions for satisfying their needs and their development. The products of functional properties containing chemical compounds from natural sources, and not as a result of chemical synthesis, have gained increasing popularity, and the demand for them among consumers with ecological awareness has been growing. The consumers demonstrate an increasing number of requirements and expect that the products or materials they buy will be simultaneously "bio" and "natural" but, on the other hand, will also be competitive with those products obtained by chemical synthesis. Biosynthesis of useful chemical compounds under natural conditions is expected to be superior to the synthesis of organic compounds in chemical plants.

12.2 Algae Biomass as Raw Product for Industry

Because of the difficult environmental conditions of living, including periodic deficiencies in nutrients, the lack of access to light, space, limitations or temporal salinity, the organisms that live in water in order to survive have developed the ability to produce different chemical compounds through metabolic transformations. Among the water organisms of particular interest are algae. They are thalloid plants that show the ability to photosynthesize (Craigie et al. 2008; Craigie 2011). They are divided into two groups, depending on their biological structure: macroalgae and microalgae. The biomass of these organisms has been used in many countries, those in Asia in particular, as a source of chemical compounds and nutrition for humans and animals, as well as a natural fertilizer in agriculture. These organisms are an excellent source of proteins, vitamin E, and beta-carotene. The ones most often used are the green algae, or chlorophytes, whose thallus may contain chlorophyll (green pigment), xanthophyll (yellow), and carotene (orange pigment); the red algae (rhodophytes), which contain red phycoerythrin, blue-green phycocyanin, or green chlorophyll; and the brown algae (phaeophytes), whose thalli can be green, yellow, or brown (xanthophyll, fucoxanthin).

In the natural environment, algae are continuously exposed to stress related to access to light, temperature fluctuations, osmotic stress, and lack of water or water salinity. In response, the plants synthesize a number of organic compounds whose task is to counteract the stress. The biologically active compounds synthesized in the algal cells in order to protect the plants against adverse conditions can be isolated from the biomass and be used as valuable raw products for industry. The organic compounds obtained in this way are both "bio" and "natural."

Biomass has been used for many years as a food component for humans and animals, as well as a raw product for industry. Presently, about 15 million Mg products are obtained from algae annually. A small fraction of these products is used for obtaining extracts, the annual production of extracts is 25,000 Mg, and the production shows a tendency toward dynamic growth. In the understanding of pharmaceutical law (Pharmaceutical Law 2001), algae are classified as medical plants. Similarly, the Fodder Law (Minister of Agriculture 2005) defines fresh and processed algae as fodder substances of plant origin. The extracts from algae are classified differently. According to the EU laws (Community Register 2003), algae extracts are animal feed additives.

Consumers from Europe, in particular, those from Central and Eastern Europe, treat algae as a food component with reserve. In its unprocessed form, algae biomass is marketed as a diet supplement and is mainly made up of microalgae – *Spirulina*. Algae extracts are easier to accept for consumers, so methods of algae processing are always eagerly sought. The use of algae extracts is more easily approved, not only because of the psychological barrier but also because of the known composition and longer stability of the products, especially the certified ones, and the possibility of adding extracts – rich in valuable bioactive compounds – to different formulations. The algae extracts in which it was possible to isolate metabolites and secondary metabolites were found to contain antioxidants, bactericides, vitamins, mineral components, and polysaccharides (Alassali et al. 2016; Bedoux and Bourgougnon 2015). In the cosmetic industry, algae extracts are used as ingredients in creams, tonics, and shampoos, while dried and powdered algae are used for face masks and slimming baths.

Algae biomass can be obtained for industry from:

- Natural water bodies (seas, lakes, rivers)
- Algae cultivation under natural conditions in selected water reservoirs (ponds, algae farms)
- Algae cultivation for industrial purposes under the controlled access of light, nutrients (N, P), and pH of the medium

Each of these methods provides raw products of different components, has a different impact on the natural environment, requires different costs, and shows different production capacities and different seasonal characters for the obtainment of biomass.

The composition of biomass strongly depends on the species of alga, which determines a given species' use. Table 12.1 presents the contents of the three main components of biomass in algae from different species.

Table 12.1 Compositions of different species of alga (Vazquez-Duhalt and Arredondo-Vega 1991; Renaud et al. 2002; Demirbas 2011; Chisti 2007; Pruvost et al. 2011; Singh et al. 2012)

Algae species	Concentration (%)		
	Protein	Carbohydrate	Lipid
Anabaena cylindrical Lemm.	43–56	25–30	4–7
Botryococcus braunii Kütz.	8–17	8–20	25–75
Chlamydomonas reinhardtii Dang.	48	17	21
Chlorella pyrenoidosa Chick	57	26	2
Chlorella vulgaris Beyer.	51–58	12–17	14–22
Dunaliella bioculata Butcher	49	4	8
Dunaliella salina (Dunal) Teod.	57	32	6
Eucheuma cottonii Weber-van Bosse	9–10	26	1
Euglena gracilis Klebs	39–61	14–18	14–20
Isochrysis sp.	31–51	11–14	20–22
Neochloris oleoabundans Chant and Bold	20–60	20–60	35–54
Porphyridium cruentum (Gray) Näg.	28–39	40–57	9–14
Prymnesium parvum Carter	28–45	25–33	22–38
Scenedesmus dimorphus (Turpin) Kütz.	8–18	21–52	16–40
Scenedesmus obliquus (Turpin) Kütz.	50–56	10–17	12–14
Scenedesmus quadricauda (Turpin) Brèb. s. Chodat	48	17	21
Spirogyra sp.	6–20	33–64	11–21
Spirulina maxima (Setch., Gard.) Geitler	60–71	13–16	6–7
Spirulina platensis (Gomont) Geitler	46–63	8–14	4–9
Synechococcus sp.	63	15	11
Tetraselmis maculata Butcher	52	15	3
Ulva lactuca L.	17	59	3–4
Undaria pinnatifida (Harvey) Suringar	24	43	3–4

According to Jung et al. (2013), global production of macroalgae has shown that macroalgae can be mass cultivated with currently available farming technology. Their various carbohydrate compositions imply that new microorganisms are needed to effectively saccharify macroalgae biomass. Up-to-date macroalgae conversion technologies for biochemicals and biofuels show that molecular bioengineering would contribute to the success of macroalgae-based biorefinery. The industrial use of algae is the decisive factor in the rate of mass gain over time. Due to the high photosynthetic ability of macroalgae, they have the potential to generate and store sufficient carbon resources as needed for biorefinery. Table 12.2 shows the photosynthetic rates of macroalgae.

Photosynthetic rates vary highly depending on the species. *Enteromorpha* (green alga) and *Porphyra* (red alga) have the highest photosynthetic rates, which are 1–2 orders of magnitude higher than those of brown algae. Biomass of micro- and macroalgae is used for:

Table 12.2 Photosynthetic rates of macroalgae (Estimation in µmol CO_2 h^{-1} on the basis of dry weight (g dry^{-1}) or wet weight (g wet^{-1}) (Jung et al. 2013)

Species	Photosynthetic rate (µmol CO_2 h^{-1})
Green algae	
Acrosiphonia centralis (Lyng.) Kjellman	468 g dry^{-1}
Cladophora rupestris (L.) Kütz.	30.5 g dry^{-1}
Codium fragile (Suring.) Hariot	68.3 g dry^{-1}
Enteromorpha sp.	1786 g wet^{-1}
Monostroma grevillei (Thuret) Wittrock	1466 g dry^{-1}
Ulva sp.	48.7 g dry^{-1}
Red algae	
Asparagopsis taxiformis (Delile) Trevisan	174 g dry^{-1}
Chondrus crispus Stackhouse	21.2 g dry^{-1}
Delesseria sanguinea (Hudson) Lamour	37.9 g dry^{-1}
Gracilaria sp.	85 g wet^{-1}
Iridaea cordata (Turn.) Bory de Saint-Vincent	29.4 g dry^{-1}
Porphyra sp.	1808.7 g dry^{-1}
Brown algae	
Alaria marginata Post. and Rupr.	109.3 g dry^{-1}
Cymathere triplicata (Post. and Rupr.) Ag.	58.5 g dry^{-1}
Dictyopteris sp.	221 g dry^{-1}
Fucus sp.	561 g dry^{-1}
Laminaria sp.	124 g dry^{-1}
Macrocystis sp.	171.8 g dry^{-1}
Sargassum sp.	415 g dry^{-1}

- Energy production (methane, ethanol, butanol, biodiesel, and hydrogen)
- Production of non-energy materials for the chemical industry (saturated and unsaturated acids, antioxidants, and pigments) and to be used as agrochemicals (natural fertilizers and plant growth promoters)

12.3 Biomass Algae as a Raw Product for the Production of Biofuels

All countries must take some precautions against the depletion of fossil fuels. The need to provide energy security, maintain economic growth, and restrict the effects of climate change has triggered an interest in biofuels. Production of sustainable bioenergy is of global interest. Algae offer great potential as a sustainable raw product. Much attention has been paid to the production of bioenergy from the algae biomass, taking into account the potential benefit for the natural environment when compared with the effects of conventional cultivations of bioenergy production and conventional fossil fuels. Algae are very promising, as they absorb considerable amounts of carbon from the atmosphere, as well as industrial gases, and can effectively use the nutrients from postindustrial processing waste and municipal waste. After the development of economically acceptable and environmentally justified technologies, algae cultivation could provide large amounts of biofuel with a minimal effect on the environment (Singh et al. 2012; Ruiz et al. 2016).

In order to economically use algae biomass for biofuel production, first of all, a continuous supply of large amounts of dry biomass is needed. In the temperate climatic zone, the harvest of biomass from natural water bodies shows a seasonal character, with variations in algae species composition and different yields of the crop depending on the weather, species composition and diversity, and the considerable amount of pollution with toxic metals, pesticides, and toxins absorbed from the environment. Various species of filamentous alga prefer different conditions for a temporary increase in their vegetation season. Growth patterns for common filamentous algae indicate the occurrence of both an early spring species and, typically, a summer species in freshwater ecosystems in Poland (Pikosz et al. 2017). Thus, the harvesting of algae biomass for industrial purposes will be characterized by seasonality. Among the filamentous algae, which commonly occur in water ecosystems, *Cladophora rivularis* (L.) Hoek and *C. fracta* (Müll. ex Vahl) Kütz. show higher aqueous water temperatures than *C. glomerata* (L.) Kütz., *Tribonema vulgare* Pascher, and *Ulothrix variabilis* Kütz, with an optimum of occurrence in water below 18 °C. Taking into account all macroalgae that form the densest patches, *C. glomerata* (L.) Kütz. occupies a distinct niche in Polish water ecosystems (Table 12.3). Moreover, *C. glomerata* (L.) Kütz., as the only species of homogenous macroalgae mass development, is able to achieve high biomass in freshwaters in a short time. This species forms its highest biomass concentration in two periods: spring (April–June) and autumn (September–October), which confirms its strong response to colder water and also the possibility of harvesting biomass for several months each year. For other freshwater species of macroalga, optimum development is limited to only a few weeks, making it difficult to accurately plan the collection of biomass. It seems that the impact of environmental factors on the size of freshwater macroalgae biomass is higher than that of the marine ecosystem, which adds an additional factor of high variability and seasonality of harvest. A lack of repeatability in the development of the same amount of biomass in natural aquatic ecosystems can negatively affect subsequent economic aspects. Table 12.3 Presence (light gray color) and growth intensity of macroalgae species determined by their ecological optimum (dark gray color) in freshwater ecosystems in Poland.

An important economic argument for development of algae biomass use is related to the necessity of improving recreational conditions of water reservoirs. Collection of algae biomass from beaches attractive to tourists has become a necessity, irrespective of the cost and manner of biomass utility, as it is vital for maintenance of the recreational value of the relevant regions. This way of collecting biomass seems to be the cheapest way to utilize waste that can become a raw product. The amount of biomass in water bodies and along their banks depends on the process of accumulation of organic matter at a rate higher than that of its assimilation; the organic matter can be provided in municipal and industrial wastewaters. Because of the weather conditions and character of the environment, it is difficult to predict the blooming and thus plan harvesting, transportation, and fast utilization. In addition, the algae collected from beaches are polluted with sand and litter and undergo fast decomposition. Presently, if algae biomass is collected, it is usually damp.

Another method for providing biomass supply for industry is the cultivation of algae in selected water bodies. Algae are characterized by a very fast pace of growth, and their cultivation permits very effective use of space. From a unit area covered with algae, it is possible to get about 30 times more energy than from a unit of the same size in which biofuels of the first or second generation are being grown. Biofuels of the first generation are obtained from corn, sugarcane, rapeseed, sugar beet, and others. Production of biofuels of the first generation is in competition with food production and is treated as a possible limitation for food production. Biofuels of the second generation are obtained from materials that are not used for food production, such as wood, straw, and other wastes from agricultural production (Carriquiry et al. 2011; Brennan and Owende 2010).

Table 12.3 Presence (light gray color) and growth intensity of macroalgae species determined by their ecological optimum (dark gray color) in freshwater ecosystems in Poland

Taxon of freshwater macroalgae	Months											
	1	2	3	4	5	6	7	8	9	10	11	12
Cladophora glomerata (L.) Kütz.												
Tribonema vulgare Pascher												
Ulothrix variabilis Kütz.												
Rhizoclonium spp.												
Ulva flexuosa Wulfen												
Cladophora fracta (Müll. ex Vahl) Kütz.												
Cladophora rivularis (L.) Hoek												

In 2013, biofuel obtained from algae was offered on the market for the first time. The Propel network of gas stations in San Francisco (USA) offered biodiesel with a 20% admixture of the component obtained from algae (Cardwell 2013). Production of algae biodiesel on a mass scale must be evaluated, taking into account the size of resources, their availability and stability of supply, the effect on the environment and economy, the yield of technology applied, and the life cycle of the product (Quinn and Davis 2015).

Microalgae are considered to be one of the most feasible options with the potential to serve as a major feedstock for biofuels and bioproduct production. However, the economic viability of commercial-scale production continues to be called into question by many researchers and investors. Biomass from micro- and macroalgae contains significant amounts of water. Microalgae generally contain only 10–15% dry matter (Chen et al. 2015). Therefore, there is a need to carry out parametric analyses so as to identify the influence of system configuration and process on their economic viability. The results show that the most important cost-driving parameters are the pond, the harvesting, and the biomass drying process (Madugu and Collu 2016).

The biggest challenge for any biofuel technology is the drying out of microalgae. This process is the most energy consumptive (Bennion et al. 2015). Research on the economics of using algae for biofuel production is primarily based on experiences and calculations from other climates and adaptation to European conditions. Malic et al. have established that, for the climatic zone of Australia, bio-crude production is environmentally, economically, and socially sustainable. To this end, an economic multi-region input-output model of Australia was completed through processing of the engineering data on algal bio-crude production. This model was used to undertake hybrid life-cycle assessment for direct measurement, as well as measuring the indirect impacts of producing bio-crude. Overall, the supply chain of bio-crude is more sustainable than that of conventional crude oil. The results indicate that producing 1 million Mg of bio-crude will generate almost 13,000 new jobs and 4 billion dollars' worth of economic stimulus. Furthermore, bio-crude production will offer carbon. Sequestration opportunities of the production process are net carbon-negative (Malic et al. 2015).

Slade and coworkers have shown three aspects of microalgae production that will ultimately determine its future economic viability and environmental sustainability: the energy and carbon balance, environmental impacts, and production cost (Slade and Bauen 2013). We find that achieving a positive energy balance in the temperate climate of Europe will require technological advances and highly optimized production systems (Ghadiryanfar et al. 2016).

Biomass of microalgae is also considered as an alternative source of bioenergy. The techno-economic characteristics of macroalgae utilization in European temperate zones was evaluated by Dave et al. (2013) in a selected anaerobic digester using the chemical process modeling software ECLIPSE. The assessment covered the mass and energy balance of the entire process, followed by an economic feasibility study, which included the total cost estimation, net present value calculation, and sensitivity analysis. The selected plant size corresponded to a community-based AD of 1.6 MW with a macroalgae feed rate of 8.64 Mg per day (dry basis). The produced biogas was utilized in a combined heat and power plant generating 237 kW of electricity and 367 kW of heat. The break-even electricity-selling price in this study was estimated at around €120/MWh. On the grounds of different national and regional policies, this study did not account for any government incentives (Dave et al. 2013).

Aitken presented a life-cycle assessment (LCA) that considered the energy return and environmental impacts of the cultivation and processing of macroalgae (seaweed) into bioethanol and biogas with a particular focus on specific species (*Gracilaria chilensis* Bird, McLachlan and Oliveira and *Macrocystis pyrifera* (L.) Agardh) and cultivation methods (bottom planting and long-line cultivation). The study was based mainly upon data obtained from a research conducted in Chile, but the results can be applied to other locations where similar cultivation is feasible (Aitken et al. 2014).

The cost-effectiveness of production of energy from macroalgae depends on the yield of macroalgae production and the cost of raw product acquisition and processing. The yield of macroalgae production varies from 150 to 600 Mg of fresh biomass per hectare per year and is much higher than the typical value for sugarcane, which varies from 70 to 170 Mg fresh biomass per hectare per year. This means that under the optimum conditions of cultivation, macroalgae could be a more effective source of energy than biomass from land plants. The most promising species for biofuel production are the algae containing large amounts of oil (triglycerides) (Table 12.4).

The estimated cost of raw product can vary within a wide range, depending on the method of cultivation, with typical values being from 5 to 60 USD for 1 GJ of energy obtained from biodiesel, biogas, or bioethanol (Aresta et al. 2003).

From the point of view of societal development, the production of biofuels from algae is attractive, taking into account the following facts:

- Harvesting of algae biomass from water reservoirs does not compete with the production of farmland crops, as it does not affect agricultural land.
- Production of biomass from algae does not interfere with food production.
- Growing algae absorb large amounts of carbon dioxide from the atmosphere.

Table 12.4 Oil (triglycerides) content of some microalgae (Chisti 2007)

Microalgae	Oil content (% in dry weight biomass)
Botryococcus braunii Kütz.	25–75
Chlorella sp.	28–32
Crypthecodinium cohnii (Seligo) Javornicky	20
Cylindrotheca sp.	16–37
Dunaliella primolecta Butcher	23
Isochrysis sp.	25–33
Monallanthus salina N.	20
Nannochloris sp.	20–35
Nannochloropsis sp.	31–68
Neochloris oleoabundans Chant. and Bold	35–54
Nitzschia sp.	45–47
Phaeodactylum tricornutum Bohlin	20–30
Schizochytrium	50–77
Tetraselmis suecica (Kylin) Butcher	15–23

– Cultivation of algae under natural conditions has modest environmental requirements.
– Removal of algae from beaches and lakes considerably enhances the touristic attractiveness of the relevant regions.

The cost-effectiveness of biofuel production from algae on an industrial scale depends on the pro-ecological policy of individual governments and their support of energy production from renewable sources.

Recently, algae biomass has been intensively studied as a raw product for biofuel production (Gunaseelan 1997; Fukuda et al. 2001; Sharma et al. 2001; Demirbas 2005; Meher et al. 2006, Sahena et al. 2009; Huang et al. 2010; Mubarak et al. 2015). The method of transformation of algae biomass into fuels is presented in Table 12.5.

The algae species that can be used for bioethanol production include the microalgae, *Chlorococcum infusionum* (Schrank) Menegh., *Chlamydomonas reinhardtii* UTEX 90, and *Chlorella vulgaris* Beyer., and macroalgae: green algae, *Ulva lactuca* L. and *Ulva pertusa* Kjellman; red algae, *Kappaphycus alvarezii* (Doty) Doty ex Silva, *Gelidium amansii* (Lamour.) Lamour., *Gelidium elegans* Kütz., and *Gracilaria salicornia* (Ag.) Dawson; and brown algae, *Laminaria japonica* Areschoug, *Laminaria hyperborea* (Gun.) Foslie, *Saccharina latissima* (L.) Lane, Mayes, Druehl, and Saunders, *Sargassum fulvellum* (Turner) Ag., *Undaria pinnatifida* (Harvey) Suringar, and *Alaria crassifolia* Kjellman.

Bioethanol can be obtained from algae through two technologies: acid hydrolysis and fermentation by microorganism, with the economic criteria being the yield of the process (Jambo et al. 2016).

According to the energy directive of the European Council, the level of use of generation I biofuels in transportation can reach, at most, 7%, while the rest of the demand (3%) should be provided by generation II and III biofuels (biomass of waste and algae). The European Union (EU) has recently begun withdrawing from promotion and funding of energy plant cultivation, in particular, from cultivation of crops for potential food production. Large areas of monocultural crops imply greater amounts of nitrates, phosphates, and pesticides in the environment. It also means restricted biodiversity, so algae and their waste products are currently preferred as biomass for energy production in Europe.

Table 12.5 Potential pathways from algae biomass to fuel (Posten and Schaub 2009; Amin 2009; Chisti 2007)

Technologies	Methods	Results
Thermochemical technologies		
	Gasification of biomass	Biogas
	Liquid transformation	Bio-oil
	Pyrolysis	Biogas or bio-oil
	Burring of biomass	Energy and heat
Biochemical conversion		
	Fermentation	Biomethane and bioethanol
	Photobiographical production	Biohydrogen

Under optimal conditions, algae are able to double their mass within 24 h, sometimes even within 3.5 h. In comparison with other plants, the period of their growth – from 1 to 10 days – is very short. This fact allows for frequent harvesting, far more frequent than the one to two times a year for traditional crops. Algae absorb CO_2 (1 kg of dry mass of algae absorbs about 1.8 kg CO_2), which considerably restricts the emission of carbon dioxide into the atmosphere. The algae need access to sources of carbon, nitrogen, and phosphorus for development, as the CO_2 from the air is insufficient for them, taking into account a very fast increase in biomass. Thus, the idea of localization of algae farms in the vicinity of industrial plants and conventional power stations. The carbon dioxide obtained during the combustion of fuel is directed directly to the water tank with algae. In this way, the emission of CO_2 into the atmosphere is further restricted. Open ponds providing optimum conditions for growth of microorganisms and plants are sensitive to pollution by other microorganisms, such as other algae species, bacteria, or water plants. In open ponds, control of temperature or light is not possible. The vegetation period depends to a high degree on localization, and apart from the tropical regions, it is restricted to 3–6 warm months.

Open pond cultivation is the cheapest. The access to water from the sea, lake, or river and intensive sunlight determine the low exploitation cost, in comparison with the cultivation of algae in closed bioreactors, which, however, guarantees the purity of the product and control of growth processes. Covering of natural ponds with a semitransparent foil or fabric generates hothouse conditions. This technology solves many of the problems of an open pond. It permits cultivation of a greater number of species, ensures domination of selected species, and extends the vegetation period; if the pond is heated, the vegetation period can last all year round. Idle lands are perfect areas for algae cultivation. It is also possible to use wastewater for algae production, with the wastewater then also being purified of nitrogen and phosphorus compounds. This is a very promising solution for reduction of cost related to water and nutrient supply. Algae biomass can be collected from the water by the following methods:

- With the use of a special platform
- With the use of foam flotation
- By filtration through cellulose membranes with the use of a vacuum pump
- With the use of flocculation and centrifugation
- With the use of magnetite (iron oxide) particles of micrometric size

The process of biomass production of final energy and non-energy products can be divided into several stages:

- Collection of algae biomass
- Cleaning of coatings and diatoms, washing with water, and air-drying
- Storage or milling of biomass
- Extraction of algae biomass by various techniques
- Extract processing, chemical isolation of bioactive compounds, and management of the extracted biomass

In this compilation, transport of the biomass and the products obtained through its processing is not included.

The alternative to algae cultivation in open ponds is the use of closed photobioreactors. The best seem to be the photobioreactors that use solar radiation. These reactors are made of light-transmitting materials (Molina et al. 1999; Pulz 2001). There are three main types of photobioreactor: vertical column, cylindrical, and flat panel. The fast growth of algae requires long exposure to intensive sunlight and a temperature no lower than 20 °C; under other conditions, the growth of the algae will be inhomogeneous and ineffective. In some algae cultivation plants, water is heated, and additional light sources are used, which increases the cost of biomass production. Such reactors are used for cultivation of algae to be used for the production of cosmetics and medical products. The additional cost of lighting is calculated into the price of the special product (Saeid and Chojnacka 2015).

Photobioreactors offer high rates of productivity, with the potential to yield 5000–15,000 gal of microalgal oil per acre per year. If an average yield of 10,000 gal of oil /acre/year could be achieved, the 63 billion gallon diesel demand of the USA could be produced on just 6.3 million acres (Oil seed crops 2016).

The yield of biodiesel production from algae is 15–300 times higher than that of oil produced from traditional crops. The yield of various plant oils (gallons per hectare) is soy 118, safflower 206, sunflower 251, castor 373, coconut 605, palm 605, and algae 26,417. The yield of oils per hectare of algae crops is high, higher than that for oilseeds (Biodigester 2016).

Bioproducts can be produced from different types of alga and through the use of different technologies. Microalgae contain the oil that is suitable for biodiesel production. Production of bioethanol mainly involves macroalgae, which are characterized by a high content of starch and cellulose, including brown algae, red algae, and chara (charophytes). Depending on the type of alga and the type of metabolism, these organisms accumulate different storage materials in their cells: lipids, carbohydrates, and proteins. Oil can be obtained from algae by pressing or by extraction, while the species rich in proteins or saccharides are used for alcoholic fermentation of methane production. Recently, algae have been frequently mentioned as a potential source of hydrogen or methanol for fuel cells.

The cost-effectiveness of biodiesel production from algae is not competitive with the cost of production of traditional fuels, which hinders the dynamic development of this technology. That is why the production of algae biomass on an industrial scale, irrespective of the final product (biodiesel, ethanol), is reasonable only if based on natural sunlight and water that can have no other use. At present, genetically modified organisms from which it is easier to extract energetically valuable substances are being tested.

Genetically modified algae are those whose genetic material has been altered using genetic engineering techniques. These techniques are generally known as recombinant DNA technology. Using this technology, DNA molecules from different sources are combined in vitro into one molecule to create a new gene. This DNA is then transferred into an organism and causes the expression of modified or novel traits. The interest in obtaining biofuels from GMO algae is growing year by year (Rosenberg et al. 2008; Radakovits et al. 2010; Sayre 2010; Enzing 2012; Snow and Smith 2012; Kumar 2015). In order to obtain the algae species for industrial bioproducts, the following genetic strategies are discussed (Glass 2015):

Enhance photosynthesis; improve carbon fixation
  Enhance pathway proteins
  Introduce new carbon fixation pathways
Enhance or alter lipid biosynthesis for improved diesel, jet fuel production
Enable secretion of lipids to improve harvesting and separation

Express, enhance transporter proteins
  Alter cell wall composition for easier cell lysis
Utilize metabolic engineering to enhance existing pathways
  Maximize carbon flow to desired product(s)
  Eliminate competing pathways
  Remove harmful, toxic compounds
Introduce new pathways for desired products
  Utilize ethanol
  Utilize butanol
Improve production of hydrogen for fuel use

Implementation of technologies in regard to genetically modified algae must comply with GMO laws, be socially acceptable, and, perhaps above all, be safe (European Union 2001, 2009).

12.4 Algae Biomass as a Raw Product for the Cosmetic Industry

Biomass of marine and freshwater algae is a good raw product for extraction of the bioorganic compounds used in the cosmetic industry (Pereira and Meireles 2010; Li et al. 2014; Michalak and Chojnacka 2014, 2015; Goto et al. 2015; Michalak and Chojnacka 2015; Messyasz et al. 2015a, b). Green extraction technologies for high-value metabolites from algae are presented by Esquivel-Hernández et al. (2017). The current list of algae products is extensive. Table 12.6 presents exemplary products obtained from different algal types.

Therapeutic supplements from microalgae comprise an important market involving compounds such as β-carotene and astaxanthin, polyunsaturated fatty acids (PUFA) such as DHA and EPA, and polysaccharides such as β-glucan. The dominant species of microalga used in commercial production include *Isochrysis*, *Chaetoceros*, *Chlorella*, *Arthrospira* (*Spirulina*), and *Dunaliella* (Priyadarshani and Rath Biswajit 2012). The production of unsaturated algae DHA from biomass algae is an innovative approach to this problem. The optimum algae species and technology have been selected to ensure the production of unsaturated acid in amounts profitable for the producers. DHA algal oil is produced via an algal fermentation process using *Schizochytrium*. Up to 50% of this species' dry cell weight can be made up of the fatty acids; approximately 30% of the total fatty acid content is DHA. It has been shown that *S. limacinum* can provide approximately 4 g of DHA for every liter of the medium, which is greater than the amounts obtained from the other species studied. Production of DHA oil from *Schizochytrium* is realized through the four-step technology (Oilgae 2016):

Table 12.6 List of algae products (Comprehensive Report 2015)

High value	Medium-high value	Low to medium value
Nutraceuticals (astaxanthin, beta-carotene, omega-3 fatty acid (DHA and EPA), and coenzyme Q10)	*Nutraceuticals* (*Spirulina* and *Chlorella*)	*Fertilizer and animal feed* (aquaculture feed (shrimp feed, shellfish feed, marine fish larvae cultivation), animal feed, and fertilizer)
Pharmaceuticals (antimicrobials, antivirals, antifungals, and neuroprotective products)	*Hydrocolloids* (agar, alginate and, carrageenan)	*Substitutes for synthetics* (biopolymers and bioplastics – lubricants)
Cosmetics (anti-cellulite, skin antiaging, and sensitive skin treatment – alguronic acid)	*Chemicals* (paints, dyes, and colorants)	*Bioremediation* (wastewater treatment and nutrient credits $CO2$ capture and carbon credits)

- Fermentation (fermentation carried out under controlled conditions using carbon and nitrogen bulk nutrients, vitamins, and trace mineral sources)
- Intermediate product (algae cell concentrate and dried algae transferred to oil extraction process)
- Oil extract (oil extract from dried algae by solvent (hexane) extraction. De-oiled biomass is separated by centrifugation or filtration. Solvent phase is crystallized (winterized) and oil is extracted.)
- Oil purification (Winterized oil is heated and pretreated with acid, neutralized, and centrifuged to get refined oil.)

The search for biostimulants of plant growth (by extraction) for ecological agriculture is another rapidly developing field. The compounds obtained this way are expected to replace synthetic biostimulants and become natural products for agriculture consistent with the idea of "plant for plants." Algae biomass contains small amounts of plant growth stimulants (Michalak et al. 2015, Michalak et al. 2016a, b; Godlewska et al. 2016; Michalak and Chojnacka 2016a, b; Górka et al. 2016; Michalak et al. 2017). The extraction of bioactive compounds from algae biomass requires the use of different extraction techniques (Michalak and Chojnacka 2016b; Rój et al. 2015; Schroeder et al. 2015a, b; Fabrowska et al. 2015a, b; 2016). A cost-effective analysis of these processes has been conducted by Kozłowski (Kozłowski et al. 2016).

12.5 Life-Cycle Assessment of Algae Production

Life-cycle assessment (LCA) is a tool for identifying and comparing the whole life cycle, or cradle-to-grave, environmental impacts of production, marketing, transport, and distribution. Life cycle impact assessment (LCIA) converts inventoried flows into simpler indicators. In LCIA, essentially two types of methods are followed: problem-oriented methods and damage-oriented methods (end points). In the problem-oriented approaches, flows are classified into the environmental themes to which they contribute. The damage-oriented methods also start with classification of a system's flows into various environmental themes, but they model the damage of each environmental theme according to its effect on human health and ecosystem health or the damage to resources. LCA is currently being used to assess the large-scale feasibility and environmental impact of alternative processing technologies being explored for processing microalgae as a feedstock into biofuels (Aiken et al. 2014; Aresta et al. 2005; Kothari et al. 2008; Beer et al. 2009; Brune et al. 2009; Batan et al. 2010; Clarens et al. 2010; Jorquera et al. 2010; Brentner et al. 2011; Campbell et al. 2011; Khoo et al. 2011; Razon and Tan 2011; Shirvani et al. 2011; Frank et al. 2011, 2013; Grierson et al. 2013; Liu et al. 2013; Passell et al. 2013; Slade and Bauen 2013; Zhang et al. 2013; Gao et al. 2013; Adesanya et al. 2014; Ponnusamy et al. 2014; Quinn et al. 2014; Bennion et al. 2015; Malik et al. 2015; Voort et al. 2015).

Ruiz has presented analyses of different markets for products from microalgae and the techno-economic evaluation of the whole process chain, including cultivation, biorefinery, and market exploitation for a 100 ha facility in six locations. The calculation has shown a current cost per unit of dry biomass of 3.4 euro kg^{-1} for microalgae cultivation in Spain, with an expected reduction to 0.5 euro kg^{-1} in 10 years. Production of high-value products (e.g., pigments) would currently be profitable, with a net present value of 657 million euro in 15 years. Markets aimed at food and chemical commodities require further cost reductions for cost competitiveness, reachable in the next decade (Ruiz et al. 2016).

LCA is the compilation and evaluation of the inputs and outputs and the potential impacts of a product system throughout its life cycle. Input and output parameters for production of biodiesel and its coproducts, such as succinic acid, and cost estimation have been presented by Gnansounou (Gnansounou and Raman 2016). The data inventories of the input/output parameters are given in Table 12.7.

LCA analysis of production from algae biomass very strongly depends on the region and country for which it is conducted. Globalization of results yields a significant difference from actual production costs and environmental impacts (Taelman and Sfez 2015).

Subcritical water extraction (SWE) of algae biomass demonstrates a significant capacity for generating liquid transportation fuels from algae with minimal environmental impacts. The SWE process expends pressurized water to produce bio-crude or bio-oil, as well as aqueous, solid, or gaseous by-products. Danquah has comprehensively reviewed the process principles, optimal conditions, engineering scale-up, and product development and has shown that the LCA analysis ascertains the viability of an industrial-scale SWE process for biofuel production from algae (Thiruvenkadam et al. 2015).

Environmental impact is an essential aspect for the introduction of algae production systems. As information on large-scale algae production is only sparsely available, process simulation is the only way to evaluate environmental sustainability in the early phase of process design. Simulation

Table 12.7 Data inventory of algae biodiesel production per kg and 1.94 kg protein production (Gnansounou and Raman 2016)

Stages	Input/output	Utilities/materials
Cultivation	Input	Water
		Urea
		Diammonium phosphate
		Concrete
		Steel
		Plastic
		Cast iron
		Electricity
	Output	Algal broth
Harvest	Input	Chitosan
		Electricity
		Algal broth
	Output	Dry algae
Oil extraction	Input	Oil extraction
		Electricity
		Heat
	Output	Algal oil
Protein extraction	Input	De-oiled algae biomass
		Ethanol
		Methanol
		Electricity
		Heat
	Output	Protein
		Algae biomass
Anaerobic digestion	Input	Electricity
		Heat
		Algae biomass
	Output	Electricity
		Heat
Biodiesel production	Input	Algal oil
		Methanol
		Sodium hydroxide
		Sulfuric acid
		Electricity
		Heat
		Water
	Output	Biodiesel
		Glycerol
Transport biodiesel	Transport biodiesel – from industry to outlet	
	Protein – from industry to outlet	

results allow for the evaluation of production and design scenarios and reveal the potential to improve the life-cycle performance of algae production systems. Van Boxtel has discussed how choices in the process design of algae production systems (cultivation, biorefinery, and the supply chain) advance LCA results (Van Boxtel et al. 2015).

Sho has found that the starch contained in algae can be converted into alkyl lactate and alkyl levulinate, important chemicals in the production of pharmaceuticals, additives, and polyesters. Two molecules of lactic acid can be dehydrated into lactone lactide and subsequently polymerized into either atactic or syndiotactic polylactide, which are biodegradable polyesters. Levulinic acid, on the other hand, is used as a precursor for pharmaceuticals and additives like plasticizers, as well as a starting material for a wide variety of compounds (Sho et al. 2017).

Micro- and macroalgae are the source of biomass used for the production of biomethane, hydrogen, and biogas, oils used for the production of biodiesel, and sugars used for the production of bioethanol (Richa et al. 2008). The production of algae biofuels is very effective, especially with the use of enzymes, but requires high financial investment. An additional problem in the production of algae biogas is the high cost of supplying the raw product and the odor that accompanies the biogas plant. From the economic point of view, companies would be much more profitable were they to utilize better products for the extraction of fine chemicals and ingredients for pharmaceuticals, cosmetics, and high-end nutraceuticals. On the basis of their market values, algae products are classified into three types: high-, medium-, and low-value products (Table 12.8).

In accordance with the principles of sustainable development, the acquisition and processing of algae biomass for industrial purposes must also be considered in its environmental and social aspects. For this reason, the following facts are relevant to sustainability: algae do not compete with farmland crops, algae do not pose a threat to food crops, algae are fast growing with low environmental requirements, algae are involved in the absorption of carbon dioxide with very high yield, and algae removed from the beaches of seas, lakes, or rivers enhance the landscape and tourism values of these areas. The objective economic, social, and environmental assessment of the algae biomass used for industrial applications needs to be based on the life-cycle assessment analysis for each region or country, taking into account not only climate but also legal and social conditions.

Table 12.8 List of products from biomass algae and their market values (Oilgae 2016)

Category	Price range ($/kg)
Biofuels	<1
Fine chemicals/food ingredients	10–50
Ingredients for pharma, cosmetics, high-end nutraceuticals	150 and higher (>$1000/kg for astaxanthin)

References

Adesanya VO, Cadena E, Scott SA, Smith AG (2014) Life cycle assessment on microalgal biodiesel production using a hybrid cultivation system. Bioresour Technol 163:343–355

Agenda 21 (1992) United Nations Division for Sustainable Development. https://sustainabledevelopment.un.org/content/documents/Agenda21.pdf. Accessed 10 May 2017

Aitken D, Bulboa C, Godoy-Faundez A, Turrion-Gomez JL, Antizar-Ladislao B (2014) Life cycle assessment of macroalgae cultivation and processing for biofuel production. J Clean Prod 75:45–56

Alassali A, Cybulska I, Brudecki GP, Farzanah R, Thomsen MH (2016) Methods for upstream extraction and chemical characterization of secondary metabolites from algae biomass. Adv Tech Biol Med 4:163. https://doi.org/10.4172/2379-1764.1000163

Amin S (2009) Review on biofuel oil and gas production processes from microalgae. J Biotechnol 142:64–69

Antizar-Ladislao B (2014) Life cycle assessment of macroalgae cultivation and processing for biofuel production. J Clean Prod 75:45–56

Aresta M, Angela D, Tommasi I (2003) Energy from macro-algae. Fuel Chem Preprints 48:260–261

Aresta M, Dibenedetto A, Barberio G (2005) Utilization of macro-algae for enhanced CO2 fixation and biofuels production: development of a computing software for an LCA Study. Fuel Process Technol 86:1679–1693

Atkinson G, Dietz S, Neumayer E (eds) (2009). Handbook of sustainable development. Edward Elgar Publishing, Cheltenham

Batan L, Quinn J, Willson B, Bradley T (2010) Net energy and greenhouse gas emission evaluation of biodiesel derived from microalgae. Environ Sci Technol 44:7975–7980

Bedoux G, Bourgougnon N (2015) Bioactivity of secondary metabolites from Macroalgae. In: Sahoo D, Seckbach J (eds) The algae world, Cellular Origin, Life in Extreme Habitats and Astrobiology, vol 26. Springer, Dordrecht, pp 391–401

Beer LL, Boyd ES, Peters JW, Posewitz MC (2009) Engineering algae for biohydrogen and biofuel production. Curr Opin Biotechnol 20:264–271

Bennion EP, Ginosar DM, Moses J, Agblevor F, Quinn JC (2015) Lifecycle assessment of microalgae to biofuel: Comparison of thermochemical processing pathways. Appl Energy 154:1062–1071

Biodigester (2016) https://pl.pinterest.com/outsidethebox2/biodigester/. Accessed 10 May 2017

Brennan L, Owende P (2010) Biofuels from microalgae – a review of technologies for production, processing, and extractions of biofuels and co-products. Renew Sust Energ Rev 14:557–577

Brentner LB, Eckelman MJ, Zimmerman JB (2011) Combinatorial life cycle assessment to inform process design of industrial production of algal biodiesel. Environ Sci Technol 45:7060–7067

Brune DE, Lundquist TJ, Benemann JR (2009) Microalgal biomass for greenhouse gas reductions: potential for replacement of fossil fuels and animal feeds. J Environ Eng 135:1136–1144

Campbell PK, Beer T, Batten D (2011) Life cycle assessment of biodiesel production from microalgae in ponds. Bioresour Technol 102:50–56

Cardwell D. (2013) Unilever to buy oil derived from algae from solazyme. The New York Times

Carriquiry MA, Du X, Timilsina GR (2011) Second generation biofuels: Economics and policies. Energy Policy 39:4222–4234

Chen H, Zhou D, Luo G, Zhang S, Chen J (2015) Macroalgae for biofuels production: progress and perspectives renewable and sustainable. Energy Rev 47:427–437

Chisti Y (2007) Biodiesel from microalgae. Biotechnol Adv 25:294–306

Clarens AF, Resurreccion EP, White MA, Colosi LM (2010) Environmental life cycle comparison of algae to other bioenergy feedstocks. Environ Sci Technol 44:1813–1819

Community Register of Feed Additives pursuant to Regulation (2003) No 1831/2003. Council Directive 70/524/EEC concerning additives in feedingstuffs – list of authorised additives in feeding stuffs (2004/C 50/01)

Comprehensive Report on Attractive Algae Product Opportunities – Preview (2015) https://secure.clixoo.com/purchase/oilgae/alg_pdt/report.html Accessed 10 May 2017

Craigie JS (2011) Seaweed extract stimuli in plant science and agriculture. J Appl Phycol 23:371–393

Craigie JS, MacKinnon SL, Walter JA (2008) Liquid seaweed extracts identified using ^{1}H NMR profiles. J Appl Phycol 20:665–671

Dave A, Huang Y, Rezvani S, McIlveen-Wright D, Novaes M, Hewitt N (2013) Techno-economic assessment of biofuel development by anaerobic digestion of European marine cold-water seaweeds. Bioresour Technol 135:120–127

Demirbas A (2005) Potential applications of renewable energy sources, biomass combustion problems in boiler power systems and combustion related environmental issues. Prog Energ Combust 31:171–192

Demirbas MF (2011) Biofuels from algae for sustainable development. Appl Energy 88:3473–3480

Enzing C (2012) Algae and genetic modification, Research, production and risks. Technopolis Groups, www.technopolis-group.com Accessed 10 May 2017

Esquivel-Hernández DA, Ibarra-Garza IP, Rodríguez-Rodríguez J, Cuéllar-Bermúdez SP, Rostro-Alanis MJ, Alemán-Nava GS, García-Pérez JS, Parra-Saldívar R (2017) Green extraction technologies for high-value metabolites from algae: a review. Biofuels Bioprod Biorefin 11:215–231

European Union (2001) Directive 2001/18/EC of the European Parliament and of the Council of 12 March 2001 on the deliberate release into the environment of genetically modified organisms.http://eur-lex.europa.eu/legal-content/EN/TXT/?uri=CELEX:32001L0018. Accessed 10 May 2017

European Union (2009) Directive 2009/41/EC of the European Parliament and of the Council of 6 May 2009 on the contained use of genetically modified micro-organisms. http://eur-lex.europa. eu/LexUriServ/LexUriServ.do?uri=OJ:L:2009:125:0075:0097:EN:PDF. Accessed 10 May 2017

Fabrowska J, Łęska B, Schroeder G (2015a) Freshwater Cladophora glomerata as a new potential cosmetic raw material. Chemik 69:491–497

Fabrowska J, Łęska B, Schroeder G, Messyasz B, Pikosz M (2015b) Biomass and extracts of Algae as material for cosmetics. In: Kim S-K, Chojnacka K (eds) Marine Algae extracts: processes, products, and applications. Wiley-VCH Verlag GmbH & Co. KGaA, Weinheim, pp 681–706

Fabrowska J, Ibañez E, Łęska B, Herrero M (2016) Supercritical fluid extraction as a tool to valorize underexploited freshwater green algae. Algal Res 19:237–245

Frank ED, Han J, Palou-Rivera I, Elgowainy A, Wang MQ (2011) Life-cycle analysis of algal lipid fuels with the greet model. Center for Transportation Research, Energy Systems Division, Argonne National Laboratory, Oak Ridge

Frank E, Elgowainy A, Han J, Wang Z (2013) Life cycle comparison of hydrothermal liquefaction and lipid extraction pathways to renewable diesel from algae. Mitigat Adapt Strat Glob Chang 18:137–158

Fukuda H, Kondo A, Noda H (2001) Biodiesel fuel production by transestryfication of oils. J Biosci Bioeng 92:405–416

Gao X, Yu Y, Wu H (2013) Life cycle energy and carbon footprints of microalgal biodiesel production in Western Australia: a comparison of byproducts utilization strategies. ACS Sustain Chem Eng 1:1371–1380

Ghadiryanfar M, Rosentrater KA, Keyhani A, Omid M (2016) A review of macroalgae production, with potential applications in biofuels and bioenergy. Renew Sust Energ Rev 54:473–481

Glas DJ (2015) Government regulation of the uses of genetically modified Algae and other microorganisms in biofuel and bio-based chemical production. In: Prokop A, Bajpai RK, Zappi ME (eds) Algal biorefineries, products and refinery design, vol 2. Springer, International Publishing, Cham, pp 23–61

Gnansounou E, Raman JK (2016) Life cycle assessment of algae biodiesel and its co-products. Appl Energy 161:300–308

Godlewska K, Michalak I, Tuhy Ł, Chojnacka K (2016) Plant growth biostimulants based on different methods of seaweed extraction with water. Biomed Res Int 2016:5973760

Górka B, Kucab K, Lipok J, Wieczorek PP (2016) Biologically active compounds in Algae and their application in plant growth stimulation. In: Chojnacka K, Michalak I (eds) Innovative bio-products for agriculture: Algal extracts in products for humans, Animals and Plants. Nova Science Publishers Inc, New York, pp 101–128

Goto M, Wahyudiono KH, Siti M (2015) Extraction of carotenoids and lipids from algae by supercritical CO2 and subcritical dimethyl ether. J Supercrit Fluids 96:245–251

Grierson S, Strezov V, Bengtsson J (2013) Life cycle assessment of a microalgae biomass cultivation, bio-oil extraction and pyrolysis processing regime. Algal Res 2:299–311

Gunaseelan VN (1997) Anaerobic digestion of biomass for methane production: a review. Biomass Bioenergy 13:83–114

Huang G, Chen F, Wei D, Zhang X, Chen G (2010) Biodiesel production by microalgal biotechnology. Appl Energy 87:38–46

Jambo SA, Abdulla R, Azhar SHM, Marbawi H, Azlanansau J, Ravindra P (2016) A review on thirde generation bioethanol feedstock. Renew Sust Energ Rev 65:756–769

Jorquera O, Kiperstok A, Sales EA, Embirucu M, Ghirardi ML (2010) Comparative energy life-cycle analyses of microalgal biomass production in open ponds and photobioreactors. Bioresour Technol 101:1406–1413

Jung KA, Lim S-R, Kim Y, Park JM (2013) Potentials of macroalgae as feedstocks for biorefinery. Bioresour Technol 135:182–190

Khoo HH, Sharratt PN, Das P, Balasubramanian RK, Naraharisetti PK, Shaik S (2011) Life cycle energy and CO_2 analysis of microalgae-to-biodiesel: preliminary results and comparisons. Bioresour Technol 102:5800–5807

Kothari R, Buddhi D, Sawhney RL (2008) Comparison of environmental and economic aspects of various hydrogen production methods. Renew Sust Energ Rev 12:553–563

Kozłowski K, Rój E, Dobrzyńska-Inger A, Kostrzewa D (2016) The analysis of Technical and Economic (TEA) aspects of natural raw materials extraction with supercritical carbon dioxide. In: Chojnacka K, Michalak I (eds) Innovative bio-products for agriculture: Algal extracts in products for humans, animals and plants. Nova Science Publishers Inc, New York, pp 43–72

Kumar S (2015) GM Algae for biofuel production: biosafety and risk assessment. Coll Biosaf Rev 9:52–75

Li Y, Naghdi FG, Garg S, Adarme-Vega TC, Thurecht KJ, Ghafor WA, Tannock S, Schenk PM (2014) A comparative study: the impact of different lipid extraction methods on current microalgal lipid research. Microb Cell Factories 13:14. https://doi.org/10.1186/1475-2859-13-14.13-14

Liu X, Saydah B, Eranki P, Colosi LM, Greg Mitchell B, Rhodes J, Clarens AF (2013) Pilot-scale data provide enhanced estimates of the life cycle energy and emissions profile of algae biofuels produced via hydrothermal liquefaction. Bioresour Technol 148:63–71

Madugu F, Collu M (2016) Parametric analysis for an algal oil production process. Int J Energy Prod Mgmt 1:141–154

Malik A, Lenzen M, Ralph PJ, Tamburic B (2015) Hybrid lifecycle assessment of algal biofuel production. Bioresour Technol 184:436–443

Meher LC, Vidya SD, Naik SN (2006) Technical aspects of biodiesel production by transesterification – a review. Renew Sust Energ Rev 10:248–268

Messyasz B, Łęska B, Fabrowska J, Pikosz M, Cieslak A, Schroeder G (2015a) Effects of organic compounds on the macroalgae culture of Aegagropila linnaei. Open Chem 13:1040–1044

Messyasz B, Łęska B, Fabrowska J, Pikosz M, Roj E, Cieslak A, Schroeder G (2015b) Biomass of freshwater Cladophora as a raw material for agriculture and the cosmetic industry. Open Chem 13:1108–1118

Michalak I, Chojnacka K (2014) Algal extracts: technology and advances. Eng Life Sci 14:581–591

Michalak I, Tuhy Ł, Chojnacka K (2015) Seaweed extract by microwave assisted extraction as plant growth biostimulant. Open Chem 13:1183–1195

Michalak I, Chojnacka K (2016a) The potential usefulness of a new generation of agro-products based on raw materials of biological origin. Acta Sci Pol Hortorum Cultus 15:97–120

Michalak I, Chojnacka K (2016b) Innovative technology of Algal extracts obtained by supercritical fluid extraction useful in the products for plants, animals and human. In: Chojnacka K, Michalak I (eds) Innovative bio-products for agriculture: Algal extracts in products for humans. Animals and Plants. Nova Science Publishers Inc, New York, pp 1–28

Michalak I, Górka B, Wieczorek PP, Rój E, Lipok J, Łęska B, Messyasz B, Wilk R, Schroeder G, Dobrzyńska-Inger A, Chojnacka K (2016a) Supercritical fluid extraction of algae enhances levels of biologically active compounds promoting plant growth. Eur J Phycol 51:243–252

Michalak I, Chojnacka K, Dmytryk A, Wilk R, Gramza M, Rój E (2016b) Evaluation of supercritical extracts of Algae as biostimulants of plant growth in field trials. Front Plant Sci 7:1591

Michalak I, Chojnacka K, Saeid A (2017) Plant growth biostimulants, dietary feed supplements and cosmetics formulated with supercritical CO_2 Algal extracts. Molecules 22:66. https://doi.org/10.3390/molecules22010066

Minister of Agriculture (2005) 13 January 2005 on the category of feed materials) (Journal of Laws 25 January 2005) Regulation (EC) No 767/2009 of the European Parliament and of the Council of 13 July 2009 on the placing on the market and use of feed, amending European Parliament and Council Regulation (EC) No 1831/2003 and repealing Council Directive 79/373/EEC, Commission Directive 80/511/EEC, Council Directives 82/471/EEC, 83/228/EEC, 93/74/EEC, 93/113/EC and 96/25/EC and Commission Decision 2004/217/EC

Molina GE, Acien F, Garcia CF, Chisti Y (1999) Photobioreactors: light regime, mass transfer, and scaleup. J Biotechnol 70:231–247

Mubarak M, Shaija A, Suchithra TV (2015) A review on the extraction of lipid from microalgae for biodiesel production. Algal Res 7:117–123

Oil seed crops, Food and Energy (2016) http://www.oilseedcrops.org/algae/. Accessed 10 May 2017

Oilgae (2016) http://www.oilgae.com/non_fuel_products/non_fuel_products_from_algae.html. Accessed 10 May 2017

Passell H, Dhaliwal H, Reno M, Wu B, Amotz AB, Ivry E, Gay M, Czartoski T, Laurin L, Ayer N (2013) Algae biodiesel life cycle assessment using current commercial data. J Environ Manage 129:103–111

Pereira CG, Meireles MAA (2010) Supercritical fluid extraction of bioactive compounds: fundamentals, applications and economic perspectives. Food Bioprocess Technol 3:340–372

Pharmaceutical law (2001) Polish Pharmaceutical Law. J Laws, No. 53 pos. 533, 6 September 2001

Pikosz M, Messyasz B, Gąbka M (2017) Functional structure of algal mat (Cladophora glomerata) in a freshwater in western Poland. Ecol Indic 74:1–9

Ponnusamy S, Reddy HK, Muppaneni T, Downes CM, Deng S (2014) Life cycle assessment of biodiesel production from algal bio-crude oils extracted under subcritical water conditions. Bioresour Technol 170:454–461

Posten C, Schaub G (2009) Microalgae and terrestrial biomass as source for fuels – a process view. J Biotechnol 142:64–69

Priyadarshani I, Rath B (2012) Commercial and industrial applications of micro algae – a review. J Algal Biomass Utln 3:89–100

Pruvost J, Van Vooren G, Le Gouic B, Couzinet-Mossion A, Legrand J (2011) Systematic investigation of biomass and lipid productivity by microalgae in photobioreactors for biodiesel application. Bioresour Technol 102:150–158

Pulz O (2001) Photobioreactors production systems for phototrophic microorganisms. Appl Microbal Biotechnol 57:287–293

Quinn JC, Ryan D (2015) The potentials and challenges of algae based biofuels: a review of the techno-economic, life cycle, and resource assessment modeling. Bioresour Technol 184:444–452

Quinn JC, Smith TG, Downes CM, Quinn C (2014) Microalgae to biofuels lifecycle assessment-multiple pathway evaluation. Algal Res 4:16–22

Radakovits R, Jinkerson RE, Darzins A, Posewitz MC (2010) Genetic engineering of Algae for enhanced biofuel production. Eukaryot Cell 9:486–501

Razon LF, Tan RR (2011) Net energy analysis of the production of biodiesel and biogas from the microalgae: *Haematococcus pluvialis* and *Nannochloropsis*. Appl Energy 88:3507–3514

Renaud SM, Van Thinh L, Lambrinidis G, Parry DL (2002) Effect of temperature on growth, chemical composition and fatty acid composition of tropical Australian microalgae grown in batch cultures. Aquaculture 211:195–214

Report of the United Nations Conference on Environment and Development (1992) Rio de Janeiro, 3–14 June 1992 (United Nations publication, Sales No. E.93.I.8 and corrigenda), vol. I: Resolutions Adopted by the Conference, resolution 1, annex I

Richa K, Buddhi D, Sawhney RL (2008) Comparison of environmental and economic aspects of various hydrogen production methods. Renew Sust Energ Rev 12:553–563

Rój E, Dobrzyńska-Inger A, Dębczak A, Kostrzewa D, Stępnik K (2015) Algae extract production methods and process optimization. In: Kim S-K, Chojnacka K (eds) Marine Algae extracts: processes, products, and applications. Wiley-VCH Verlag GmbH & Co. KGaA, Weinheim, pp 101–121

Rosenberg JN, Oyler GA, Wilkinson L, Betenbaugh MJ (2008) A green light for engineered algae: redirecting metabolism to fuel a biotechnology revolution. Curr Opin Biotech 19:430–436

Ruis JS, Olivieri G, de Vree J, Bosma R, Willems P, Reith JH, Eppink MHM, Kleinegris Dorinde MM, Wijffels RH, Barbosa MJ (2016) Towards industrial products from microalgae. Energy Environ Sci 9:3036–3043

Saeid A, Chojnacka K (2015) Toward production of microalgae in photobioreactors under temperate climate. Chem Eng Res Des 93:377–391

Sahena F, Zaidul ISM, Jinap S, Karim AA, Abbas KA, Norulaini NAN, Omar AKM (2009) Application of supercritical CO_2 in lipid extraction – a review. J Food Eng 95:240–253

Sayre R (2010) Microalgae: the potential for carbon capture. Bioscience 60:722–727

Schroeder G, Leska B, Messyasz B, Pikosz M, Fabrowska J (2015a) Extraction of macroalgae biomass for cosmetics industry. Przem Chem 94:405–407

Schroeder G, Łęska B, Fabrowska J, Messyasz B, Pikosz M (2015b) Analysis of green algae extract. In: Kim S-K, Chojnacka K (eds) Marine Algae extracts: processes, products, and applications. Wiley-VCH Verlag GmbH & Co. KGaA, Weinheim, pp 81–99

Sharma R, Chisti Y, Banerjee UC (2001) Production, purification, characterization and applications of lipases. Biotechnol Adv 19:627–662

Shirvani T, Yan XY, Inderwildi OR, Edwards PP, King DA (2011) Life cycle energy and greenhouse gas analysis for algae-derived biodiesel. Energy Environ Sci 4:3773–3778

Sho Y, Yuuki K, Hidetaka Y, Kan T, Sousuke I (2017) Development of new carbon resources: production of important chemicals from algal residue. Sci Rep 7(1). https://doi.org/10.1038/s41598-017-00979-y

Singh A, Pant D, Olsen SI, Nigam PS (2012) Key issues to consider in microalgae based biodiesel production. Energy Educ Sci Technol A: Energy Sci Res 29:687–700

Slade R, Bauen A (2013) Micro-algae cultivation for biofuels: cost, energy balance, environmental impacts and future prospects. Biomass and Bioenergy 53:29–38

Snow AA, Smith VH (2012) Genetically engineered Algae for biofuels: a key role for ecologists. Bioscience 62:765–768

Sustainable development goals – United Nations (2016) http://www.un.org/sustainabledevelopment/sustainable-development-goals/. Accessed 10 May 2017

Taelman SE, Sfez S (2015) Environmental Life Cycle Assessment (LCA) of algae production in North West Europe (NWE), Public Output report of the EnAlgae project, Swansea, December 2015, 1–35, Available online at http://www.enalgae.eu/public-deliverables.htm. Accessed 10 May 2017

Thiruvenkadam S, Izhar S, Yoshida H, Danquah MK, Harun R (2015) Process application of Subcritical Water Extraction (SWE) for algal bio-products and biofuels production. Appl Energy 154:815–828

Transforming our word (2015) The 2030 Agenda for sustainable development. United Nations – sustainable development knowledge platform. Retrieved 23 Aug 2015

United Nations (2015) Global Sustainable Development Report

United Nations Development Programme (2015).World Leaders Adopt Sustainable Development Goals. Retrieved 25 Sept 2015

United Nations Official Document (2016). Un Org. Retrieved 2016-10-18

Van Boxtel AJB, Perez-Lopez P, Breitmayer E, Slegers PM (2015) The potential of optimized process design to advance LCA performance of algae production systems. Appl Energy 154:1122–1127

Vazquez-Duhalt R, Arredondo-Vega BO (1991) Haloadaptation of the green alga Botryococcus braunii (race a). Photochemistry 30:2919–2925

Voort MPJ, Vulsteke E, Visser CLM (2015) Marco-economics of Algae products, Public Output report WP2A7.02 of the EnAlgae project. Swansea 2015:1–47

Zhang Y, Hu G, Brown RC (2013) Life cycle assessment of the production of hydrogen and transportation fuels from corn stover via fast pyrolysis. Environ Res Lett 8:025001. https://doi.org/10.1088/1748-9326/8/2/025001

MIX
Papier aus verantwortungsvollen Quellen
Paper from responsible sources
FSC® C105338
FSC
www.fsc.org

If you have any concerns about our products,
you can contact us on
ProductSafety@springernature.com

In case Publisher is established outside the EU,
the EU authorized representative is:
Springer Nature Customer Service Center GmbH
Europaplatz 3, 69115 Heidelberg, Germany

Printed by Libri Plureos GmbH
in Hamburg, Germany